The New African American Urban History

The New African American Urban History

edited by

Kenneth W. Goings
Raymond A. Mohl

SAGE Publications
International Educational and Professional Publisher
Thousand Oaks London New Delhi

For information address:

SAGE Publications, Inc.
2455 Teller Road
Thousand Oaks, California 91320
E-mail: order@sagepub.com

SAGE Publications Ltd.
6 Bonhill Street
London EC2A 4PU
United Kingdom

SAGE Publications India Pvt. Ltd.
M-32 Market
Greater Kailash I
New Delhi 110 048 India

Printed in the United States of America

Library of Congress Cataloging-in-Publication Data

The new African American urban history / editors, Kenneth W. Goings, Raymond A. Mohl
p. cm.
Includes bibliographical references (p.) and index.
ISBN 0-7619-0308-9 (cloth: acid-free paper). — ISBN 0-7619-0309-7 (pbk.: acid-free paper)
1. Urbanization—United States—History. 2. Afro-Americans—Social conditions—To 1964. I. Goings, Kenneth W., 1951- , II. Mohl, Raymond A.
HT384.U5N49 1996
973'.0496073'0091732—dc20 95-50234

96 97 98 99 10 9 8 7 6 5 4 3 2 1

This book is printed on acid-free paper.

Sage Production Editor: Diana E. Axelsen
Sage Typesetter: Janelle LeMaster
Cover photograph by Ernest C. Withers, 333 Beale St., Memphis, TN.

Contents

Preface

arely is a scholarly project of this kind the sole inspiration of a single person, and this book is no exception. This project began as a session at the 1992 annual meeting of the American Historical Association in Washington, D.C. At a session titled "Racial Violence in the Twentieth-Century Urban South," Goings and Smith presented a paper on racial violence in Memphis, and Mohl offered a paper on second-ghetto violence in Miami (both papers are included in this volume in revised form). Commenting on that session were Arnold R. Hirsch, University of New Orleans, and Gail W. O'Brien, North Carolina State University. Their cogent and insightful comments, along with contributions from the audience, led the presenters and David R. Goldfield, University of North Carolina at Charlotte, who chaired the session, to believe that, taken together, the papers pointed the way to some new directions in African American urban history. Goldfield, who is also editor of the *Journal of Urban History*, suggested that we guest edit a special issue of the JUH, soliciting articles that played upon and extended some of the themes developed at the AHA session. The

special issue on "The New African Ameican Urban History" appeared as the March 1995 and May 1995 issues of the JUH. This book includes all of the articles from the JUH, as well as four additional articles that also illustrate new historiographical tendencies in African American urban history.

Goings would like to acknowledge the invaluable support and contribution of Gerald L. Smith, not only to our article on Memphis but indeed to my understanding of African American urban history. Stephanie Shaw and Eugene O'Connor deserve special mention. Throughout the writing of the original Memphis article, and then as we attempted to put this volume together, they both provided encouragement and critical responses at just the right moment. Mohl would like to acknowledge careful critical readings of his Miami essay by Kenneth W. Goings, Mark H. Rose, Arnold R. Hirsch, David Schulyler, and Christopher Silver, as well as the essential support provided by Eva Franziska.

The book's cover photo—an image recorded during the Sanitation Workers' Strike in Memphis in 1968—was taken by Mr. Ernest C. Withers. Mr. Withers has been "making pictures" for close to sixty years. His most famous photo, the 1955 photograph of the battered and mutilated body of Emmett Till, was published in *Jet* magazine and shocked a nation just as civil rights issues were rising to the surface in 1950s America. Mr. Withers was a participant/observer of the Civil Rights Movement in the South. He has always been concerned with how ordinary people attempt to live their lives with dignity and self-respect. We wanted to use this particular image because it personified the theme of autonomy and agency—themes central to the new African American history. We thank Mr. Withers for permission to reproduce this photograph.

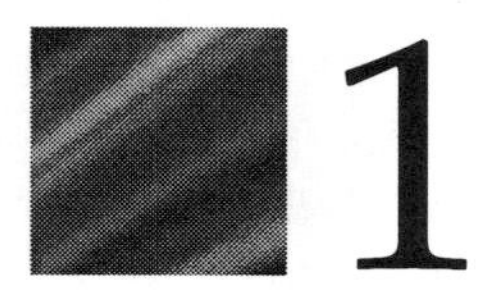

1

Toward a New African American Urban History

Kenneth W. Goings
Raymond A. Mohl

Few areas of American historical scholarship have produced as much exciting new work in recent decades as African American history. An outpouring of scholarship on slavery, emancipation, Reconstruction, sharecropping, late-nineteenth-century race politics, and southern segregation dominated work in the field through the 1970s. But as scholars moved forward to explore twentieth-century African American history, they encountered the city and the urbanization of the black population. The great modern migrations of African Americans to the city, the creation and expansion of black communities, and the examination of black life and culture, especially working-class culture, have provided a central focus for recent scholars of twentieth-century African American urban history.

This change in scholarly focus has been paralleled by several shifts in interpretation and analysis. Earlier works, particularly those that pursued a race-relations perspective, tended to adopt what has been labeled the *ghetto synthesis model* of African American urban history. This approach focused heavily on the physical and institutional structure of black communities and the degree to which whites regulated and controlled black life. By contrast, later studies emphasized an *agency model,* demonstrating the extent to which African Americans in slavery and freedom shaped and controlled their own destinies. Earlier works on black urban history—Gilbert Osofsky on Harlem and Allan Spear on black Chicago, for example—concentrated on the role of institutional forces in the creation of the ghetto and on black urban life generally.[1] More recent work has emphasized an internal focus on kinship and communal networks, class and culture, and the diversity and complexity of black communities. Earlier studies of African American urban history, moreover, concentrated almost exclusively on the black ghettos of northern cities—New York, Cleveland, Detroit, Philadelphia, and Chicago (Howard Rabinowitz's *Race Relations in the Urban South, 1865-1890,* [1978], was a notable exception).[2] Northern and western cities continue to receive attention, but much of the recent scholarship reflects a shift back to black southern roots, to southern cities where blacks generally lived in closer proximity to whites rather than in sprawling ghettos as in the North. Chapters in this book, for instance, examine facets of the African American urban experience in Richmond, Memphis, Norfolk, Atlanta, Birmingham, and Miami, as well as some northern cities.

In a February 1983 review essay in the *Journal of Urban History,* Elliott Rudwick surveyed the state of the field in a piece titled "Black Urban History: In the Doldrums." At the end of the 1960s, when urban problems and racial conflict dominated the public consciousness, Rudwick noted that "black urban history appeared to be headed for an unbeatable future." Policymakers, scholars, and Americans generally, he contended, sought "some historical depth about how these troubles began." In 1983, however, Rudwick concluded that African American urban history had become bogged down in narrowly focused, poorly conceived, or weakly researched studies. Rudwick ended on a pessimistic note, implying that the early promise of the field remained unfulfilled.[3] Four years later, in the

August 1987 issue of the *JUH,* Kenneth L. Kusmer reexamined the scholarly landscape in a review essay titled "Urban Black History at the Crossroads." Kusmer evaluated a spate of new books published in the mid-1980s, suggesting "a renewal of interest in black urban history as a research field." Important new community studies offered a variety of new ways of understanding the urbanization of the black population. Yet for Kusmer, the very diversity of the new scholarship in African American urban history raised "questions about where the field is going."[4]

If African American urban history stood poised at a crossroads in the mid-1980s, uncertain of future directions, those uncertainties surely have been banished by the mid-1990s. In fact, Kusmer's crossroads metaphor may have been too confining conceptually, suggesting only one or two directions that scholars might choose. The virtual explosion of newly published research, by contrast, has demonstrated that African American urban historians have been marching boldly down numerous highways, some moving in parallel directions, others converging, and still others diverging. The very diversity of this recent work provides the best evidence of the vitality of the field. Taken as a whole, this outpouring of scholarship represents a "new African American urban history."

What are the defining interpretive characteristics of this new history? One significant and overarching line of investigation emphasizes the importance of agency among African Americans—an interpretive thrust that has shaped new writing in the field for a decade or more. Earlier studies often portrayed African Americans as passive or powerless, as victims of white racism or slum pathologies. The new African American urban history conveys a sense of active involvement, of people empowered, engaged in struggle, living their lives in dignity and shaping their own futures. Indeed, some new work, such as that of Robin Kelley, has unearthed new forms of agency that demonstrate wider patterns of resistance. In a study of blacks in the urban South, for example, Kelley has described a tradition of "infrapolitics"—a pattern of daily behavior, an oppositional culture, more or less overt, in which African Americans demanded respect and recognition in uncounted daily encounters in the streets, on streetcars and buses, on the job, in the courts, and elsewhere.[5]

Earlier scholarship in African American urban history was heavily weighted toward the study of black elites and elite organizations; it often tended to focus on segregation, race relations, and the role of political and economic institutions (white and black) in establishing the parameters of black life. The new African American urban history has abandoned the exclusive emphasis on prominent African Americans to portray a more diverse community with sharply etched divisions of class and culture. Earlier work depicted the working class as an undifferentiated mass that followed the lead of prominent black citizens. The new history has provided a deeper and more textured sense of the black working class—of its transformation from southern agricultural roots to urban industrial labor, from peasantry to proletariat.[6] At the same time, there is a recognition that many aspects of the rural, African American culture and tradition persisted or were adapted to life in the big city.[7] Black working-class community and culture were made and remade in response to urban and industrial life.

Consequently, an exciting new focus has emerged on black working-class life, culture, and community, on black people in labor and radical movements, on African American sports, and on black festival behavior and the use of streets and public spaces.[8] New studies of the migrations of southern blacks have been written from the perspective of the migrants themselves.[9] New research on twentieth-century urban uprisings (or race riots) now focuses more centrally on the working-class African Americans who participated.[10] The new history also has demonstrated greater appreciation of class division and contestation within the black community. How, for instance, did elite and middle-class blacks in northern cities deal with waves of southern migrants? How did black churches and black organizations respond to the Great Migration? How did black individuals and families navigate their lives in segregation or deal with the new conditions of northern urban life? How did internal conflict within Marcus Garvey's Universal Negro Improvement Association, as well as attacks from other African Americans, undermine the movement's mass appeal? Scholars are now working on these and other questions.[11]

The new African American history has initiated a reexamination of earlier conceptualization and periodization. For instance, some scholars have now rejected the argument of earlier historians, such as Rayford W. Logan, that the period from the end of Reconstruction through the Progressive Era represented the "nadir" for African Americans. According to Logan's analysis, the high rates of lynching during that period served as an expression of white power and black victimization. Revisionist scholars now question the proposition that African Americans were powerless and deferential. They suggest that the rising incidence of lynching was not a symptom of white power but a reaction to black insurgency and the job competition in southern towns and cities that stemmed from African American urbanization. As early as 1923, sociologist Charles S. Johnson noted that the areas with the highest number of lynchings were the same places that attracted the most black rural migrants. White terrorism had much less to do with stimulating the migration of African Americans than did the economic opportunities that beckoned in New York, Chicago, and Detroit, or in Atlanta, Birmingham, Memphis, and New Orleans. There is considerable evidence that rape charges (traditionally, the standard explanation for lynching in the South) were involved in only a minority of lynchings, as well as that African Americans often were armed and defended themselves against lynching when possible. As scholars have reconsidered notions of African American agency during the "nadir" period, they have had to rethink the history and significance of lynching.[12]

The role of religion and of women in the African American community also has come under renewed scholarly examination. The new African American urban history has begun to explore the importance of church and theology in black communities. Scholars are beginning to recognize the degree to which religious conviction empowered African Americans and moved them to action. For example, black preachers actively involved in the late-nineteenth and early-twentieth-century social gospel movement pushed for social justice and racial reform.[13] As early as 1900, some African American denominations, such as the Church of God in Christ, pursued a theology of resistance to segregation, producing militant behavior among congregants.[14] The militant Robert Charles, killed in a vicious police hunt in New Orleans in 1900, was a follower of Bishop Henry M. Turner of the African Methodist Episcopal Church, who preached not only a back-to-Africa message but African American pride and self-defense as well.[15] More recently, the secular application of theology played an important role in the early years of the civil rights movement.[16]

Similarly, scholars are beginning to appreciate the important role of women in black organizations and churches, as well as in family and community. To take one recent example, Evelyn Brooks Higginbotham's new book *Righteous Discontent* (1993) provides an examination of the women's movement in the black Baptist Church between 1880 and 1920: She contends that women "were crucial to broadening the public arm of the church and making it the most powerful institution of racial self-help in the African-American community." Individually and collectively, Baptist women contested segregation, demanded antilynching legislation, advocated voting rights and women's rights, supplied needed social services, and promoted a feminist theology of resistance. They pursued, Higginbotham suggests, "everyday forms of resistance to oppression and demoralization." Other recent studies have provided overwhelming evidence of the powerful and shaping role of women in early-twentieth-century interracial reform, as well as in the mid-twentieth-century civil rights movement.[17]

The civil rights movement itself is undergoing some dramatic reinterpretation as a result of new work in African American history. Much of the published work on the civil rights movement has focused on national elite leaders and organizations and on the push for passage of national legislation for civil rights and voting rights. This top-down history may have presented a misleading picture of the struggle for civil rights. The new history offers a distinctly different view—a view from the grass roots, from the community level, a view most fully articulated in Aldon D. Morris's *The Origins of the Civil Rights Movement* (1984), but also evident in other recent work.[18] This emerging new interpretation contends that the freedom struggle at the local community level took place independently of national activities in response to local conditions, suggesting many different civil rights *movements* rather than a single unified movement dominated by a few elite leaders. Such a community-oriented interpretation provides a new appreciation of the role of individual struggle in achieving civil rights; it also turns a generation of scholarship upside down, or inside out.

The new academic interest in historic memory has also reflected an inward turn, although in more self-conscious ways. Some recent historians have begun exploring the ways in which individuals, groups, and cultures have used the construction of memories in the shaping and reshaping of identities.[19] For African Americans, as for others, the social construction of historic memory was both a collective effort and an individual endeavor. Collectively, African Americans established a sense of the past that conveyed cultural meaning, that supported group traditions, holidays, parades, festivals, and similar public rituals.[20] Such socially constructed and symbolic histories also aided in guiding current conduct, such as navigating in a world bound by segregation, discrimination, and racism. As Elsa Barkley Brown and Gregg D. Kimball have suggested in their essay in this book, collective public memories served as important historical forces shaping the behavior of those doing the remembering. Individually, African Americans created and used notions about the past in understanding and locating their own lives in home, neighborhood, and community. Historic memories conveyed to people a distinct sense of themselves, a sense of understanding about their place and position, and about their relations with others. Race, class, religion, family, and gender all were involved in this process, serving as filters for individual and group conceptions about the past and present. Whether in the public sphere or in private life, historic memories were often contested, which meant that the process of "inventing tradition" was a matter of continuous discourse. Although application of these conceptual approaches remains at an early stage, some intellectually exciting studies, including several in this book, have already been completed. The cultural history of African Americans in the city shows great promise.

Two final areas of new research—on the "second ghetto" and on the "underclass"—have also proven productive. The initial ghetto model of African American urban history, dominant in the 1960s and early 1970s, has given way to new approaches and new conceptions. As historians have advanced their focus beyond the first Great Migration of 1915 to 1930 and into the post-1940 period, a second and more massive black migration from the South, that from 1940 to 1970, has come under examination. So, too, has the newer, second ghetto, first explored in Arnold R. Hirsch's *Making the Second Ghetto: Race and Housing in Chicago, 1940-1960* (1983). A consequence of the new migration of blacks to northern and western cities, and of the simultaneous migration of urban whites to the new postwar suburbs, second ghettos sprouted throughout big-city America. Urban renewal, highway building, school desegregation, and other government policies speeded the process. Uneasy residential transitions took place as neighborhoods turned over, usually facilitated by the real estate industry, the hidden hand shaping neighborhood changes. Behind the shift to the second ghetto, however, was the driving force of African Americans seeking improved housing for their families and better schooling for their children. Black agency had powerful consequences in the postwar era not just for civil rights but in the struggle for decent housing, schooling, and other quality-of-life issues.[21]

The underclass issue is a matter of extensive academic debate at the moment. The term first entered popular usage in the late 1970s in an article in *Time* magazine; by the 1980s, it had come to serve as a shorthand for the high levels of crime, poverty, unemployment, and social disorganization in the mostly black central cities. In his influential 1987 book, *The Truly Disadvantaged: The Inner City, the Underclass, and Public Policy*, sociologist William Julius Wilson explained the emergence of this black underclass as a function of deep structural shifts in the American economy since about 1970. These changes included the decline of manufacturing, the automation of many production processes, the rise of a high-skill, high-tech economy (mostly based in the suburbs and the Sunbelt) and a parallel low-skill, low-pay service and sweatshop economy (mostly in the central cities), and an educational and training mismatch that left most inner-city African Americans unprepared for the postindustrial job market. While these economic changes were taking place, middle-class blacks were moving to the suburbs or more distant second-ghetto neighborhoods, leaving the working class and the poor behind in the inner cities.[22]

The Wilson thesis recently has been challenged by several scholars, some contesting the social breakdown interpretation, others blaming the plight of the inner cities on the persistence of racism and segregation. The recently published *American Apartheid: Segregation and the Making of the Underclass* (1993), authored by sociologists Douglas S. Massey and Nancy A. Denton, offers an especially powerful antidote to the Wilson interpretation.[23] Still other challenges to the Wilson underclass thesis have been made by the historians and social scientists who contributed to Michael B. Katz's edited collection, *The "Underclass" Debate: Views From History* (1993). In examining the underclass issue from many angles, Katz and his colleagues found both continuity with the past and great disjunctions between past and present experience. Katz argued the case for "the centrality of history to the troubling and urgent questions that underlie the underclass debate."[24] The new African American urban history, in short, provides important insight and perspective on contemporary urban issues.

The chapters in this book illustrate in concrete ways several of the new directions scholars have been exploring in African American urban history. Shane White's essay on black parades and festivals in eighteenth- and early-nineteenth-century northern cities provides a close rereading of African American cultural practice and public behavior. He demonstrates a shift from more or less private holiday festivals oriented around music, dance, feasting, and socializing, common in the eighteenth century, to more public and political parades typical of the early nineteenth century, reflecting an emerging African American sense of pride and citizenship. Essentially, free blacks in the urban North were claiming the rights of American citizens, but they were also demanding their rights to the streets and public spaces of the city.

Two chapters engage issues involving communal and collective memory. In their study of Richmond, Elsa Barkley Brown and Gregg D. Kimball provide a new cultural history of the city's black community, weaving several separate historical strands into a text on festival behavior, historic memory, and power, class, and gender. Urban space, public ritual, and private lives are linked in this imaginative piece of black urban history. Earl Lewis, too, applies conceptions of historic memory, collective and individual, to the power of place in his article on Norfolk. The social construction of memory guided the behavior of African Americans and shaped their conceptions of self and community. Taken together, these creative essays on two Virginia cities reveal what a newer African American urban history might look like and what might be achieved through the use of new conceptual approaches.

A focus on patterns of black resistance to white domination and control has provided the interpretive framework for articles by Kenneth W. Goings and Gerald L. Smith, by Tera W. Hunter, and by Robin D. G. Kelley. In their study of racial violence in Memphis, Goings and Smith analyze the effect of migration on a southern community and on race relations generally. They report a pattern of black resistance and the use of violence in response to segregation and daily harassment. Tera Hunter's essay similarly advances new arguments about African American agency. She begins with a discussion of a massive strike by black household workers in Atlanta in 1881, using this event as a means to explore the dimensions of covert and overt resistance to racial and class oppression. Challenging exploitation by employers, African American domestic workers pursued a variety of options in articulating their grievances and asserting their demands, and they also developed a high degree of autonomy and political consciousness in new south Atlanta. Finally, Robin Kelley's essay provides a full elaboration of the "infrapolitics" interpretation mentioned earlier, demonstrating the oppositional culture that emerged in the Jim Crow South and that sustained or empowered African Americans in their daily confrontations with segregation and racism. In the workplace and in the white-dominated public spaces of the cities, the daily business of racial contestation took place, exhibiting patterns of resistance that challenge earlier interpretations of black passivity and acceptance.

Another thrust of the new African American urban history deals with migration—both regional migration of blacks from the South to the North, Midwest, and West, and intracity migration as African Americans moved to new neighborhoods in search of better housing. In each case, migration can be interpreted as a purposeful strategy to achieve greater freedom and better opportunities. Darlene Clark Hine's essay follows this line of argument, but applies it especially to African American women who may have chosen migration to Detroit or Cleveland or Chicago for noneconomic reasons, such as the pursuit of personal autonomy or to escape sexual exploitation by black and white men. In any case, black women had different migratory patterns than black men, and, for various reasons, they also assimilated to northern urban life less completely and thus contributed more extensively to the "southernization of the Midwest," as southern black culture (foodways, religious practices, family structures, and music) penetrated the urban North.

A second form of black migration serves as the basis of Raymond A. Mohl's essay on Miami. Mohl explores the multiple impulses behind the creation of the second ghetto in the two decades after 1940—the consequence of the white migration to the suburbs during those years and the simultaneous arrival of massive numbers of black urban migrants. Clearly, southern cities were experiencing the second-ghetto phenomenon as well as northern cities like Detroit, Chicago, and Philadelphia. Particularly notable as driving forces were the real estate and building interests while public policies such as urban renewal, highway building, and school desegregation intensified the racial turnover of neighborhoods. Not to be discounted, however, were the aspirations of black Americans for new and better housing, attested to by their willingness to confront whites who used violence and terror in defense of their segregated neighborhoods. The second-ghetto battles in postwar Miami mirrored the racial patterns that typified northern and midwestern cities.

We bring the book to a close with a scholarly assessment of the changing historiography of African American urban history. Joe W. Trotter's essay on African Americans in the industrial city, 1900 to 1950, provides an analytical overview of scholarly writings in history and the social sciences. He traces the contemporary historiographic patterns of black urban history back to their roots in late-nineteenth- and early-twentieth-century sociology, especially the work of W. E. B. Du Bois, Charles S. Johnson, Robert E. Park, and E. Franklin Frazier. These early social scientists, as well as such midcentury scholars as St. Clair Drake, Horace R. Cayton, Robert C. Weaver, and Gunnar Myrdal, pioneered the urban community study. They also asserted the importance of theory while demonstrating how it might be applied in specific situations. The black urban history that emerged in the 1960s built on the foundations of that earlier scholarship, but it also pushed far beyond the work of the pioneering social scientists of race and urban community. Trotter brings some conceptual order to the scholarship of the past thirty years, suggesting as well the many new lines of analysis and interpretation. The focus throughout is on the building of African American community within the industrial city, a process that derived from the adaptation of agricultural workers to the new demands of urban and industrial life.

Kenneth L. Kusmer's essay provides a detailed assessment of the current state of knowledge about African Americans in the city since World War II. In the largest sense, this era was shaped by the economic and social transition from an industrial to a postindustrial nation. For black Americans, the new forces of change had powerful consequences. These included the Great Migration to northern and western cities between 1940 and 1970; the emergence of the civil rights movement; the drive toward black political empowerment; periodic outbreaks of urban racial violence, especially but not exclusively in the 1960s; the persistence of economic inequality and of segregation in housing and schooling; and the ghettoization of the African American underclass. Kusmer's essay presents a complex but compelling conceptual framework for understanding the momentous changes that have affected African Americans during the second half of the twentieth century. At the same time, there is an appreciation for "the aspirations and struggles of urban African Americans" that "continually shaped and reshaped the social, cultural, and political life of black communities." Taken together, the essays of Trotter and Kusmer demonstrate the richness and vitality of the historical and social science literature on African Americans in the city.

We present this book, then, as a testament to the new African American urban history. It is a historiographic thrust that focuses heavily (although not exclusively) on the black experience in southern cities; on the diversity of class, culture, and gender within the black community; and on the shaping force of African American agency. The research presented here on African American resistance strategies, on black culture and festival behavior, on the black working class, on regional and intraurban migration, and on historic memory should help to open up new lines of investigation for students and researchers in the field. Much work remains for scholars in African American urban history, to be sure, but the essays presented in this book should help mark the way.

Notes

1. Gilbert Osofsky, *Harlem: The Making of a Ghetto, Negro New York, 1890-1930* (New York, 1966); Allan H. Spear, *Black Chicago: The Making of a Negro Ghetto, 1890-1920* (Chicago, 1967).

2. Seth M. Scheiner, *Negro Mecca: A History of the Negro in New York City, 1865-1920* (New York, 1965); David M. Katzman, *Before the Ghetto: Black Detroit in the Nineteenth Century* (Urbana, 1973); Kenneth L. Kusmer, *A Ghetto Takes Shape: Black Cleveland, 1870-1930* (Urbana, 1976); Vincent P. Franklin, *The Education of Black Philadelphia: The Social and Educational History of a Minority Community, 1900-1950* (Philadelphia, 1979); Thomas L. Phillpott, *The Slum and the Ghetto: Neighborhood Deterioration and Middle-Class Reform, Chicago, 1880-1930* (New York, 1978); Leonard P. Curry, *The Free Black in Urban America, 1800-1850* (Chicago, 1981). For a notable exception to the northern focus of these early ghetto studies, see Howard N. Rabinowitz, *Race Relations in the Urban South, 1865-1890* (New York, 1978). For some more recent community studies of western and southern cities, see Douglas Henry Daniels, *Pioneer Urbanites: A Social and Cultural History of Black San Francisco* (Philadelphia, 1980); George C. Wright, *Life Behind a Veil: Blacks in Louisville, Kentucky, 1865-1930* (Baton Rouge, 1985); Albert S. Broussard, *Black San Francisco: The Struggle for Racial Equality in the West* (Lawrence, 1993); Quintard Taylor, *The Forging of a Black Community: Seattle's Central District From 1870 Through the Civil Rights Era* (Seattle, 1994).

3. Elliott Rudwick, "Black Urban History: In the Doldrums," *Journal of Urban History* 9 (February 1983), 251-60. For similar sentiments, see also August Meier and Elliott Rudwick, *Black History and the Historical Profession, 1915-1980* (Urbana, 1986), especially 277-308.

4. Kenneth L. Kusmer, "Urban Black History at the Crossroads," *Journal of Urban History* 13 (August 1987), 460-70. See also Kenneth L. Kusmer, "The Black Urban Experience in American History," in Darlene Clark Hine, ed., *The State of Afro-American History: Past, Present, and Future* (Baton Rouge, 1986), 91-122.

5. Robin D. G. Kelley, " 'We Are Not What We Seem': Rethinking Black Working-Class Opposition in the Jim Crow South," *Journal of American History* 80 (June 1993), 75-112; Robin D. G. Kelley, *Race Rebels: Culture, Politics, and the Black Working Class* (New York, 1994). Kelley's work has been informed by the anthropological insights provided in two books by James C. Scott: *Weapons of the Weak: Everyday Forms of Peasant Resistance* (New Haven, 1985) and *Domination and the Arts of Resistance: Hidden Transcripts* (New Haven, 1990).

6. See, for instance, Joe W. Trotter, Jr., *Black Milwaukee: The Making of an Industrial Proletariat, 1915-45* (Urbana, 1985); Richard W. Thomas, *Life for Us Is What We Make It: Building Black Community in Detroit, 1915-1945* (Bloomington, 1992).

7. For an early study in this tradition, see James Borchert, *Alley Life in Washington: Family, Community, Religion, and Folklife in the City, 1850-1970* (Urbana, 1980).

8. On labor and radical movements, see Peter Rachleff, *Black Labor in Richmond, 1865-1890* (Philadelphia, 1984); Mark Naison, *Communists in Harlem During the Depression* (Urbana, 1983); Robin D. G. Kelley, *Hammer and Hoe: Alabama Communists During the Great Depression* (Chapel Hill, 1990); Eric Arnesen, *Waterfront Workers of New Orleans: Race, Class, and Politics, 1863-1923* (New York, 1991). On African American sports, see Rob Ruck, *Sandlot Seasons: Sport in Black Pittsburgh* (Urbana, 1987); Jules Tygiel, *Baseball's Great Experiment: Jackie Robinson and His Legacy* (New York, 1983).

9. Peter Gottlieb, *Making Their Own Way: Southern Blacks' Migration to Pittsburgh, 1916-1930* (Urbana, 1987); James R. Grossman, *Land of Hope: Chicago, Black Southerners, and the Great Migration* (Chicago, 1989); William Cohen, *At Freedom's Edge: Black Mobility and the Southern White Quest for Racial Control, 1861-1915* (Baton Rouge, 1991); Joe W. Trotter, Jr., *The Great Migration in Historical Perspective: New Dimensions of Race, Class, and Gender* (Bloomington, 1991); Alferdteen Harrison, ed., *Black Exodus: The Great Migration From the American South* (Jackson, Miss., 1993); and, for a more controversial journalistic treatment, Nicholas Lemann, *The Promised Land: The Great Black Migration and How It Changed America* (New York, 1991).

10. Dominic J. Capici, Jr. and Martha Wilkerson, *Layered Violence: The Detroit Rioters of 1943* (Jackson, Miss., 1991); Roberta Senechal, *The Sociogenesis of a Race Riot: Springfield, Illinois, in 1908* (Urbana, 1990). For earlier studies partially in this tradition, see Elliott Rudwick, *Race Riot at East St. Louis, July 2, 1917* (Carbondale, 1964); William M. Tuttle, Jr., *Race Riot: Chicago in the Red Summer of 1919* (New York, 1970).

11. On the churches, see, for example, Robert Gregg, *Sparks From the Anvil of Oppression: Philadelphia's African Methodists and Southern Migrants, 1890-1940* (Philadelphia, 1993). On the Garvey movement, see Judith Stein, *The World of Marcus Garvey: Race and Class in Modern Society* (Baton Rouge, 1986); Robert A. Hill, ed., *The Marcus Garvey and Universal Negro Improvement Association Papers*, 9 vols. (Berkeley, 1983-1995).

12. Rayford W. Logan, *The Negro in American Life and Thought: The Nadir, 1877-1901* (London, 1954); Charles S. Johnson, "How Much Is the Migration a Flight From Persecution?" *Opportunity* 1 (September 1923), 272-74. The makings of a new interpretation of lynching can be constructed from several new works, including W. Fitzhugh

Brundage, *Lynching in the New South: Georgia and Virginia, 1880-1930* (Urbana, 1993); Robert P. Ingalls, *Urban Vigilantes in the New South: Tampa, 1882-1936* (Knoxville, 1988); Herbert Shapiro, *White Violence and Black Response: From Reconstruction to Montgomery* (Amherst, 1988), 30-63; George C. Wright, *Racial Violence in Kentucky, 1865-1940 Lynchings, Mob Rule, and "Legal Lynchings"* (Baton Rouge, 1990); Edward L. Ayers, *The Promise of the New South: Life After Reconstruction* (New York, 1992), 153-59; Neil R. McMillen, *Dark Journey: Black Mississippians in the Age of Jim Crow* (Urbana, 1989), 224-53; Stewart E. Tolnay and E. M. Beck, "Rethinking the Role of Racial Violence in the Great Migration," in Harrison, ed., *Black Exodus*, 20-35; Stewart E. Tolnay and E. M. Beck, *A Festival of Violence: An Analysis of Southern Lynchings, 1882-1930* (Urbana, 1995).

13. Ronald White, *Liberty and Justice for All: Racial Reform and the Social Gospel* (New York, 1990).

14. Cheryl Townsend Gilkes, " 'Together and in Harness': Women's Traditions in the Sanctified Church," *Signs: Journal of Women in Culture and Society* 10 (1985), 678-99.

15. William I. Hair, *Carnival of Fury: Robert Charles and the New Orleans Race Riot of 1900* (Baton Rouge, 1976).

16. James F. Findlay, Jr., *Church People in the Struggle: The National Council of Churches and the Black Freedom Movement, 1950-1970* (New York, 1993); C. Eric Lincoln and Lawrence M. Mamiya, *The Black Church in the African American Experience* (Durham, 1990), 196-235.

17. Evelyn Brooks Higginbotham, *Righteous Discontent: The Women's Movement in the Black Baptist Church, 1880-1920* (Cambridge, Mass., 1993), quotations on pp. 1, 2; Vickie L. Crawford, Jacqueline Anne Rouse, and Barbara Woods, eds., *Women in the Civil Rights Movement: Trailblazers and Torchbearers, 1941-1965* (Bloomington, 1993); J. A. Robinson, *The Montgomery Bus Boycott and the Women Who Started It: The Memoir of Jo Ann Gibson Robinson*, (Knoxville, 1987).

18. Aldon D. Morris, *The Origins of the Civil Rights Movement: Black Communities Organizing for Change* (New York, 1984); Armstead L. Robinson and Patricia Sullivan, eds., *New Directions in Civil Rights Studies* (Charlottesville, 1991); Steven F. Lawson, "Freedom Then, Freedom Now: The Historiography of the Civil Rights Movement," *American Historical Review* 96 (April 1991), 456-71; Adam Fairclough, "Historians and the Civil Rights Movement," *Journal of American Studies* 24 (December 1990), 387-98. Local studies of the civil rights movement include William H. Chafe, *Civilities and Civil Rights: Greensboro, North Carolina, and the Black Struggle for Freedom* (New York, 1980); Robert J. Norrell, *Reaping the Whirlwind: The Civil Rights Movement in Tuskegee* (New York, 1985); Kim Lacy Rogers, *Righteous Lives: Narratives of the New Orleans Civil Rights Movement* (New York, 1993); Michael K. Honey, *Southern Labor and Black Civil Rights: Organizing Memphis Workers* (Urbana, 1993); James R. Ralph, Jr., *Northern Protest: Martin Luther King, Jr., Chicago, and the Civil Rights Movement* (Cambridge, Mass., 1993); John Dittmer, *Local People: The Struggle for Civil Rights in Mississippi* (Urbana, 1994); Adam Fairclough, *Race and Democracy: The Civil Rights Struggle in Louisiana, 1915-1972* (Athens, Ga., 1995).

19. David Thelan, ed., *Memory and American History* (Bloomington, 1990); David Glassberg, *American Historical Pageantry: The Uses of Tradition in the Early Twentieth Century* (Chapel Hill, 1990); George Lipsitz, *Time Passages: Collective Memory and American Popular Culture* (Minneapolis, 1990); John Bodnar, *Remaking America: Public Memory, Commemoration, and Patriotism in the Twentieth Century* (Princeton, 1992).

20. William H. Wiggins, Jr., *O Freedom: Afro-American Emancipation Celebrations* (Knoxville, 1987); Shane White, " 'It Was a Proud Day': African American Festivals and Parades in the North, 1741-1834," *Journal of American History* 81 (June 1994), 13-50;

Richard White, "Civil Rights Agitation: Emancipation Days in Central New York in the 1880s," *Journal of Negro History* 78 (Winter 1993), 16-24; William H. Wiggins, Jr., *Jubilation! African American Celebrations in the Southeast* (Columbia, 1993); Geneviève Fabre and Robert O'Malley, eds., *History and Memory in African-American Culture* (New York, 1994).

21. Arnold R. Hirsch, *Making the Second Ghetto: Race and Housing in Chicago, 1940-1960* (Cambridge, 1983); John F. Bauman, *Public Housing, Race, and Renewal: Urban Planning in Philadelphia, 1920-1974* (Philadelphia, 1987); Raymond A. Mohl, "Race and Space in the Modern City: Interstate-95 and the Black Community in Miami," in Arnold R. Hirsch and Raymond A. Mohl, eds., *Urban Policy in Twentieth-Century America* (New Brunswick, 1993), 100-58; W. Edward Orser, *Blockbusting in Baltimore: The Edmondson Village Story* (Lexington, 1994); and many of the essays in Henry Louis Taylor, Jr., *Race and the City: Work, Community, and Protest in Cincinnati, 1820-1970* (Urbana, 1993).

22. William Julius Wilson, *The Truly Disadvantaged: The Inner City, the Underclass, and Public Policy* (Chicago, 1987). See also John F. Bauman, Norman P. Hummon, and Edward K. Muller, "Public Housing, Isolation, and the Urban Underclass: Philadelphia's Richard Allen Homes, 1941-1965," *Journal of Urban History* 17 (May 1991), 264-92.

23. Douglas S. Massey and Nancy A. Denton, *American Apartheid: Segregation and the Making of the Underclass* (Cambridge, Mass., 1993). See also Gary Orfield and Carole Ashkinaze, *The Closing Door: Conservative Policy and Black Opportunity* (Chicago, 1991).

24. Michael B. Katz, "The Urban 'Underclass' as a Metaphor of Social Transformation," in Michael B. Katz, ed., *The "Underclass" Debate: Views From History* (Princeton, 1993), 3-23, quotation on p. 23. Particularly useful essays in the Katz book are Jacqueline Jones, "Southern Diaspora: Origins of the Northern 'Underclass,' " 27-54; Joe W. Trotter, "Blacks in the Urban North: The 'Underclass Question' in Historical Perspective," 55-81; and Thomas J. Sugrue, "The Structures of Urban Poverty: The Reorganization of Space and Work in Three Periods of American History," 85-117.

2

"It Was a Proud Day"

African Americans, Festivals, and Parades in the North, 1741-1834

Shane White

Dawn, in Boston Harbor, on Monday, June 2, 1817, failed to bring the hoped-for winds. It seemed likely that the *Canton-Packet*, anchored a little distance off Long Wharf and loaded with lumber for the Isle of France and four hundred thousand dollars in specie to be invested in Canton, would not be able to get underway until the following day. At-

---

Author's Note: This chapter originally appeared in the *Journal of American History, 81*(1), June, 1993, pp. 75-112, and is reprinted here by permission of the Organization of American Historians. I am extremely grateful to the Australian Research Council, the American Philosophical Society, the American Antiquarian Society, and my own department for their generous support of my work. I would also like to thank Jim Horton, Eric Lott, Gary B. Nash, Susan Armeny, Chad Berry, Meg Meneghel, Dave Thelen, Greg Dening, Donna Merwick, Bob St. George, Ian Tyrrell, and Mike Zuckerman for suggestions that have improved this piece immeasurably. I thank Richard Bosworth, Michael Fellman, Stephen Garton, Jim Gilbert, Charles Joyner, Larry Levine, Leon Litwack, Marcus Rediker, David Waldstreicher, Richard Waterhouse, Al Young, and especially Roger Abrahams, Rhys Isaac, and Graham White (no relation) for their helpful comments and splendid encouragement.

tracted by the manifold pleasures of spending the holiday time called Election in Boston, William Read, the ship's young black steward, sought permission to go ashore, but the request or, as one newspaper reported, the "demand," was refused. After a heated exchange, an angry Read stormed below deck, collected a loaded pistol from the captain's cabin, and fired a shot point-blank into several tightly packed barrels of gunpowder wrapped in woolen cloth and stacked just outside the cabin door. The ensuing blast lifted the quarterdeck to a great height in the air, drove out the ship's quarters and stern, threw the mizzenmast against the main yard, and set the *Canton-Packet* on fire. Most of the crew were forward and, surprisingly, sustained only minor injuries. Read, of course, died in the explosion.[1]

For several years now, William Read's violent death has haunted me, gently mocking my pretension that, as a historian, I can know the past. Why did he blow up the ship and kill himself? What did he think he was doing as he pulled the trigger? We know that, as Greg Dening has pointed out, being in port disturbed the usual running of a vessel, put added stress on shipboard relationships, and occasioned much punishment.[2] Perhaps there had been trouble between the young free black steward and the whites in charge of the ship, a history that transformed the refusal to be allowed to go to Election into the straw that broke the camel's back—clichés lay to rest the inexplicable or the too easily explained. William Read still eludes me, but I do have an idea of the meaning of his death in the scheme of things. So did Read's contemporaries.

Whites would not allow black Bostonians easily to forget Read. According to the historian Alice Morse Earle, for years afterward, white boys taunted African Americans on the streets of town:

> Who blew up the ship?
> Nigger, why for?
> 'Cause he couldn't go to 'lection
> An' shake paw-paw.[3]

We have no way of knowing what Boston blacks made of these gibes, or of Read's desperate act, but it would be surprising if their account did not differ from the simple narrative that was quickly constructed by whites. The *Columbian Centinel* presumed that Read had "perpetrated this act from motives of revenge—for which he foolishly sacrificed his life," a judgment echoed by the Reverend William Bentley, in the nearby town of Salem, who recorded in his diary that the explosion had originated in the "mad resentment of a black man because he was not allowed yesterday to go to election." For Bentley, the incident revealed "the true spirit of our free blacks & the still bewitching influence of what they call election." Yet the Salem clergyman was clearly baffled by what he saw as the irrational behavior of ordinary blacks, even those with no "eccentricities at any other time," when caught up in the excitement of Election: "And yet most of them without a penny have no other amusement than in a long walk & absence from home, the most fatiguing dances & the never ceasing sound of the violin, & return exhausted, dirty, ragged & often hungry & emaciated." Even in Boston and its surrounds, with their small black population and constant, close contact between Euro-American and African American cultures, much of what was important to blacks remained a mystery to whites.[4]

My account of William Read's demise—pieced together from a couple of fugitive diary entries, laconic newspaper reports, and an antiquarian's musings three quarters of a century after the event—encapsulates the problems of trying to reconstruct the history of the relatively small number of African Americans living in the North in the eighteenth and early nineteenth centuries. Town historians, antiquarians, and newspaper writers, whose works are among the main sources employed here, used a language suffused with racism and constantly displayed patronizing attitudes toward African Americans. Frequently, their accounts of black life were set in some timeless past, and the details of that life reported in vague and often contradictory terms. Yet though severe, the problem of the sources is not insurmountable (many historians nowadays habitually work in such broken terrain), and it is possible to find within Bentley's diary entry and the ditty chanted by children the seeds of an alternative account of Election. This version offers the chance to transcend the narrow-mindedness of nineteenth-century whites and to arrive at a reading of the event that incorporates African Americans as full and conscious participants and focuses on the dynamic relationship between the two Boston cultures.[5]

For Bentley, the newspaper writers, and Earle, Read's apparent fixation with going to Election and his violent death were slightly bizarre proofs of the immaturity and childishness of the Negro race, further affirmations of the superiority of white Americans. Like theirs, my purpose is to exploit this black man's suicide to form part of a larger story, but the narrative I wish to construct is different. In the broader context of African American culture in the North, Read's death can be seen as a final dramatic flourish, signaling the end of a festive event. This reading is a fiction—Election Day would drag on, in a few isolated locales, for a while longer—but it is a convenient one that contains a considerable element of truth. Election Day in Connecticut, Rhode Island, and Massachusetts and Pinkster in New York and New Jersey were at the very heart of slave culture in the eighteenth-century North. But as the slaves slowly secured their freedom, these festivals must have seemed almost dated, certainly less appropriate, and they were gradually replaced by the parade, a celebration more suited to African Americans' position in northern cities. Six weeks after Read's death, for example, several hundred black males marched through the streets of Boston in an annual procession, first held in 1808, that celebrated the abolition of the slave trade. Parades and festivals existed, for a time, side by side, but Read's explosive self-immolation coincided with a shift toward the new cultural practice. Indeed, it seems only right that so richly textured and important an African American cultural event as Election Day should not end with a whimper.

My intentions in this essay are to chart the transformation of the northern black festivals of the eighteenth century into the city parades of the first third of the nineteenth century and to demonstrate how this process reflected the changed circumstances of African northerners. In the colonial period, transplanted Africans created a ritual that if not exactly private—participants often played to, and were always aware of, the watchful gaze of whites—was conducted principally for their own benefit. But the nineteenth-century parades, with their overtly public and political cast, prominently, even intrusively, displayed the newly freed northern blacks to whites and to themselves as both African Americans and citizens of the new nation. My account begins with the festivals, which have never been considered as a whole before, and then shows how, in the early nineteenth century, the newly freed African Americans established marches through northern cities and towns. The marches gradually replaced the festivals as the preeminent form of African American public display. In the final section of the essay, I consider both the links between these two forms of celebration and the question of what was new and distinctive about the parades of northern free blacks. By examining these African American celebratory rituals in the North over nearly a century and by carefully delineating how whites and blacks negotiated their different concerns over these events, I hope to reveal something new about both slave and free black culture.

* * *

In the middle third of the eighteenth century, it seems, northern slaves created distinctively African American festivals. Those years, as Ira Berlin has pointed out, saw a transformation of African American culture in northern colonies, brought about by an increasing reliance on slave labor and, more important, by the new practice of importing slaves directly from Africa rather than from the West Indies. Negro Election Day in New England was only the best known of those festivals, which included Pinkster in New York and New Jersey and General Training in various parts of the North. Initially, slaves had merely participated in holiday activities organized by their masters, but gradually they took the festivals over, infusing them with new life and meanings, so that what had begun as white practices, relics of a European past, had by the end of the century become recognized as African American events. These festivals helped to break up the work year for northern slaves, most of whom were employed on small farms, in artisanal workshops, and in domestic service. They also provided a rare and valued opportunity for African Americans, who were few and sparsely settled, to socialize among themselves.[6]

From 1693 onward in Massachusetts, Election Day occurred on the last Wednesday in May. Gradually, this event became an important popular festival, attracting large numbers of people to the metropolis. Churches, societies, and the like held their annual meetings during what became known as Election Week, adding to the crowds. For example, on the Monday following Election Day the officers of the Ancient and Honorable Artillery Company were chosen, and they received their commissions from the governor.[7] Doubtless, early in the eighteenth century, blacks participated in the activities associated with Election Week but during the second third of the century, Election Day became particularly associated with African Americans and became known as Negro Election Day, or "Nigger 'Lection."

In Boston on Negro Election Day, blacks congregated on the common, drinking, gambling, dancing, and generally enjoying themselves without interference from whites. That they were able to do these things was a privilege. In his memoirs, published in 1868, Sol Smith, a theatrical entrepreneur, remembered that as a young lad in Boston in 1814, he had mixed with what he called the Republican boys, both Northenders and Southenders, and "chased all the niggers off the Common, as was usually done on occasions of gatherings except on what was termed 'nigger 'lection.' " Smith confessed that he had no idea why the occasion was called "nigger 'lection," but according to him, it was only then that "the colored people were permitted to remain unmolested on Boston Common."[8] By the nineteenth century, though, there was a certain looseness in the use of the terms "Negro Election Day" and "Election Day." In 1817, when Read committed suicide, the usage in both Bentley's diary and the children's ditty suggests that "election" had become a general term for the period stretching from the Wednesday to Artillery Election on the Monday, the day of Read's death.

Elsewhere in Massachusetts and New England, Smith's confusion about the meaning of "Negro Election" would scarcely have been possible. There the holiday had probably begun as it had in Boston, that is, as a general holiday for slaves. Such an origin is suggested by an entry in the diary of Benjamin Lynde, a resident of Salem, Massachusetts, for Wednesday, May 27, 1741—"Fair weather, Election; Negro's hallowday here at Salem; gave Scip 5s. and Wm 2s 6d." But within a few years, slaves in Salem were meeting on the last Wednesday in May to elect a black ruler, whose inauguration they celebrated the next two or three days. Although the day on which African American kings or governors were chosen differed from colony to colony and later from state to state—in Rhode Island it is the last Saturday in June, in Connecticut it was the day after the white election—much the same annual ritual occurred, if sporadically, during the next sixty or seventy years, in over twenty New England towns.[9]

In New York and New Jersey, the black festival was called Pinkster. Initially Pinkster had been a Dutch holiday—Pentecost—that occurred seven weeks after Easter, in late May or early June, at about the same time of year as the New England Election Day. Throughout most of the eighteenth century, Pinkster was probably a low-key event in which both the Dutch and their slaves took part—certainly, it is difficult to find, between the late seventeenth century and the 1780s, any reference at all to the holiday. In 1736, the *New York Weekly Journal* published a long account of a biracial holiday that may have been Pinkster, although I think it more likely that the description is of a General Training Day. After the Revolution, however, Pinkster became associated closely with African Americans. They celebrated this holiday in areas first settled by the Dutch—along the Hudson Valley, on the western end of Long Island, and in New Jersey—but it was in Albany in the 1790s and the first decade of the nineteenth century that Pinkster reached its apogee. Here King Charles (who, unlike the African American kings and governors of New England, was not elected) presided over the week-long festivities on Pinkster Hill.[10]

The other important holiday for many northern slaves was variously called Negro Training or General Training and usually occurred in June. Blacks would come from miles around, ostensibly to watch the black militia drill. In New England, this occasion was sometimes held on a separate day, as in Windsor, Connecticut, where, according to the town's historian, Henry Stiles, Negro Trainings were common. Elsewhere the training appears to have been incorporated into the activities of Negro Election Day. General Training was particularly important in New Jersey. It was a "great day for the Bedminster colored people" that was "always celebrated by Dick and Nance," slaves in the 1790s on the "old farm" described by Andrew D. Mellick, Jr., late in the nineteenth century. According to Mellick, slave children sat quietly "in round-eyed wonder at all the glories of the day." In 1883, Silvia Dubois, an ex-slave from New Jersey, recollected in more down-to-earth detail how the holiday was celebrated in the early part of the nineteenth century. General Training, she told C. W. Larison, "was the biggest day. The niggers were all out to General Training—little and big, old and young; and then they'd have some rum—always had rum at general trainings—and then you'd hear 'em laugh a mile. And when they got into a fight, you'd hear 'em yell more than five miles."[11]

Throughout the North, then, where slaves were concentrated, often in small pockets along the coast and up the valleys of the major rivers, African Americans appear to have created midyear festivals. Pennsylvania stands as a possible exception. John Fanning Watson's brief description of slaves in Philadelphia contains, as far as I know, the only reference to such an event. According to Watson, writing in the first part of the nineteenth century, "many can still remember when the slaves were allowed the last days of the fairs for their jubilee." Up to a thousand men and women would be present, "divided into numerous little squads, dancing and singing, 'each in their own tongue' after the customs of their several nations in Africa."[12]

There was a striking dynamism and cultural fluidity to these festivals. Pinkster, Election Day, and General Training began among the white colonists, but that past provided only the loosest of structures within which northern slaves improvised. They did so brilliantly. There was a lightness of touch, a quality simultaneously threatening and tongue-in-cheek, an unpredictability, in the ways in which northern slaves used what they found around them to reinvent these holidays. It is generally accepted that the festivals were African American events, but other influences were also at work.

Accounts of Negro Election Day, in particular, frequently mention the involvement of Native Americans. Although we know little of what looks like an important cultural contact between blacks and Indians in New England, Indians too probably helped shape the festival. Earle noted that "old squaws" were prominent in collecting the barks and roots used in brewing the "'lection beer" drunk in Boston, and an anonymous poet, after viewing the diverse crowd attracted to Boston Common, wrote,

> Thither resorts motley crew,
> Of Whites & Blacks & Indians too
> And Trulls of every sort.

Similarly, according to Thomas Williams Bicknell, "both negro and Indian slaves" participated in the election of the black governor in Narragansett, Rhode Island, at the end of the eighteenth century.[13]

In Albany, New York, in the early years of the nineteenth century, at the zenith of the reign of King Charles, Pinkster became increasingly commercial. Although Charles and the African American slaves remained at the center of the festival, an array of activities took place on the periphery. According to Dr. James Eights, writing in the 1860s, at the back of the natural amphitheater where the blacks performed were exhibitions of wild animals, ropedancing, and circus riding. Someone with an "unpronounceable name" performed on the slack rope; the rider of the famous horse Selim somersaulted through a blazing hoop; and Rickett, "the celebrated clown of the day," displayed his "stock of buffoonery on horseback." Another writer claimed that there were also such games as "egg butting" and "riding at the ring," as well as "impromptu horse races, wrestling matches and occasional 'scraps.' "[14]

But diversity—which ranged from European, Native American, and African influences to a homegrown commercialism and which pleases the multicultural sensibilities of most academic historians—was just one feature of this cultural dynamism. Even though these festivals were, for over half a century, the showy centerpiece of northern slave culture, they were transitory, particularly in the smaller towns where the slave population usually comprised only a few score individuals. In such places, the festivals seldom persisted for more than a few years at a time. Even in Albany, where hundreds, perhaps thousands, participated, Pinkster as an African American event probably lasted only a quarter of a century. In part, this perception may reflect the usual shortcomings in the sources, but more may be involved. The festivals appear as a series of discontinuous events, of traditions interrupted and recovered, of celebrations that erupted sporadically into something spectacular and obvious, even to the most inattentive whites, and then faded back into a more subdued form or even disappeared. Factors as diverse as the attitudes of local whites, occurrences in nearby towns, or the presence of a charismatic African suited to the role of king or governor may have shaped the festivals' history.

Both the contingent nature of Negro Election Day, Pinkster, and General Training and their dependence on circumstances peculiar to the North are highlighted by a comparison with southern festivals. Particularly noticeable is the extent to which the white population left northern slaves to their own devices. Whites watched—indeed, it is only through their gaze that we know of these events—but blacks were in control. In the case of Pinkster, the authority of King Charles was "absolute" and his will "law," even over whites who ventured up Pinkster Hill. Later in the nineteenth century, southern slaves created festivals, but the slave owners, ever sensitive to the precariousness of their situation, were more closely involved. In the corn shuckings, ubiquitous in the antebellum South, the planter was at the very center. He welcomed slaves from surrounding plantations, was an obvious presence during the shucking, and, at the end of the night on many plantations, was tossed in the air, carried around the big house, and had his hair and beard combed by the slaves. In the North, by contrast, the spotlight was very much on the slaves.[15]

There is another important difference. Whereas southern festivals such as Jonkonnu and corn shucking took place almost within the shadow of the plantation house, northern ones often occurred on marginal land on the edge of the cities. In Boston, Negro Election was held on the common; in Philadelphia, blacks gathered on a burying ground that later became Washington Square; in New York, Pinkster was celebrated at the market at Catharine Slip and, across the East River in Brooklyn, at the old market, until that was torn down in 1814. The use of a marketplace was particularly appropriate. In European tradition, the market operated as a sanctuary that attracted peoples of diverse cultures, and within its bounds authorities tolerated transgressions of normal laws.[16] In the smaller northern towns, the festivities typically took place outside of town. Election in Rhode Island occurred at "some suitable place near a country tavern"; at Portsmouth, New Hampshire, it was held on the nearby Portsmouth Plains; Windsor, Connecticut, blacks attended Negro Training in the countryside at Pickett's Tavern, about half a mile above Hayden's Station; and in Albany, New York, the Pinkster festivities occurred on Pinkster Hill. These grounds often resembled a European fairground or a marketplace. In 1736, "The Spy" (a pseudonymous newspaper contributor from an unidentified town, probably in the Hudson Valley) noted that the plain a little way out of town on which the festivities would occur "was partly covered with Booths," as was Pinkster Hill in Albany. The important difference was that the booths on Pinkster Hill were made of bushes and leafy branches, a practice with African origins.[17]

But for white observers, the most noticeable characteristic of these northern holidays and festivals was the extravagant style of the African American participants, an extravagance that many whites actively encouraged. Pinkster, Negro Election Day, and the Negro Trainings occurred in May or June in the brief lull after the crops were planted, when winter was well and truly over. Descriptions of Pinkster in Albany, in particular, emphasized the "verdant scenery" emerging on the "barren and dusty hills," and the way King Charles and many of the slaves covered themselves with Pinkster *blummies*, the flowers of the wild azaleas (which bloomed for the festival that bore their name).[18] It was a time for excess, for release from the rigors of a northern winter and the everyday exigencies of a slave regime, and the exuberance exhibited by the slaves—in their clothing, feasting, music, and dance—made the African American participants seem larger than life.

Clothing was an important medium for slaves' self-expression.[19] Most of the time slaves presented their bodies in a manner circumscribed by the need to wear work clothing, a lack of money, and the censorious eyes of their owners, particularly those who were Puritans. But in their free time, and particularly when they went to church on Sundays, slaves took pride in the way they looked. This was also the case at the festivals. In Salem, slaves were "attired in their best," and in Rhode Island, according to Wilkins Updike, they wore "cues, real or false," sported cocked hats on top of their "pomatuned and powdered" heads, and rode their owners' best Narragansett pacers. In both Salem and Rhode Island, slaves often wore their masters' swords, again suggesting the complicity of whites in the festivals. Indeed, according to Updike, "it was degrading to the reputation of the owner if his slave appeared in inferior apparel." Cyrus Bruce, the slave of Governor John Langdon in Portsmouth, appeared at Election flamboyantly dressed, wearing a massive gold chain, cherry-colored smallclothes, silk stockings, ruffles, and silver shoe buckles. Observers often commented on the appearance of the black kings and governors at Election, but few managed to cut a figure quite as striking as King Charles. His Pinkster ceremonial garb consisted of a British brigadier's broadcloth scarlet jacket, covered in bright gold lace and reaching almost to his heels, fresh and new yellow buckskin smallclothes, blue stockings, highly burnished silver buckles on well-blackened shoes, and a three-cornered cocked hat trimmed with gold lace. His carefully constructed, and to whites outrageous, appearance was an act of cultural bricolage, the imaginative mediation of an African-born slave in a new, European-dominated environment. As such, it aptly captured the timbre of these northern festivals.[20]

On these festive occasions, quantities of food and drink were sold and consumed. In many places in New England, election cakes, described by Earle as "a sort of rusk rich with fruit and wine," were made and sold. Pinkster and Election Day were particularly noted for gingerbread. Sidney Perley, writing a history of Salem in the 1920s, noted that although some "old-fashioned" families still made election cake, "the gingerbread of the ancient days has long ago disappeared from refreshment booths." Early in May, the herb gatherers brought barks and roots into Boston for the brewing of election beer, which was sold on the streets during Election Week. There was a commercial aspect to the festivals. Dick and Nance, slaves from Bedminster, New Jersey, attended General Training in the late 1790s in a covered wagon, which contained a barrel of root beer and a corn basket full of large round ginger cakes. At two cents a glass of beer and a penny a piece of cake, they made enough "pocket-money" to last them several months. In the week before Pinkster, Long Island blacks sold sassafras and swingled tow (cleaned flax) in New York City to raise money for their celebrations. William Pynchon, of Salem, noted in his diary in May 1788 that he had gone "to election at Primus's flag," had taken ale and pies, and had watched the dances. Two days later he recorded that "Titus and Primus and attendants are getting money apace."[21]

But it was not just whites who partook of the food and drink on such occasions. According to an anonymous poet writing of Boston around 1760:

> The blacks their forces summon.
> Tables & benches, chairs & stools
> Rum-bottles, Gingerbread & bowls
> Are lug'd into the common.

During these festivals, black slaves, freed from the rigors of their usual existence, ate and drank, often to excess. In towns where a king or a governor was elected, it was incumbent on the victor to treat the assembled blacks to a feast, a not inconsiderable expense that apparently was frequently met by the slave's owner. In Norwich, Connecticut, owners "liberally" supplied "provisions, decorations, fruits and liquors," for their slaves' enjoyment. According to an oft-told story, E. R. Potter, a state and federal legislator from Narragansett, informed his slave, Governor John, that "one or the other must give up politics, or the expense would ruin them both."[22]

Through their music and dancing, more than through the structure of the festivals, northern slaves conclusively revealed their distinctive culture. Music was provided by the fiddle, ubiquitous in eighteenth-century African American life, the Jew's harp, the banjo, drums, and even fish horns. The Guinea drum used at Pinkster was made from a log, four feet long, about fourteen inches in diameter, burnt out at one end and covered with a tightly drawn sheepskin. On this, an anonymous spectator observed disdainfully, the musician "thump[ed] with his fists a kind of barbarous ill composed or uncomposed air," accompanied by a "hoggish sort of grunting, a bawling and mumbling." The improvisatory nature of the slaves' music, their willingness to incorporate what Europeans considered novel sounds, was neatly suggested by the fact that in New York whistling was called "Negro Pinckster Music" long after the festival itself had died.[23]

As the slaves were caught up in the performance, their behavior became more African. This was most apparent at Pinkster. George Roberts, historian of Schenectady, New York, in the Hudson Valley near Albany, wrote that at Pinkster slaves "were granted unusual liberty to enjoy themselves according to their own ideas," which had clearly been carried across the Atlantic Ocean from Africa. One unknown white has left a contemporaneous description of toto, or the Guinea dance, on Pinkster Hill. The chief musician, "dressed in a horrid manner [and] rolling his eyes and tossing his head with an air of savage wildness," banged away on his drum, while on either side of him an "imp," adorned with "feathers and cows tails" matched the leader's actions on smaller drums, creating "sounds of frightful dissonance." The dancing was as alien to white eyes as the music was to their ears. It was not just the "lewd and indecent" gesticulations of the revelers and the climax of the Guinea dance, when the participants met in an embrace "which must cover even a harlot with blushes to describe," that so fascinated and repelled white observers; it was also the absence of "regular movements" and the performers' seeming loss of control. Blacks joined in the dance, each "as the spirit moved him or her to do so," surrounding the musician with "twisting, wriggling histerical slaves who, for the time, were thousands of miles away in the heart of superstitious Africa." They often danced themselves into exhaustion, and it was "not unusual for this wild dance to continue through two days."[24]

The Guinea dance provides the best documented and clearest example of the enormous impact of African culture on the music and dance of northern slaves. Elsewhere, descriptions lack detail, merely suggesting the magnetic attraction black dances held for participants. As we have already seen, the Reverend Mr. Bentley's servant girl put up with a lot, including the condescension of her owner, to participate at Election time in "the most fatiguing dances" and to lose herself, if only momentarily, in "the never ceasing sound of the violin." It was always difficult to control the behavior of northern slaves, particularly in urban areas, but many owners made special efforts to prevent their slaves regularly dancing the night away. In 1805, the Schenectady owners of Yat negotiated a contract with him that specified that, except on holidays—and Yat enjoyed three days at Christmas, two at New Year's, two at Easter, and three at Pinkster —he was not to "go a Fidling" without their specific permission.[25] Little wonder, then, that northern slaves, freed for the duration of these festivals from petty restrictions and the interference of their seemingly ever present owners, sometimes danced to exhaustion.

In Boston and New York City, and probably in Philadelphia, African American festivals were diffuse affairs, but elsewhere in New England and in Albany both whites and blacks focused much of their attention on the king or governor. A significant proportion of surviving northern black folklore concerns these men. Earle, aware of the interest in such material, centered two of her stories in *In Old Narrangansett* on a fictional black governor.[26] Her Cuddymonk is a feckless stereotype; the real kings and governors, even when viewed through condescending white eyes, were more impressive figures: often African-born, usually owned by prominent whites, and (even if in their youth) typically physically strong.

Albany's King Charles was such a larger-than-life figure. Charles was enslaved in Angola and brought to New York as a very young boy in the first half of the eighteenth century. He was owned by Volckert P. Douw of Wolvenhoeck, who was recorder and mayor of Albany for many years, as well as commissioner of Indian affairs and vice-president of the first provincial congress. It is not known when or how Charles assumed his position—I suspect it was soon after the

Revolution and that the prestige of his owner and the rumor that Charles had "noble" blood in his veins had something to do with it—but he is the only king of Pinkster Hill known to us. Stories accreted to his person. According to one tale, at a conference with the Indians at the Douw mansion a mile and a half below Albany, General Philip Schuyler offered to wager a large sum of money that his horse could beat Sturgeon, Volckert Douw's famous racehorse. It was midwinter, there was rain, and the ice was slushy, but the Indians and African Americans, under the supervision of Douw's overseer, cleared the ice and, holding lanterns, manned the course across and down the river. Sturgeon, ridden by Charles, easily won the race and the money. In 1857, an octogenarian reminisced in *Harper's Monthly* about Pinkster in the years just after the Revolution. King Charles, garbed in his full ceremonial outfit, had placed the young author on his shoulders and "leaped a bar more than five feet in height," a feat the more remarkable when it is considered that King Charles, who supposedly lived to be over one hundred twenty, must have been well past his biblical allotment of three score and ten years. Within an hour, King Charles was "gloriously drunk" on the free drinks proffered by observers of his athletic prowess. The author led him home at sunset.[27]

Athletic prowess was one thing—white Americans have usually, if grudgingly, managed to accept blacks' excelling at physical skills, which were the principal reason for their presence in the New World—but the stories adhere to clearly circumscribed and unflattering patterns. Most of the *surviving* stories about these black monarchs are not celebratory; many have a decided edge. As one story had it, King Charles liked to boast of his bravery, but to his considerable annoyance, both whites and blacks around Albany "twitted" him for his cowardice, calling out "Saratoga" as he passed. During the Revolution, Charles, as body servant, had set out with his master to join the army at Saratoga. It was a moonlit night, and Charles, observing the wind rustling the Indian salt or sumac ("which, when ripe presents a red appearance"), cried out "*Heer, ik zag een vyant*" ("Master, I saw the enemy"). Whereupon he promptly wheeled his horse, dug his spurs in, and "rode in hot haste for home, proclaiming that his master had been captured, and he, after hard fighting, had escaped." Similar stories about New England kings and governors abounded. As a young boy in Wethersfield, London, a native of Africa and a black governor in Connecticut had mistaken a skunk for a "very pretty young puppy," but the smell soon convinced him otherwise. This story was memorialized for a time by citizens of Simsbury referring to a skunk as a "Wethersfield puppy."[28]

The selection of such stories that has survived (and perhaps even more noticeably Earle's portrait of the fictional Cuddymonk) reveals a white discomfort with the idea of African Americans as leaders, even of their own people, an inability to take black aspirations seriously. It also demonstrates how such attitudes had become inextricably tied up with the minstrel show, one of the most important cultural developments of the nineteenth century. The intermingling happened in two ways. First, many of the constituent elements of the minstrel show may have had their origins in white observations of northern black culture, particularly as displayed in the slave festivals and, later in the nineteenth century, in the cities. Consider, for example, the minstrel show's stump speech, one of the genre's central attractions and distinguishing characteristics, in which the star performer gave a disquisition, loaded with malapropisms and ludicrous verbosity, often on some contemporary issue. This recalls the *parmateering,* or parliamenteering, that preceded Election Day outside of Boston, as, sometimes for weeks, candidates for king or governor solicited votes. In Norwich, according to Frances Caulkins, "great electioneering prevailed, parties often ran high, stump harangues were made, and a vast deal of ceremony [was] expended in counting the votes, proclaiming the result, and inducting the candidate into office." As we shall see, these goings-on were not entirely serious; they were tinged with burlesque, as blacks lampooned whites. Minstrelsy's stump speech may have been a white parody of blacks who were already parodying whites, a pattern replicated in other cultural crossovers.[29]

Second, even if the speculation above is not correct, the minstrel show was such a ubiquitous and important institution that it shaped the way whites looked at blacks. That shaping was particularly important in the present case because most testimony used in this analysis of the northern black festivals was written long after the event. Even an artist's rendition of the governor's parade in Connecticut, the only view of these festivals known to me and one completed years after its demise, shows the influence of conventions governing the representation of African Americans in minstrel show posters and the like. Moreover, some stories told about the festivals could have come straight from a minstrel show script. For instance, at the election dinner at Hartford, Connecticut, the deputy governor, seated at the opposite end of the table from the governor, commented, "Mr. Gubnor, seems to me dere ort to be sumthin' said on dis 'casion." The governor replied, "Will Mr. Deputy say sumthin?" The deputy then stood up and began: "Tunder above de Hebens. Litnin' on de earth, Shake de tops of de trees. Table spread afore us, no eat a'yet, eat a-bimeby, for Christ's sake, Amen." Upon which the governor exclaimed, "Well done, well done, Mr. Deputy, I no idee you such able man in prayer." This form of demeaning humor existed long before the minstrel show, but it must have been strengthened and broadcast by the success of that institution. It is hardly surprising, then, that many of the stories about these festivals remembered by whites are of this kind.[30]

Whites who viewed African Americans electing kings and governors or taking part in Negro Trainings had little doubt about what they were watching. The words they most commonly applied to such events were "imitation," "mock," and "sham." Henry Stiles, writing about Negro Election Day in Windsor, Connecticut, attributed it to that "spirit of emulation and imitation which is peculiar to their race, and the monkey tribe." Orville Platt, one of the first to give serious historical consideration to the practice of electing Negro governors, cautiously agreed that the "negro is an imitator and a mimic." For most white observers, the festivals were not only exotic and entertaining but also at least slightly ludicrous; the spectacle helped cement their stereotype of African Americans. Caulkins described "Sam Hun'ton," elected to this "mock dignity" for many years in Norwich, as a "sham dignitary." She found it "amusing" to see him riding through town "puffing and swelling with pomposity"; the "Great

Moghul" never "assumed an air of more perfect self-importance than the negro Governor."[31]

Treating African Americans as little more than a source of public entertainment stigmatized them as a marginal group. In European culture, as Roger Abrahams has pointed out, the role of the performer is one of the few available to outsiders, because it does not challenge stereotypes. Many whites, slave owners in particular, encouraged the slaves in their celebratory activities, cheerfully donating food, drink, and money and lending clothing and horses, as well as giving them time off. For these whites, their slaves were like children—difficult and demanding, but also capable of being immensely entertaining and providing considerable pleasure. If a slave became a king or a governor, it reflected well on the owner, no matter in how deprecating a fashion he or she might choose to tell friends that news. On these festive occasions, drunkenness, which slave owners usually censured, was tolerated or even benevolently condoned. In Brooklyn, the morning after Pinkster usually saw twenty-five or thirty slaves brought up before Squire Nicolls on charges of disorderly conduct. According to Henry Stiles, the squire, knowing that Pinkster came only once a year and "appreciating the peculiar weakness of the Negro character," contented himself with fining them whatever money they had left in their pockets after their "spree."[32]

We may be relatively confident about white reactions, but working out what slave participants thought they were doing as they celebrated these festivals is another matter altogether. Penetrating the tangled skein of condescension and racism that envelops white accounts of these festivals—and virtually all the accounts are from whites—is fraught with difficulty. But it seems clear that blacks did not view these festivals as their owners did. That slaves elected prominent blacks, who were often African-born, and that on occasion black elected officials may have performed a governing or policing function within the local slave community (in a few cases even meting out corporal punishment to black wrongdoers) certainly suggests this.[33] There was a disjuncture between white and slave views of what was going on, and in this space, the festivals flourished.

Manifestly, these black festivals were not solemn occasions weighed down with ponderous ritual. Using the license granted them, slaves satirized, parodied, and poked fun at themselves and at whites. James Newhall pointed out that, at Election in Lynn, Massachusetts, "the masters did not interfere till the utmost verge of decency had been reached, good-naturedly submitting to the hard hits leveled against themselves, and possibly profiting a little by some shrewd allusion."[34] Perhaps the allusions were sometimes a little too shrewd for many whites to pick up. What onlookers regarded as the almost slapstick attempts of blacks to negotiate the customs of their masters, often signaled by whites' use of such words as "mimic" or "imitation," contained meanings that eluded them.

Consider, for example, Stiles's account of the Negro Trainings held near Windsor, Connecticut, just after the Revolution, "a source of no little amusement to whites, who often visited them to witness the evolutions and performances of their sable competitors." The central figure was General Ti, "the observed of all the observers," resplendent in an outfit mostly borrowed from his master, which included a fob watch. Stiles related a couple of anecdotes. Knowing General Ti could not tell time, "Esquire Bissell," a local white notable, asked him the time. Ti held out the watch and with "scornful dignity" told Bissell to "Look for yourself, genmen, by ______." The black troops then formed ranks and drilled in what onlookers thought a ludicrous fashion. But even through the miasma of Stiles's racist assumptions, it is not difficult to detect a strain of parody, the slaves were almost certainly sending up the white militia. Furthermore, Ti's response to Bissell is notable more for its verbal agility and for Ti's presumption in addressing the "squire" in this manner than for his inability to tell the time. But most interesting of all is General Ti's name. During the revolutionary war, Colonel Tye (short for Titus) was the noted African American leader of a band of Loyalist guerrillas who created havoc among the New Jersey Patriots. Although he was killed in 1780, even his white opponents long remembered him as a "brave and courageous man." Because Windsor blacks, too, probably remembered Colonel Tye's exploits, perhaps the name

"General Ti" suggested to the slaves that they too had a military tradition and reminded them of compatriots who, in an attempt to become free, had sided with the enemies of the Windsor slaves' Yankee masters. Naming the black troops' leader "General Ti" gave the proceedings an edge missed by white onlookers, who perceived little more in the General Training than farce and black inferiority.[35]

Slaves came to these festivals to drink, to gamble, to dance, to talk, to mingle among their own people, and in some cases to make money by selling food and drink. Permeating such gatherings was a sense of exhilaration at being free, if only for a few brief hours. White onlookers, in control of their own lives, might look disdainfully at what they saw as puerile sociability, but, in a society where rural slaves were dispersed and average slaveholdings minuscule, these face-to-face interactions between African Americans performed a crucial function. Whether the participants were elected king or drank themselves insensible while telling stories about their owners' foibles or gambled away their meager savings at paw-paw or just danced for hours with other young black men and women did not really matter. What was important about Negro Election Day, Pinkster, and General Training was that slaves were for a while in control of their lives and could transcend their normally relentless and humdrum existence.

Why, then, did these festivals, so valued a part of slave life, end? In part, the answer lies with the attitude of northern whites. From the outset, some whites were not enamored of the development of African American festivals. In Salem in the 1760s, an unsuccessful attempt was made "to prevent slaves especially on election days (so called) from wearing swords, beating drums and making use of powder." New England ministers, in particular, objected: Earle quoted one, Urian Oakes, as complaining that Election Day had become a time "to meet, to smoke, carouse, and swagger and dishonor God with the greater bravery." At first, the opposition to the festivals came only from a disgruntled minority, and many whites encouraged their slaves to take part. But by the nineteenth century, opposition had increased and had become subsumed within the more powerful currents of moral reform.[36]

What seemed particularly to annoy commentators was the involvement of whites. After noting that whites, as well as blacks and Indians, were in attendance at Election, the anonymous poet who described the festival in Boston about 1760 penned the following lines:

> There all day long they sit & drink,
> Swear, sing, play paupau, dance and stink

Another anonymous observer, this time of Pinkster in 1803, noted that "blacks and a certain class of whites" formed "a motley group of thousands" and presented "to the eye of the moral observer, a kind of chaos of sin and folly, of misery and fun." Later, he took a walk across Pinkster Hill to see the results of the dissipations of "depraved nature": "Here lies a beastly black and here lies a beastly white," he lamented, "sleeping or wallowing in the mud or dirt." In the early nineteenth century, the good burghers of Albany became increasingly concerned at the way Pinkster was growing and, it seemed, getting out of hand. In 1811, the common council passed an ordinance forbidding persons to "collect in numbers for the purpose of gambling or dancing, or any other amusements" or "to march or parade, with or without any kind of music" on "the days commonly called Pinkster," effectively ending the festival in the area. Authorities in New England moved at a slower pace, but by 1831 the official election day in Massachusetts was moved to the first Wednesday in January, hardly a propitious time for an outdoor festival.[37]

But probably more important than the attitude of whites was that of northern African Americans. As slavery gradually wound down in the North, the slave festivals no longer seemed as relevant. Also, as we have seen, the history of the festivals was sporadic. In Albany, in the early 1800s, Pinkster was flourishing—there were still slaves in the area, although the passage of the Gradual Manumission Act was beginning to take effect. But in 1811, when the Albany Common Council ended the holiday, there apparently was remarkably little concern among African Americans. Within a few years, Pinkster in Albany was little more than fodder for the reminiscences of old men. Slavery met its demise in New England long before it did in New

York and New Jersey, and the action of the Massachusetts authorities two decades later in ending Negro Election Day was of little moment. By that time, the newly freed African Americans were expending their considerable energies elsewhere.

* * *

In colonial times, the small black population of Salem—in 1764 there were only 117 slaves and free blacks in a population of 4,469—elected a governor in the last week of May, the festivities taking place on a clearing or plain adjacent to the Collins farm in nearby Danvers. Here Election survived the Revolution, but the practice of choosing a black governor appears to have died out around the turn of the century. In 1805, some free blacks in Salem formed an African Society, and a year later they celebrated that event with a march to Washington Hall, where they listened to a sermon. About thirty well-dressed black males, wearing the insignia of their brotherhood and marching to the accompaniment of music, proceeded through town. A parade of thirty might strike us as risible, but drama is hardly a function of size. The *Salem Gazette's* comment, that this was a "novel exhibition," and the paper's hasty reassurance that "we are informed that their characters are respectable," underlined the unprecedented action of free blacks in impinging on what whites had unthinkingly assumed to be their prerogative and their ceremonial space. It also hinted at an unease on the part of the town's whites, an unease that boded ill for any attempt by blacks to move from Salem's periphery to its center.[38]

The changes that occurred in Salem were repeated elsewhere in the North, particularly in the cities. The postrevolutionary period appeared to offer a brave new world for northern free blacks, and they were determined to ascertain the boundaries of their freshly won liberty. For recently freed blacks from the countryside, the easiest way to shuck off the stigma of the hated institution was to move: All across the North, ex-slaves headed for urban centers, further accentuating the urban bias in the black population distribution and contributing to substantial free African American communities, most obviously in Boston, New Haven, New York, and Philadelphia.[39] The urban environment—and in the quarter century after the Revolution those cities were rapidly expanding—gave these blacks a chance, but it also brought them face-to-face with a white population that nourished a deep-seated and intensifying racism and resented any sign of African American achievement.

White distaste for the growing population of urban free blacks would have been readily apparent to any observer of the burgeoning number of civic ceremonies that dotted the northern calendar.[40] If the processions and parades of the 1790s and early decades of the nineteenth century were, to quote Clifford Geertz, "stories a people tell about themselves," then blacks were rarely part of the early-nineteenth-century American story. To be sure, in the half century after the Revolution, African Americans were sometimes integrated into the ritual occasions of northern cities.

Many inhabitants turned out to watch sundry blacks being escorted in imposing processions to the gallows. In Philadelphia in 1800, the black Masons were given a position at the rear of the parade marking George Washington's birthday. But the apogee of this tepid integrative impulse occurred during the war that began in 1812. In August and September 1814, parades of up to a thousand blacks marched through the streets of New York and Philadelphia on their way to help fortify those cities. These were splendid moments for the recently freed slaves, holding out to ordinary black men and women the possibility of acceptance into the life of their cities, but they were also just moments, chimeras that passed with the end of the crisis. Normally, African Americans were not included in the ceremonial events of the new republic. Parades and processions to mark Evacuation Day (celebrating the British retreat from New York City during the Revolution), Washington's Birthday, the Fourth of July, and special events such as the ratification of the Constitution in 1788 or the opening of the Erie Canal in 1825 could be viewed by blacks, but only whites could participate.[41]

Thrown back on their own meager resources, free blacks engaged, in the 1790s and the early decades of the nineteenth century, in a burst of associational activity. They formed African American societies, whose first actions almost invariably included holding a procession. For city dwellers, perhaps even more than for the inhabitants of smaller towns such as Salem, the processions were novel sights. In the colonial period, slave festivals usually took place on the outskirts of the city, but after the Revolution free blacks made their presence much more visible and organized funeral processions and celebratory marches that traversed major thoroughfares. In Philadelphia in 1797, the Quaker diarist Elizabeth Drinker witnessed two parades of Masons—one white, the other black—on the same day and noted with some surprise in her diary that "tis the first I have heard of negro masons—a late thing I guess." Drinker's surmise was correct; she had watched the first instance of what became an annual procession in her city.[42]

Drinker's quizzical reaction to the black Masons presaged the response of other whites to the spectacle of African Americans parading through northern cities, although many would not exercise the same restraint as this upper-class matron. Two decades later, David Claypoole Johnston, scion of one of Philadelphia's first Quaker families, produced a caricature of the march of that city's black Masons. As Gary Nash has pointed out, it was probably the first of the torrent of racist prints in the 1820s and 1830s (including, most notably, Edward Clay's series, *Life in Philadelphia*) that skewered the perceived pretensions of African Americans. In New York in 1809, members of the African Society for Mutual Relief marked their first anniversary with a procession. They had prepared some "very handsome silk banners," including a full-length depiction of an African American with the motto, "Am I Not A Man And A Brother?" According to James McCune Smith, who had spoken with participants in the 1809 march, white "friends" attempted to dissuade the society from being so "foolhardy" as to march. That attempt was in vain, and the blacks went ahead with their celebration, "easily thrusting aside by their own force the small impediments that blocked their way."[43]

In the early nineteenth century, northern blacks were increasingly interested in asserting their right to the streets, a practice that, as Smith's account suggests, almost inevitably involved jostling for space with whites. Rumors, in New York in 1801, that a Madam Volunbrun was about to dispatch twenty slaves to the South caused a volatile crowd of several hundred blacks to gather outside her house. Force had to be used to disperse the blacks, twenty-three of whom were arrested and sentenced to sixty days. In Philadelphia on July 4, 1804, about a hundred blacks armed with clubs and swords and apparently organized in semimilitary formation took control of the streets in the southwest part of the city for several hours and roughed up several whites. The following night, a larger group "committed similar if not greater excesses," but what was most unsettling for white Philadelphians was African Americans' "damning the whites and saying they would show them St. Domingo," thus raising the dreaded specter of the great West Indian rebellion. A year later, on July 4, 1805, the usual crowd that gathered on the square in front of Independence Hall to enjoy Philadelphia's festivities was disrupted when whites turned on the free blacks present and drove them from the square with a mixture of pushing, shoving, and abuse. This was probably a response to the events of the previous year, but the import of defining a certain space as for "whites only" was not lost on anyone. (Most recently, it has provided John Edgar Wideman with richly symbolic material for his novel *Philadelphia Fire.*) How whites and an increasing free black population in northern cities would accommodate to one another and what limits whites would impose on African Americans as they attempted to establish their own right to public urban space were issues worked out piecemeal, through a pattern of black challenge and white response. That process is most clearly revealed in the controversies surrounding the marches celebrating the abolition of the slave trade in Boston and the abolition of slavery in New York City.[44]

On Thursday, July 14, 1808, about two hundred of Boston's "African and colored people" celebrated the "late *Abolition of the Slave Trade,* by governments of the United States of *Great Britain* and *Denmark,*" by marching from Elliot Street to the African Meeting House on Belknap Street, where they participated in a service. July 14 quickly established itself as a major event in the calendar of African Bostonians, with the march, rather than the service in the

African church (invariably featuring a prominent white minister), becoming its major focus. In 1816, for instance, blacks were told to gather at 9 a.m. for the march, though they were not due at the African church until noon.[45]

Negro Election with African Americans dancing and drinking was one thing, but the sight of a well-organized group of free blacks seeming to flaunt their recently won freedom by parading all over town was another. Jokes and stories about what quickly became known as "Bobalition" circulated, and beginning in the late 1810s, broadsides savagely caricaturing the march were plastered all over Boston in advance of the event. At least once, however, a broadside disputed the accuracy and fairness of this derogatory image. A "Reply to Bobalition" conforms to the genre of Bobalition" posters, using black "dialect" and malapropisms (though more sparingly than the hostile broadsides). But it also criticizes the behavior of white Bostonians toward the city's African American celebration. Thus, Scipio Smilax tells Mungo Meanwell, "De folks no let use alone to joy ourself peaceably, but dey muss do juss as dey do for number year pass—dey disrepresent all our proceedum, and publish such a set of tose as de old boy heself wouldent be consider de autor of." The African American duo highlighted the hypocrisy of white Bostonians, who would not have tolerated any interference with their July 4 festivities, asked why newspapers virtually ignored the celebration of July 14, and gave a scathing account of whites' pretentious behavior at the ceremonial opening of the Mill Dam.[46]

The "Reply to Bobalition" may have caused a few people to consider what they were doing, but the anonymous pamphleteer had little impact on most Bostonians. They were left with an image of the city's blacks as less than human, as objects of derision, and, most important, as a group who were getting ideas beyond their station. The result was probably inevitable. While Sol Smith and the "Republican boys" of Boston may have been willing to let blacks put on a spectacle on Election Day and may even have joined the appreciative audience, they tried to shut down any other African American public activities. By the later 1810s and early 1820s, increasingly hostile crowds were turning out to watch and to harass the marchers verbally and physically, although at least once, Boston blacks fought back, and a running brawl from the Common to Belknap Street resulted.[47]

In the end, what is most impressive about the July 14 celebration is that each year two or three hundred black Boston males were prepared to assemble and, quite literally, to run the gauntlet of white Boston. Twenty-five years later, a writer in the *Liberator* remembered graphically that the marches in the early 1820s "were followed by the rabble; hissed, hooted and groaned at every turn, and one would suppose that Bedlam had broken loose." And there was the ever present threat that the spectacle would erupt into serious violence.[48] Their liberty had been hard won, and these African Americans were determined to commemorate the abolition of the slave trade and their own freedom, even though an intolerant Boston would neither allow them to do so peaceably nor acknowledge their right to participate in the wider political discourse.

Much the same pattern occurred in New York nearly a decade later. The end of the slave trade in 1808 was celebrated in New York City as well as in Boston, but these marches were very quickly discontinued. Such events probably seemed inappropriate to free blacks while so many of their compatriots remained enslaved. In 1817, the legislature passed a law that would free the remaining New York slaves a decade later, and as 1817 approached there were some celebrations. On October 3, 1825, Bernhard, duke of Saxe-Weimar, observed a procession organized by the Wilberforce Benevolent Society to mark the "anniversary of the abolition of slavery in New York." According to the German visitor, "during a quarter of an hour, scarcely any but black faces were to be seen in Broadway." That evening a dinner and a ball were held.[49]

In 1827, the public demeanor of African New Yorkers became a matter of general debate. Because of the establishment, in New York in 1827, of *Freedom's Journal*, the first African American newspaper in the United States, we not only can follow the controversy in detail but also, for the first time, can hear unmediated and distinct black voices. Ten years earlier, the legislature, prompted by Governor Daniel D. Tompkins, and with none too subtle symbolism, had picked July 4, 1827, as the date on which those New York slaves born before July 4, 1799, would be set free.[50] In fact, most of the slaves affected by the legislation had already negotiated their way out of bondage, but Independence Day 1827 was still a milestone for New York blacks, ending two centuries of slavery in the state. Yet Tompkins's choice of the Fourth of July posed some awkward questions for African Americans: Should they accept the all-too-easy assertion of a link between the founding principles of the Republic and their freedom, particularly given the persistence of the hated institution in the South? On a more practical level, how could blacks celebrate safely when much of the state's white population, hardly friendly to blacks at the best of times, was intoxicated and in a mood to be rowdy and boisterous, or worse?

Preparations for the celebrations began early. On March 27, Albany blacks held a meeting at the African Meeting House and agreed to celebrate publicly the abolition of slavery in the state's capital. Lewis Topp successfully moved that "whereas the 4th day of July is the day that the National Independence is recognized by the white citizens of this country, we deem it proper to celebrate the 5th." A month later, in New York, a "very large and respectable" gathering of African Americans meeting in the Mutual Relief Hall on Orange Street came to a different conclusion. They decided to celebrate, not the fifth, but the fourth day of July as "Jubilee of Emancipation From Domestic Slavery." For these more conservative African New Yorkers, the point of marking the Fourth of July was "to express our gratitude for the benefits conferred on us by the honorable Legislature," and to that end they agreed to "do no act that may have the least tendency to disorder" and therefore to "abstain from all processions in the public streets on that day."[51]

Not all New York City blacks agreed. By late June, a rival group had organized its own celebration, and *Freedom's Journal,* a staunch supporter of the decision not to parade, was forced to acknowledge that two separate events would take place in the city: "One party will celebrate the Fourth of July, without any public procession, and the other, the Fifth, with a Grand Procession, Oration and Public Dinner." This division among African Americans—which one correspondent of *Freedom's Journal* labeled "disgraceful"—aroused comment as to the propriety of a public display. The committee of arrangements for the celebration on July 4, the Mutual Relief Society, and the Asbury and Presbyterian churches condemned the decision to hold a procession on July 5, but this action had no effect.[52]

On July 4, 1827, a "large and respectable" gathering of blacks attended the Zion Church and listened to an oration by the black clergyman William Hamilton. The hall was decked out with banners from several of New York's black societies and portraits of white benefactors, including John Jay and Matthew Clarkson. Not surprisingly, *Freedom's Journal* was impressed by the "discriminating taste" of those who had decorated the church, by the fact that "no public parade added to the confusion of the day," indeed, by the whole occasion. The next day, according to the *New York Daily Advertiser,* between three and four thousand blacks (*Freedom's Journal* put the figure at "near two thousand") marched through "the principal streets, under their respective banners, with music and directed by a marshal on horseback," and assembled at the Zion Church, where they listened to an oration by John Mitchell. Even *Freedom's Journal* had to admit that "not withstanding the great concourse" watching the spectacle, "the day passed off without disturbance."[53]

The success of the parade on July 5, 1827, settled the issue for the following year. On July 5, 1828, blacks assembled by 9 a.m. for a march that would cover a good portion of New York City—the marchers were not due at the church until 2:30 p.m.—and the parade, rather than the church service, was clearly the centerpiece of the day's proceedings. *Freedom's Journal* conceded that the procession was "quite an imposing spectacle, and elicited approbation from the most prejudiced, wherever they appeared"; what aroused the black newspaper's ire in 1828 was a march in Brooklyn. Brooklyn blacks, having participated in the New York events, had held a parade of their own, "and a pretty large one too it was, extending over half a mile, as we are informed." As far as *Freedom's Journal* was concerned, with the single exception of crime, "nothing serves more to keep us in our present degraded condition, than these foolish exhibitions of ourselves." The newspaper went on to castigate blacks for spending money on costumes rather than on food, clothing, and fuel, expenditure that rendered them "complete and appropriate laughing stocks" to citizens. Furthermore, the writer had heard of leaders of the march whose condition left them "scarcely able to bear their standards," of "the insolence of certain Coloured females," and of "debasing excesses." All confirmed him in his conclusion "that nothing is more disgusting to the eyes of a reflecting man of colour than *one of these grand processions*, followed by the lower orders of society."[54]

White New Yorkers reacted to the July 5 festivities, particularly the march, with the same derision and violence that white Bostonians had shown toward the July 14 procession. In covering the festivities, the *New York American* could not resist a "diverting" anecdote mocking the pretensions of blacks. On Greenwich Street, a sudden downpour caused the band and marchers to scramble for shelter under the awnings and stoops of the shops. The marshal and his aides were left in the front "without a single follower," until the marshal's appeal—"For shame, gentlemen—for shame! you behave like boys! form, and move on!"—persuaded the marchers to reassemble. The newspaper's claim that in relating this incident, "we by no means wish to throw ridicule on the ceremony" was hardly convincing. Similarly, the cutting observations on the 1833 parade by two foreign observers were probably refractions of the views of their American acquaintances. James Boardman, an Englishman, thought the event a "half ludicrous procession," and Carl David Arfwedson, a Swede, wrote that "the whole was a perfect farce." New York also had its equivalents of Sol Smith and the Boston "Republican boys." Boardman noted that the "insulting behaviour of many of the coachmen and carters was unblushingly displayed in their driving their vehicles so as to interrupt the progress and order of the procession," even though there had been no provocation from the black New Yorkers.[55]

The petty harassments and derision of white New Yorkers and the ambivalent attitude of some of the African American elite could not diminish the enthusiasm of the marchers. The struggle to end slavery in New York had been longer and harder than that in any other northern state, and it had been sustained largely by ordinary black men and women. Now they had their chance to take to the streets and to demonstrate pride in their achievement. The annual parade became a popular event with black New Yorkers and remained so until 1834, when emancipation in the West Indies gave them the opportunity to rid themselves of Governor Tompkins's legacy. From that year, August 1 became the date of the major African American celebration in New York and Boston.

When, in 1828, Prince Abduhl Rahaman of Footah Jallo, who had been a slave in the South, visited Boston, the African American community held a public dinner for him at the African Masonic Hall. It was a fascinating occasion, particularly revealing of Boston blacks' attitudes toward Africa. The proposer of the opening toast expressed the hope that Africa would soon "stretch forth her hands unto God," that is to say, become Christianized. "May the sons and daughters of Africa," another speaker intoned, "soon become a civilized and christian-like people, and shine forth to the world as conspicuous as their more highly favoured neighbours." Other toasts looked to the ending of slavery. One offered support to Haiti, "the only country on earth where the man of color walks in all the plenitude of his rights." Boston blacks may have expressed the customary allegiance to "old Africa" the place of their grandparents' origin—but they seemed decidedly more at home in the "civilized" New World. Rahaman was a prince and a Muslim, but those who had gathered to honor him viewed him simply as an ex-slave from the South, whose family was still in captivity there.[56]

The occasion was replete with ironies. For over half a century, African Americans had sustained festivals centered on the election of kings, and earlier in the eighteenth century at least, the successful candidates had often claimed African royal blood. But by the time of Rahaman's visit in 1828, Negro Election Day in Massachusetts had become a shadow of its former self, an anachronism about to be put out of its agony by the legislature. It would be hard to imagine a function further removed from Election Day than that put on for the African prince. If celebrations at the election of a black king or governor were characterized by excess, exuberance, and a leveling spirit, that for the prince was a formal occasion of high seriousness. The male hierarchy of African American Boston was proudly on display. Toasts were given in order, "regular" ones first, followed by the "volunteer" toasts from the president, two vice-presidents, the marshals, and members of the Committee of Arrangements. Doubtless, seating was also by rank. Although the participants were black, and the toasts different in content, the occasion resembled nothing so much as the formal dinners white Bostonians held for visiting dignitaries. There was, however, one more layer of irony; wittingly or not, the spirit of the festivals did intrude, not at the dinner, but in the procession to it. At four o'clock, a parade—comprising numerous marshals, presidents and vice-presidents in charge of the arrangements, "young men," "elder citizens," and a band, as well as Prince Rahaman—set off from the African School for the African Masonic Hall. In this and other processions of the early nineteenth century, African Americans showed, often in subtle ways, that they were the heirs of the traditions of their slave parents and grandparents.[57]

In the nineteenth century, the African American parade rapidly developed into an important cultural form. It did have antecedents; in most, but not all, of the eighteenth-century festivals, processions were important. Portsmouth, New Hampshire, slaves "went up from town in procession" to Portsmouth Plains, the site of their Election Day festivities. Following their election of a black governor at a country tavern near Providence, Rhode Island, blacks would "march up and down the road." In Albany, Pinkster week began with King Charles leading a procession from State Street to Pinkster Hill. But increasingly, in the half century following the Revolution, one element of these festivals, the parade, replaced the festival itself as the primary vehicle for the public display of African American culture.[58]

This transformation both reflected and helped shape important changes in northern black life. Most obviously, by the 1830s virtually all blacks in the North were free, and, given their tendency to migrate to urban areas, the focus of African American life had shifted from the countryside and smaller towns to the major cities. Freedom weakened the apparent unity of slavery, a unity ritualized in events such as Pinkster and Negro Election Day, but it also broadened black horizons, fostering the development of black organizations and the emergence of a black elite. African Americans now spoke with many voices, as an event in Philadelphia in 1817 demonstrated. Many leading black citizens there were impressed by the colonization movement, and they called a mass meeting to consider it. Nearly three thousand blacks crammed into the Bethel Church on Sixth Street and listened to Absalom Jones, Richard Allen, and John Gloucester lauding colonization. But when James Forten, chair of the meeting, called for a vote, not a single person agreed with these prominent spokesmen for the black community. And when Forten asked those opposed to colonization to make their view known, the unanimous response seemed, he remembered, as if "it would bring down the walls of the building."[59] The leaders took heed of mass opinion—black Philadelphians became leading opponents of colonization—but this extraordinary meeting shows that "race leaders" concerned with the uplift of African Americans did not always have the support of ordinary black men and women.

In the years after the Revolution, the increasing concentration of blacks in the cities and the growing complexity of their lives brought new cultural tensions. Ever since many slaves had begun to secure their freedom at the end of the eighteenth century, prominent blacks had been convinced that, if they and their compatriots were to have any influence in the new republic, particularly on the crucial issue of slavery, they would have to lead exemplary lives. Even had the emerging black leadership wavered in their conviction, the point was continually emphasized by whites in the antislavery movement. In 1796, for example, the Convention of Deputies from the Abolition Societies in the United States had issued a widely distributed appeal to free Africans, imploring them to be religious, to refrain from spirituous liquors, to marry legally, and to behave in a civil and respectful manner. In these ways, blacks could refute some white objections to freeing slaves and promote the cause of the "general emancipation of such of your brethren as are yet in bondage."[60]

In this context, then, many black leaders can hardly have lamented the demise of Pinkster and Negro Election Day. The injunction of the 1796 convention to "avoid frolicking and amusements that lead to expense and idleness; [because] they beget habits of dissipation and vice," was echoed over the years by prominent blacks, notably the editor of *Freedom's Journal,* as they fulminated against African Americans whose visible high spirits might jeopardize black achievements and aspirations. But these men did not have things all their own way. In New Haven in the 1820s and 1830s, the leadership struggle between Amos Beman, pastor of the Temple Street Church, an abolitionist and organizer of the African Improvement Society, and William Lanson, who had been elected king in 1825 and owned a hotel that, according to Colonel Gardner Morse, operated as a "house of resort and entertainment for guests of his kind," neatly encapsulated the divergent outlooks in the free black communities.[61] Moreover, because there were significant continuities between the festivals and the parades (despite the efforts of African American reformers), the later became an important site on which cultural tensions within the African American community were worked out.

It is very difficult to find detailed descriptions of African American parades that would permit a full analysis of the genre and of the continuities and discontinuities between festival and parade. Newspapers were saturated with coverage of Fourth of July celebrations—dozens of accounts, including many from the most obscure parts of the United States, appeared every year—but only fleeting attention was paid to these important black rituals that often occurred at about the same time. Nevertheless, by piecing together the little material available, we can picture an African American parade in the first three decades of the nineteenth century.

The most immediately noticeable feature of the large public processions was the presence of marshals. Observing the Wilberforce Society's march in 1825, the duke of Saxe-Weimar noted that "marshals with long staves walked outside the procession." The July 5 marches followed much the same pattern, but the marshals were on horseback. In 1833, Carl Arfwedson described the marshals as "dressed in white, with epaulettes, sword and cocked hat." Most attention focused on the chief marshal, who rode at the front. In New York, Samuel Hardenburgh was usually appointed to this position. Forty years later, James McCune Smith could still remember him leading the first march on July 5, 1827. Hardenburgh, Smith recalled, was "a splendid looking black man, in cocked hat and drawn sword, mounted on a milk-white steed." In Boston in the late 1820s, Isaac Woodland, a grain inspector in his forties, often acted as chief marshal. The African American historian William C. Nell remembered that Woodland, too, had a commanding physical presence: "His towering and manly form was always the observed of all observers." These descriptions could just as easily have been applied to the African American king on Negro Election Day or Pinkster—Hardenburgh's horse was even the same color as King Charles's Pinkster steed.[62]

Other aspects of the marches also recalled the festivals. In June 1827, an editorial in the *New York Morning Chronicle* anticipated that this "Jubilee nonsense" would contribute to the increase of the city's "criminal calendar, pauper list and *dandy* register." The paper was right to emphasize the continued importance attached by the participants to clothing; other observers made the point in less acerbic fashion. In 1827, the *New York Daily Advertiser* thought the marchers "remarkably well dressed," and in 1833 James Boardman described them as "extremely well dressed" and looking striking in their "sashes and ribands." Smith remembered that in 1827 the various black societies—the New York African Society for Mutual Relief, the Wilberforce Benevolent Society, and the Clarkson Benevolent Society—were "splendidly dressed in scarfs of silk with gold edgings." In 1825, in the Wilberforce Society's own march celebrating the forthcoming abolition of slavery in New York members wore "ribands of several colours, and badges like the officers of free masons." The society's funds were carried in the procession in a "sky-blue box," and the treasurer held in his hand "a large gilt key." Such societies usually formed contingents, often at the front of the large processions, and carried banners advertising their presence. In 1827, the *New York American* commented that the numerous banners were "neatly executed, and appropriate to the occasion." The marchers also carried paintings of Daniel Tompkins, John Jay, and Wilberforce. There were in all this distinct echoes of the march in Albany at Pinkster, at whose front marchers bore a standard displaying "significant colours" and a portrait of King Charles, specifying the years of his reign.[63]

To white observers, more than the color of the marchers and the paraphernalia they carried distinguished black processions, whether they were large and public, involving several thousand people, or small and private. In Philadelphia in 1798, an African American funeral procession passing Elizabeth Drinker's door caught her eye; it was, she recorded in her diary, "in different order from any I have ever before seen." Six men, one carrying a book in his hand, went in front of the coffin and all the mourners were singing psalms in "a very loud and discordant voice." Drinker assumed that the marchers were Methodists. Just over three decades later, Boardman, the traveler from England, was also taken with the appearance of a black funeral procession in Philadelphia. After the hearse "about fifty blacks of both sexes, extremely well dressed, followed, walking two and two, at a slow pace. The effect to Europeans was striking."[64]

Specific customs, such as the pace of the marchers or their arrangement in twos, cumulated to create a parade that was different from those of white citizens. Something about the organization and appearance of black parades caught white onlookers' attention. Unfortunately, accounts of such marches, while often commenting on their strangeness, seldom include details. Nevertheless, we know that marchers in the Wilberforce Society's 1825 procession walked in pairs, and an 1819 caricature of the black Masons in Philadelphia depicted them in pairs. On the other hand, the great mass of blacks marching behind the societies in the July 5, 1827, procession in New York were, according to Smith, five or six abreast. African Americans walking in pairs may well have conjured up images of a coffin, in which slaves were usually chained in pairs, but the practice also harked back to the slave festivals. At Election Day in Hartford, if the troop of blacks did not have horses they marched "two and two on foot." As to the tempo of black marches, according to Frances Caulkins, in Norwich, Samuel Huntington, the black governor, and his procession moved "with a slow majestic pace, as if the universe was looking on." It is even possible that such customs had their origins in Africa.[65]

Some of the spirit of the festivals may have filtered through to the parades. Boardman, an Englishman and an outsider, having closely watched white New Yorkers in their July 4 celebrations, concluded that African Americans involved in the parade on July 5, particularly those wearing uniforms and dashing about on horseback, were "a parody upon the shopkeeper colonels of the previous day." There is support for Boardman's observation buried in the tirades against processions in *Freedom's Journal.* One of the editor's particular complaints was that blacks wasted money by buying "the cast off garments of some field officer, or the sash and horse trappings of some dragoon serjeant," in order to appear as "Generals, or Marshals, or Admirals," making them "complete and appropriate laughing stocks for thousands of our citizens, and to the more considerate of our brethren, objects of compassion or shame." The writer, who had little knowledge of, or interest in, the traditions of northern blacks, couched his plea for sober and staid behavior in generational terms. Aware that he would have no influence on "the stiff necks of 45 or 55," precisely the group most likely to remember earlier customs, he pinned his hopes on the "younger members of our Colour, from whose discretion and knowledge we expect more." This was clearly a cultural struggle, and one in which time was on the editor's side.[66]

Although some of the traditions of Pinkster, Negro Election Day, and General Training continued to animate the black parades, the continuities between the festivals and the parades are less significant than the differences. Inevitably, much of African American culture was lost when the multifaceted festival was reduced to the parade. The latter, typically organized and controlled by black societies or a committee of arrangements composed of prominent blacks, differed most obviously from the festivals, in their downgrading of music and dancing. To be sure, at least one band usually accompanied the marchers. Perhaps there was dancing at the formal dinners (the sources are silent on this point). But what was missing was the spectacular and wild dancing, to a distinctly African beat, that had entranced white spectators and been integral to Negro Election Day and Pinkster.

In the eighteenth and nineteenth centuries, dancing and music were the most important forms of self-expression available to ordinary black men and women. Regardless of the wishes of whites in the antislavery movement and of self-appointed black spokesmen, they were not going to disappear. With the demise of the festivals these crucial cultural forms became separated from public African American rituals, but they made a vigorous reappearance among the rapidly developing concentrations of free blacks in the cities. There had, of course, been a tradition of secular black music and dancing under slavery, but this was strengthened immensely as the institution began to crumble and more blacks gained control of their lives. Drinker and her circle had to cope with the attraction Philadelphia's night life held for their African American servants. In an 1806 entry in her diary, she recorded, "Sally much teas'd with their Negros, Harry and Dan going to dances &c." As Drinker had noted a few years earlier in 1799, "Times is much altered with the black folk." In New York City, music and dancing went underground—literally: Beginning in the 1790s and the early years of the nineteenth century, numerous dancing cellars sprang up, most noticeably in the area around Bancker Street. At these black-run establishments, thirty or forty African Americans would jam into a small, crowded room to eat, drink, gamble, meet members of the opposite sex, and, most important, dance. We can occasionally glimpse this vibrant world through court cases, but the most graphic account comes from the pen of Charles Dickens. In 1842, Dickens was taken by a guide to a stifling underground dancing hall in the Five Points, not far from Bancker Street. There he saw "the greatest dancer known" (William Henry Lane, or "Juba") perform.

> Single shuffle, double shuffle, cut and cross-cut; snapping his fingers, rolling his eyes, turning in his knees, presenting the backs of his legs in front, spinning about on his toes and heels like nothing but the man's fingers on the tambourine; dancing with two left legs, two right legs, two wooden legs, two wire legs, two spring legs—all sorts of legs and no legs—what is this to him?[67]

For the most part, this dancing was deliberately hidden from white New Yorkers, which helps explain why it is so hard for historians to find material on the world of which it was part and why everyone grasps so gratefully at Dickens's marvelous description.

There was also, however, a more public venue for black dancing in New York that helped satisfy continuing white curiosity. At Catharine Market, blacks held competitions in which individual dancers vied for the applause of whites and perhaps for money or "a bunch of eels or fish." Here such noted performers as Ned, Jack, and Bobolink Bob would do a jig or a breakdown, often while confined to a board about five or six feet long and of "large width." This tradition, which probably dated back to colonial times, seems to have waxed in the 1790s and the very early nineteenth century.[68] I have managed to turn up a drawing of a black dancing for eels at Catharine Market in 1820. Unlike the vicious caricatures of blacks created by well-known contemporaries such as Edward Clay, this image, by an unknown artist, appears to be a relatively straightforward attempt to represent an aspect of city life and black culture. The dancer is not confined to a board or shingle. A black off to the left is patting juba, or keeping time, which African Americans achieved by stamping their heels and beating their hands against their legs or by clapping. Dancers undoubtedly gained personal satisfaction from the performance, but the drawing also makes clear the extent to which blacks were objects for the gaze of white spectators. Regardless of what the myths may say about the southern roots of the minstrel show, the origins of parts of this most popular form of nineteenth-century entertainment are probably to be found in Catharine Market.

The other major cultural difference resulting from the increased importance of the parade relates to the position of women. Although women took part in funeral processions, African American parades were typically all-male affairs. According to James McCune Smith, writing about New York in 1827, "the side-walks were crowded with the wives, daughters, sisters and mothers of the celebrants," who ranged from "grown men to small boys." The women too were well dressed—Smith commented on the "gay bandanna handkerchiefs" betraying the West Indian birth of many, and the English traveler Boardman more archly referred to "sombre innamortas" in the "gossamer of Parisian modes" but they were reduced to being spectators. Although slave women had not assumed the prominent positions at Pinkster, Negro Election Day, or General Training, they had played an important role, particularly in the dancing. Such festivals had encompassed the entire slave community, including women and children. In effect, then, the transformation from festival to parade had narrowed the public role of African American women.[69]

It has not been my intention in this essay either to lament the passing of a "traditional" custom or to promulgate a whiggish version of African American progress and assimilation into a northern white world, although elements of both paradigms can be discerned easily enough in the transformation of slave festivals into the African American parade of the nineteenth century. Rather, this is a story of African American cultural adaptation and flexibility in the face of the changing circumstances of northern blacks.

The northern slave festivals had existed, had indeed flourished, on the ambiguous ground that lay between the white and African American worlds. Multifaceted and dynamic, Pinkster, Negro Election Day, and General Training had meant different things to different cultural constituencies. For whites, they were entertainment and confirmation of all their stereotyped views of the capabilities of blacks, views that, ironically, allowed the relatively small number of northern slaves a considerable latitude in celebrating their holidays. For the slaves, these festivals presented an opportunity to put on a performance that drew on their African past and that became the centerpiece of northern slave culture, a culture more complex than most white observers were prepared to allow. But the festivals were a product of slavery, and as that institution gradually wound down in the North, the festival yielded to the parade, a new and less ambiguous cultural form. Now that they were free citizens, it was, northern blacks believed, no longer so necessary to mask their feelings. This was their opportunity to participate in the discourse of the American republic. Although initially it contained important features that harked back to Pinkster and Negro Election Day, the parade clearly registered black leaders' attempts to shuck off customs that provoked white disdain. Dancing, which had entertained whites but had held overwhelming significance for African Americans, was banished from the realm of public ritual to the dancing cellars of the cities. The white-held stereotypes that had allowed blacks room to create the slave festivals were now an embarrassment. That African Americans would fail to dislodge these pervasive and entrenched ideas from the crabbed, inflexible minds of whites was hardly their fault.

Some authors have claimed that Negro Election Day, in particular, provided a framework for the development of black politics.[70] Perhaps so. But the African American parades of the early nineteenth century seem a much more important and conscious attempt by northern blacks to enter public life. Certainly, in attempting to break the stereotypes, northern urban blacks gave the parades a confrontational edge not present in the festivals. Some of the exuberance of these liberating times is captured in Smith's recollection, from the vantage point of the 1860s of the first parade to celebrate abolition in New York, a description whose exhilarating language reverberates down to us through the history of the civil rights movement: "That was a celebration!" Smith wrote, "A real, full-souled, full-voiced shouting for joy, and marching through the crowded streets, with feet jubilant to songs of freedom!" He continued,

> It was a proud day in the city of New York for our people, that 5th day of July 1827. It was a proud day for Samuel Hardenburgh, Grand Marshal, splendidly mounted, as he passed through the west gate of the Park, saluted the Mayor on the City Hall steps, and then took his way down Broadway to the Battery &c. It was a proud day for his Aids, in their dress and trappings; it was a proud day for the Societies and their officers; it was a proud day, never to be forgotten by young lads, who, like Henry Garnet, first felt themselves impelled along that grand procession of liberty; which through perils oft, and dangers oft, through the gloom of midnight, dark and seemingly hopeless, dark and seemingly rayless, but now, through God's blessing, opening up to the joyful light of day, is still "*marching on.*"[71]

Probably the most important aspect of the parades, then, concerns their function in the rapidly expanding metropolises of the new republic, their public representation of African Americans to themselves and to whites. The parades were attempts to foster unity among an increasingly disparate black population; they proclaimed to a skeptical and often hostile white audience that blacks were no longer slaves and that as American citizens they, too, had a right to the streets.

Notes

1. *New England Palladium & Commercial Advertiser*, June 3, 1817; *Boston Columbian Centinel*, June 4, 1817; *Boston Intelligencer*, June 7, 1817.

2. Greg Dening, *Mr. Bligh's Bad Language: Passion, Power, and Theatre on the Bounty* (New York, 1992), 121. I have been influenced by both Rhys Isaac, "On Explanation, Text, and Terrifying Power in Ethnographic History," *Yale Journal of Criticism* 6 (Spring 1993), 217-36; and, more strangely, a novel: Paul Auster, *Leviathan* (New York, 1992).

3. Alice Morse Earle, *Customs and Fashions in Old New England* (New York, 1893), 225-26. Paw-paw was a gambling game played with four seashells.

4. *Boston Columbian Centinel*, June 4, 1817; William Bentley, *The Diary of William Bentley, D.D.* (4 vols., Salem, 1914), vol. 4, 457.

5. For a helpful discussion of the problem of using white sources to study the African American past, see Roger D. Abrahams, *Singing the Master: The Emergence of African American Culture in the Plantation South* (New York, 1992), xv-xxvi. Writing from Australia, I have in mind the use of sources by the so-called Melbourne group (including Inga Clendinnen, Greg Dening, Rhys Isaac, and Donna Merwick). See, in particular, a wonderful work: Inga Clendinnen, *Aztecs: An Interpretation* (New York, 1991).

6. Ira Berlin, "Time, Space, and the Evolution of Afro-American Society on British Mainland North America," *American Historical Review* 85 (February 1980), 51-54. On Negro Election Day, see Orville H. Platt, "Negro Governors," *New Haven Colony Historical Society Papers* 6 (1900), 315-35; Hubert H. S. Aimes, "African Institutions in America," *Journal of American Folklore* 19 (January-March 1905), 15-32; Samuel E. Morison, "A Poem on Election Day in Massachusetts About 1760," *Transactions of the Colonial Society of Massachusetts* 18 (1917), 54-62; Lorenzo J. Greene, *The Negro in Colonial New England* (1942; New York, 1974), 249-55; Joseph P. Reidy, " 'Negro Election Day' & Black Community Life in New England, 1750-1860," *Marxist Perspectives* 1 (Fall 1978), 102-17; Melvin Wade, " 'Shining in Borrowed Plumage': Affirmation of Community in the Black Coronation Festivals of New England, ca. 1750-1850," in *Material Life in America, 1600-1860*, Robert Blair St. George, ed. (Boston, 1988), 171-82; and William D. Piersen, *Black Yankees: The Development of an Afro-American Subculture in Eighteenth-Century New England* (Amherst, 1988), 117-40. On Pinkster, see David Steven Cohen, "In Search of Carolus Africanus Rex: Afro-Dutch Folklore in New York and New Jersey," *Journal of the Afro-American Historical and Genealogical Society* 5 (Fall-Winter 1984), 147-68; A. J. Williams-Myers, "Pinkster Carnival: Africanisms in the Hudson River Valley," *Afro-Americans in the New York Life and History* 9 (January 1985), 7-17; and Shane White, *Somewhat More Independent: The End of Slavery in New York City, 1770-1810* (Athens, Ga., 1991), 95-106. Little has been written about African Americans and General Training, but more generally on General Training, see H. Telfer Mook, "Training Day in New England," *New England Quarterly* 11 (December 1938), 675-97. Recently, two outstanding studies of African American festivals in the South have been published. See Abrahams, *Singing the Master*; and Samuel Kinser, *Carnival, American Style: Mardi Gras at New Orleans and Mobile* (Chicago, 1990).

7. Morison, "Poem on Election Day in Massachusetts About 1760," 56-59.

8. Earle, *Customs and Fashions in Old New England*, 225-26; Sol Smith, *Theatrical Management in the West and South for Thirty Years* (1868; New York, 1968), 12.

9. On the differentiation between Boston and the rest of New England, see Earle, *Customs and Fashions in Old New England*, 227; *The Diaries of Benjamin Lynde and of Benjamin Lynde, Jr.* (Boston, 1880), 109; Joseph B. Felt, *Annals of Salem* (2 vols., Salem, Mass., 1849), vol. 2, 419-20; Sidney Perley, *The History of Salem, Massachusetts* (3 vols., Salem, 1926), vol. 2, 37; James Duncan Phillips, *Salem in the Eighteenth Century* (Boston, 1937), 272; *The Life of William J. Brown of Providence, R.I.* (Providence, R.I., 1883), 12-14; Henry R. Stiles, *The History of Ancient Windsor, Connecticut* (New York, 1859), 491. For a listing of twenty-one sites, see Piersen, *Black Yankees*, 175.

10. White, *Somewhat More Independent*, 95-106. For the celebration of the holiday by the Dutch in the seventeenth century, see Donna Merwick, *Possessing Albany, 1630-1710: The Dutch and English Experiences* (New York, 1990), 75. *New-York Weekly Journal*, March 7, 1736. I am grateful to Graham Hodges for this reference. Shane White, ed., "Pinkster in Albany, 1803: A Contemporary Description," *New York History* 70 (April 1989), 191-99.

11. Stiles, *History of Ancient Windsor*, 494-94; Jane DeForest Shelton, "The New England Negro: A Remnant," *Harper's New Monthly Magazine* 88 (March 1894), 537; Andrew D. Mellick, Jr., *The Story of an Old Farm: or, Life in New Jersey in the Eighteenth Century* (Somerville, N.J., 1889), 607; C. W. Larison, *Silvia Dubois, A Biografy of the Slav Who Whipt Her Mistres and Gand Her Fredom*, Jared C. Lobdell, ed. (1883; New York, 1988), 67.

12. John F. Watson, *Annals of Philadelphia and Pennsylvania, In the Olden Time; Being a Collection of Memoirs, Anecdotes, and Incidents of the City and Its Inhabitants* (2 vols., Philadelphia, 1845), vol. 1, 265. The historiography of African Americans in Philadelphia is excellent—largely due to Gary B. Nash's influence—but no one appears to have uncovered much more than Watson's comment. See Gary B. Nash, *Forging Freedom: The Formation of Philadelphia's Black Community, 1720-1840* (Cambridge, Mass., 1988); and Gary B. Nash and Jean R. Soderlund, *Freedom by Degrees: Emancipation in Pennsylvania and Its Aftermath* (New York, 1991).

13. Earle, *Customs and Fashions in Old New England*, 225-26; Morison, "Poem on Election Day in Massachusetts About 1760," 61. A *trull* was a prostitute. Thomas Williams Bicknell, *The History of the State of Rhode Island and Providence Plantations* (5 vols., New York, 1920), vol. 2, 485.

14. James Eights, "Pinkster Festivities in Albany Sixty Years Ago," in *Collections on the History of Albany, From Its Discovery to the Present Time*, Joel Munsell, ed. (4 vols., Albany, 1865-1871), vol. 2, 324; George S. Roberts, *Old Schenectady* (Schenectady, N.Y., n.d.), 288-90.

15. White, ed., "Pinkster in Albany, 1803," 197; Abrahams, *Singing the Master*, 3-21. See also Elizabeth A. Fenn, " 'A Perfect Equality Seemed to Reign': Slave Society and Jonkonnu," *North Carolina Historical Review* 65 (April 1988), 127-53.

16. Smith, *Theatrical Management*, 12; Watson, *Annals of Philadelphia*, vol. 1, 265; Thomas F. De Voe, *The Market Book: A History of the Public Markets of the City of New York* (1862; New York, 1970), 344; Henry Stiles, *A History of the City of Brooklyn* (2 vols., Brooklyn, 1869), vol. 2, 39-40. On the special nature of the marketplace, see M. M. Bakhtin, *Rabelais and His World*, Helene Iswolsky, trans. (Cambridge, Mass., 1968). For a particularly helpful account, see Roger D. Abrahams, "The Discovery of Marketplace Culture," *Intellectual History Newsletter* 10 (April 1988), 23-32.

17. *Life of William J. Brown*, 12-14; Charles W. Brewster, *Rambles About Portsmouth* (Portsmouth, N.H., 1859), 210; Stiles, *History of Ancient Windsor*, 492; "Albany Fifty Years Ago," *Harper's Monthly* 14 (March 1857), 453; *New-York Weekly Journal*, March 7, 1736; Whited, ed., "Pinkster in Albany, 1803," 196-97. The method of construction and probably the layout of the structures at Pinkster appear to be of African origin. See

Mechal Sobel, *The World They Made Together: Black and White Values in Eighteenth-Century Virginia* (Princeton, 1987), 71-73, especially 73.

18. White, ed., "Pinkster in Albany, 1803," 195; "A Glimpse of an Old Dutch Town," *Harper's Monthly Magazine* 63 (March 1881), 526. On the azaleas, see also Alice Morse Earle, *Colonial Days in Old New York* (New York, 1896), 200-1.

19. See White, *Somewhat More Independent,* 185-206; Shane White and Graham White, "Every Grain Is Standing for Itself: African-American Style in the Nineteenth and Twentieth Centuries," *Australian Cultural History* 13 (1994); and Jonathan Prude, "To Look Upon the 'Lower Sort': Runaway Ads and the Appearance of Unfree Laborers in America, 1750-1800," *Journal of American History* 78 (June 1991), 124-59.

20. Felt, *Annals of Salem,* vol. 2, 419-20; Wilkins Updike, *History of the Episcopal Church, in Narrangansett, Rhode-Island* (New York, 1847), 117-79; Thomas Bailey Aldrich, *An Old Town by the Sea* (Boston, 1893), 78; "Glimpse of an Old Dutch Town," 256.

21. Earle, *Customs and Fashions in Old New England,* 225-26; "Albany Fifty Years Ago," 453; Perley, *History of Salem,* vol. 2, 37; Mellick, *Story of An Old Farm,* 607; De Voe, *Market Book,* 344; Fitch Edward Oliver, ed., *The Diary of William Pynchon of Salem* (Boston, 1890), 308-9.

22. Morison, "Poem on Election Day in Massachusetts About 1760," 60; Frances Manwaring Caulkins, *History of Norwich, Connecticut, From Its Possession by the Indians to the Year 1866* (n.p., 1874), 330-31; Updike, *History of the Episcopal Church, in Narragansett,* 178-79.

23. White, ed., "Pinkster in Albany, 1803," 196; Gabriel Furman, *Antiquities of Long Island* (New York, 1875), 266. On slave music, see Dena J. Epstein, *Sinful Tunes and Spirituals: Black Folk Music to the Civil War* (Urbana, 1981); and Eileen Southern, *The Music of Black Americans: A History* (New York, 1983). On dance, see Lynne Fauley Emery, *Black Dance in the United States From 1619 to 1970* (Palo Alto, 1972); and Katrina Hazzard-Gordon, *Jookin': The Rise of Social Dance Formations in African-American Culture* (Philadelphia, 1990).

24. Roberts, *Old Schenectady,* 289-90; White, ed., "Pinkster in Albany, 1803," 198.

25. Bentley, *Diary of William Bentley,* vol. 4, 457. For the contract, see Percy M. Van Epps, "Slavery in Early Glenville, NY," in Percy M. Van Epps, *Contributions to the History of Glenville, New York* (Glenville, 1932), 101-2.

26. Alice Morse Earle, *In Old Narragansett: Romances and Realities* (New York, 1898), 79-101, 183-96.

27. "Glimpse of an Old Dutch Town," 525-26, 535-36; "Albany Fifty Years Ago," 453. On the supposed royal blood of King Charles, see William Henry Johnson, *Autobiography of Dr. William Henry Johnson* (Albany, 1900), 60-61. On his age, see "Theatrical Reminiscences," in *Collections on the History of Albany,* Munsell, ed., vol. 2, 56.

28. "Glimpse of an Old Dutch Town," 526; Sherman W. Adams, *The History of Ancient Wethersfield, Connecticut* (New York, 1904), 702.

29. The standard account is Robert C. Toll, *Blacking Up: The Minstrel Show in Nineteenth-Century America* (New York, 1974). A particularly useful tonic for those caught up in the usual American exceptionalist rhetoric is Richard Waterhouse, *From Minstrel Show to Vaudeville: The Australian Popular Stage, 1788-1914* (Kensington, 1990). The minstrel show was an international phenomenon. See also the particularly suggestive Eric Lott, *Love & Theft: Blackface Minstrelsy and the American Working Class* (New York, 1993). Toll, *Blacking Up,* 56; Updike, *History of the Episcopal Church in Narragansett,* 177-78; Caulkins, *History of Norwich,* 330-31. On cultural crossovers, see, for example, Brooke Baldwin, "The Cakewalk: A Study in Stereotype and Reality," *Journal of Social History* 15 (Winter 1981), 205-18.

30. Stiles, *History of Ancient Windsor*, 492.

31. Ibid., 491; Platt, "Negro Governors," 318; Caulkins, *History of Norwich*, 331.

32. Roger Abrahams, "Afro-American Cultural Patterns in Plantation Literature," in *The Man-of-Words in the West Indies: Performance and the Emergence of Creole Culture*, Roger Abrahams, ed. (Baltimore, 1983), 48. This chapter was particularly influential in writing this essay. Stiles, *History of the City of Brooklyn*, vol. 2, 40.

33. Other historians have emphasized the policing function of those chosen on Negro Election Day. See Reidy, " 'Negro Election Day' "; and Piersen, *Black Yankees*, 129-40.

34. James Newhall, *History of Lynn, Massachusetts, 1864-1890* (Lynn, Mass., 1890), 236.

35. Stiles, *History of Ancient Windsor*, 492-94; Graham Hodges, *African-Americans in Monmouth County During the Age of the American Revolution* (Lincroft, 1990), 15-23, especially 23; Gary B. Nash, *Race and Revolution* (Madison, 1990), 61.

36. Phillips, *Salem in the Eighteenth Century*, 272; Earle, *Customs and Fashions in Old New England*, 225-26. On these broader currents, see Paul G. Faler, *Mechanics and Manufacturers in the Early Industrial Revolution: Lynn, Massachusetts, 1780-1860* (Albany, 1981), 109-38, especially 127-30. See also the excellent study: David R. Roediger, *The Wages of Whiteness: Race and the Making of the American Working Class* (London, 1991), especially 101-4.

37. Morison, "Poem on Election Day in Massachusetts About 1760," 61; White, ed., "Pinkster in Albany, 1803," 199. The emphasis on the dirt and smell of the lower classes—both black and white—was not accidental. As Peter Stallybrass and Allon White have pointed out, in England the development of the bourgeois public sphere was closely linked to the desire to be distant from what, after M. M. Bakhtin, they label the grotesque bodies of the working class. See Peter Stallybrass and Allon White, *The Politics and Poetics of Transgression* (London, 1986), especially 94-96. For a pioneering and particularly suggestive account of space, sound, and smell in the cities of the early republic, see Dell Upton, "The City as Material Culture," in *The Art and Mystery of Historical Archaeology: Essays in Honor of James Deetz*, Anne Elizabeth Yentsch and Mary C. Beaudry, eds. (Boca Raton, 1992), 51-74. The Albany ordinance is reprinted in Cuyler Reynolds, *Albany Chronicles* (Albany, 1906), 409. Roediger, *Wages of Whiteness*, 103.

38. Phillips, *Salem in the Eighteenth Century*, 272; *Salem Gazette*, March 21, 1806.

39. White, *Somewhat More Independent*, 153-56.

40. On the civic ceremonies, see Alfred Young, "English Plebeian Culture and Eighteenth-Century American Radicalism," in *The Origins of Anglo-American Radicalism*, Margaret Jacob and James Jacob, eds. (London, 1984), 185-212, especially 200-4; Sean Wilentz, "Artisan Republican Festivals and the Rise of Class Conflict in New York City, 1788-1837," in *Working Class America: Essays on Labor, Community and American Society*, Michael H. Frisch and Daniel J. Walkowitz, eds. (Urbana, 1983), 37-77; Susan G. Davis, *Parades and Power: Street Theatre in Nineteenth-Century Philadelphia* (Philadelphia, 1986); Mary Ryan, "The American Parade: Representations of the Nineteenth-Century Social Order," in *The New Cultural History*, Lynn Hunt, ed. (Berkeley, 1989), 131-53; and Mary Ryan, *Women in Public: Between Banners and Ballots, 1825-1880* (Baltimore, 1990).

41. For the statement by Clifford Geertz, see Ryan, "American Parade," 132. For an account of whites turning out to see the execution of a black, see Henry Laurence, *Centinel of Freedom*, October 8, 1805. And see *Account of the Last Moments of the Unfortunate Sinclair and Johnson With the Particulars of Their Execution* (New York, 1811).

Johnson was an African American. Nash, *Forging Freedom*, 218; White, *Somewhat More Independent*, 150-51.

42. Leonard Curry, *The Free Black in Urban America, 1800-1850: The Shadow of the Dream* (Chicago, 1981), 196-215; Elaine Forman Crane, ed., *The Diary of Elizabeth Drinker* (3 vols., Boston, 1991), vol. 2, 935.

43. Nash, *Forging Freedom*, 254-55; *A Memorial Discourse by Reverend Henry Highland Garnet With an Introduction by James McCune Smith, M.D.* (Philadelphia, 1865), 20-21. See also Robert J. Swan, "John Teasman: African-American Educator and the Emergence of Community in Early Black New York City, 1787-1815," *Journal of the Early Republic* 12 (Fall 1992), 331-56.

44. White, *Somewhat More Independent*, 144-45; Timothy J. Gilfoyle, *City of Eros: New York City, Prostitution and the Commercialization of Sex, 1790-1920* (New York, 1992), 77; *Philadelphia's Freeman's Journal*, reprinted in *Albany Centinel*, July 20, 1804; Nash, *Forging Freedom*, 177; John Edgar Wideman, *Philadelphia Fire: A Novel* (New York, 1990). Historians examining the marches have typically been more interested in the orations than in the events themselves. In many ways, the best work remains that of Benjamin Quarles. See Benjamin Quarles, *Black Abolitionists* (New York, 1969) and Benjamin Quarles, "Antebellum Free Blacks and the Spirit of '76," *Journal of Negro History* 61 (July 1976), 229-42. See also Leonard I. Sweet, "The Fourth of July and Black Americans in the Nineteenth Century: Northern Leadership Opinion Within the Context of the Black Experience," *Journal of Negro History* 61 (July 1976), 256-75; William B. Gravely, "The Dialectic of Double-Consciousness in Black American Freedom Celebrations, 1808-1863," *Journal of Negro History* 61 (July 1976), 67, (Winter 1982), 302-17. For a broader, if slightly idiosyncratic, study, see William H. Wiggins, *O Freedom! Afro-American Emancipation Celebrations* (Knoxville, 1990).

45. *Boston Columbian Centinel*, July 16, 1808; *Boston Gazette*, July 15, 1816; *New England Palladium & Commercial Advertiser*, July 16, 1816.

46. For examples of the jokes, see *Boston Daily Advertiser*, July 17, 1817. "Grand Bobalition, or Great Annibersary Fussible," 1821, Broadside Collection, American Antiquarian Society (Worcester, Mass.); "Reply to Bobalition," 1821, Broadside Collection, American Antiquarian Society (Worcester, Mass.).

47. On "Signifyin(g)," see Henry Louis Gates, Jr., *The Signifying Monkey: A Theory of African-American Literary Criticism* (New York, 1988). On the hostile reception, see *Boston Daily Advertiser*, July 15, 1820 and *Boston Columbian Centinel*, July 13, 1822. For Lydia Maria Child's account of black retaliation, see William C. Nell, *The Colored Patriots of the American Revolution, With Sketches of Several Distinguished Colored Persons: To Which Is Added a Brief Survey of the Condition and Prospects of Colored Persons* (Boston, 1855), 26-27.

48. *Liberator*, August 13, 1847.

49. Bernhard, Duke of Saxe-Weimar, *Travels Through North America During the Years 1825 and 1826* (2 vols., Philadelphia, 1828), vol. 1, 133.

50. See White, *Somewhat More Independent*, 53-55.

51. *Freedom's Journal*, April 20, April 27, 1827.

52. Ibid., June 22, June 29, 1827.

53. Ibid., July 6, 1827; *New York Daily Advertiser*, reprinted in *Long Island Star*, July 12, 1827; *Freedom's Journal*, July 13, 1827. On the celebration of the event elsewhere in the state, see ibid.

54. See the notice of the route of the march in *Freedom's Journal*, July 4, 1828. Ibid., July 11, July 18, 1828. For a letter attacking *Freedom's Journal's* account of the Brooklyn march, see ibid., August 1, 1828.

55. *New York American,* July 10, 1827; James Boardman, *America and the Americans* (London, 1833), 309-11; Carl David Arfwedson, *The United States and Canada in 1832, 1833, and 1834* (1834; 2 vols., New York, 1969), vol. 2, 252-54.

56. On the dinner, see *Freedom's Journal,* September 6, October 24, 1828. See also Sterling Stuckey, *Slave Culture: Nationalist Theory and the Foundations of Black America* (New York, 1987), 119.

57. *Freedom's Journal,* October 24, 1828.

58. Brewster, *Rambles About Portsmouth,* 210-11; *Life of William J. Brown,* 12-14; White, ed., "Pinkster in Albany, 1803," 197. In addition to the traditions of parading among northern blacks, "respectable" blacks had in front of them the example of the respectable mechanics and tradesmen. The movement from the festival to the parade among African Americans paralleled the trajectory of white mechanic celebrations. In colonial Boston, for example, there was a tradition of Pope's Day activities and of crowds tarring and feathering those who incurred their contempt. But in the 1780s a distinctively new tradition of mechanics marching in processions by trade emerged. The camp meeting also appears to have been important to northern African Americans, although I have uncovered little material on it. For a report that over 3,700 "people of colour" came from New York City to Flushing for a camp meeting, see *Freedom's Journal,* September 12, 1828. This form of black gathering also suffered physical interference from whites. On a riot at a camp meeting near Gloucester, New Jersey, see ibid., August 29, 1828. On a camp meeting at Crosswicks near Trenton, New Jersey, and the "white trash around here" who broke up meetings near Rock Mills, see Larison, *Silvia Dubois,* Lobdell, ed., 96-97.

59. Nash, *Forging Freedom,* 237-38.

60. *American Minerva,* January 20, 1796.

61. Reidy, "Negro Election Day," 113; Colonel Gardner Moore, "Recollections of the Appearance of New Haven and of Its Business Enterprises and Movements in Real Estate Between 1825 and 1837," *Papers of the New Haven Colony Historical Society* 5 (1894), 98-99.

62. Bernhard, *Travels Through North America,* vol. 1, 133; Arfwedson, *United States and Canada in 1832, 1833, and 1834,* vol. 2, 252-54; *Memorial Discourse by Reverend Henry Highland Garnet,* 24-26; Nell, *Colored Patriots of the American Revolution,* 98-99.

63. *Morning Chronicle,* reprinted in *Freedom's Journal,* June 29, 1827; *Daily Advertiser,* reprinted in *Long Island Star,* July 12, 1827; Boardman, *America and the Americans,* 310; *Memorial Discourse by Reverend Highland Garnet,* 24-26; Bernhard, *Travels Through North America,* vol. 1, 133; *New York American,* July 10, 1827; White, ed., "Pinkster in Albany, 1803," 197.

64. Crane, ed., *Diary of Elizabeth Drinker,* vol. 2, 1043; Boardman, *America and the Americans,* 268.

65. Bernhard, *Travels Through North America,* vol. 1, 133; *Memorial Discourse by Reverend Henry Highland Garnet,* 24-26. On the "odious spectacle of 'the drove,' tied two and two," passing through Philadelphia, see Watson, *Annals of Philadelphia,* 265. Scaeva [I W. Stuart] *Hartford in the Olden Time: Its First Thirty Years* (Hartford, 1853), 38; Caulkins, *History of Norwich,* 331.

66. Boardman, *America and the Americans,* 310; *Freedom's Journal,* July 18, 1828.

67. Crane, ed., *Diary of Elizabeth Drinker,* vol. 3, 1906, vol. 2, 1127; White, *Somewhat More Independent,* 179; Charles Dickens, *American Notes and Pictures From Italy* (1842; New York, 1966), 90-91. This dancing house was run by Peter Williams. Ironically, after the visit of Charles Dickens, it became notorious as "Dickens's Place." For a description written in 1850, see George G. Foster, *New York by Gas-Light and Other Urban Sketches,* Stuart M. Blumin, ed. (Berkeley, 1991), 140-43.

68. De Voe, *Market Book*, 344-45.

69. *Memorial Discourse by Reverend Henry Highland Garnet*, 24-26; Boardman, *America and the Americans*, 310. On the position of African American women, see James Oliver Horton, *Free People of Color: Inside the African American Community* (Washington, D.C., 1993), 98-121.

70. See, for example, Berlin, "Time, Space, and the Evolution of Afro-American Society on British Mainland North America," 54.

71. *Memorial Discourse by Reverend Henry Highland Garnet*, 24-26.

3

Mapping the Terrain of Black Richmond

Elsa Barkley Brown
Gregg D. Kimball

he "city" stands at the intersection of several strands of current historiography. The new cultural history seeks the meaning of parades, ceremonies, and all manner of public ritual in the urban landscape, and the social geographer looks for the articulation of race, class, and gender in urban space. What is the importance of these trends in urban and cultural history for African American history? We explore this question by mapping the physical and social terrain of one southern industrial

Authors' Note: Our appreciation to Kenneth Goings for encouraging us to write this; Barbara Batson, Richard Love, and Nataki H. Goodall for helping us find the forest amidst all the trees: Susan Johnson, Robin D. G. Kelley, and Earl Lewis for providing helpful commentaries in incredibly quick time; David Watson for the graphics; and Jimeequa Harris and Teresa Roane for sharing work space, computer, and good humor. For Elsa Barkley Brown, the writing of this essay was facilitated by a research leave from the University of Michigan and fellowships at the W. E. B. Du Bois Institute for Afro-American Research at Harvard University, and the Virginia Center for the Humanities, Charlottesville, Virginia.

city: Richmond, Virginia. In doing so, we open up issues and debates specific to African American urban history and others that resonate throughout contemporary historical scholarship. For example, led by scholars in subaltern and cultural studies, there is a lively debate over questions of hegemony and resistance,[1] and recent works have begun to explore the ways in which historians might benefit from the poststructuralist emphasis on text and meaning while at the same time remaining focused on material reality.[2] And while much of African American urban history has emphasized external spatial relationships through segregation and ghettoization, more recently historians have focused on intracommunity relations, raising new questions about the dynamics of spatial relations among African Americans.[3] Our aim in this essay is not to provide a full discussion of these issues in Richmond, but to suggest the ways in which a closer reading of the spatial dimensions of the city may aid our exploration of the dynamics of power and culture, providing more nuanced ways of discerning the development of a discourse of class and gender among black Richmonders, and complicating our understanding of the changing racial discourse between black and white Richmonders. To do this, we look at the "black city," focusing on three areas: civic space and public ritual, conceptualizations of the city, and the moral dimensions of urban spaces.[4]

The Geography of Richmond

Visitors to Richmond from other east coast urban areas in 1860 would not have been surprised by what they saw in the general outline of the city. Richmond was a classic mid-Atlantic "walking city" on the eastern fall line, with neighborhoods clustered tightly around a central core of industrial and commercial activity (see Figure 3.1). Industries, large and small, hugged the riverfront, canal, and creeks, drawing water power from and moving raw materials and finished goods along these waterways. Industries not dependent on waterpower, such as tobacco manufacture, spread along the eastern end of Cary, Main, and Franklin Streets. In the center city,

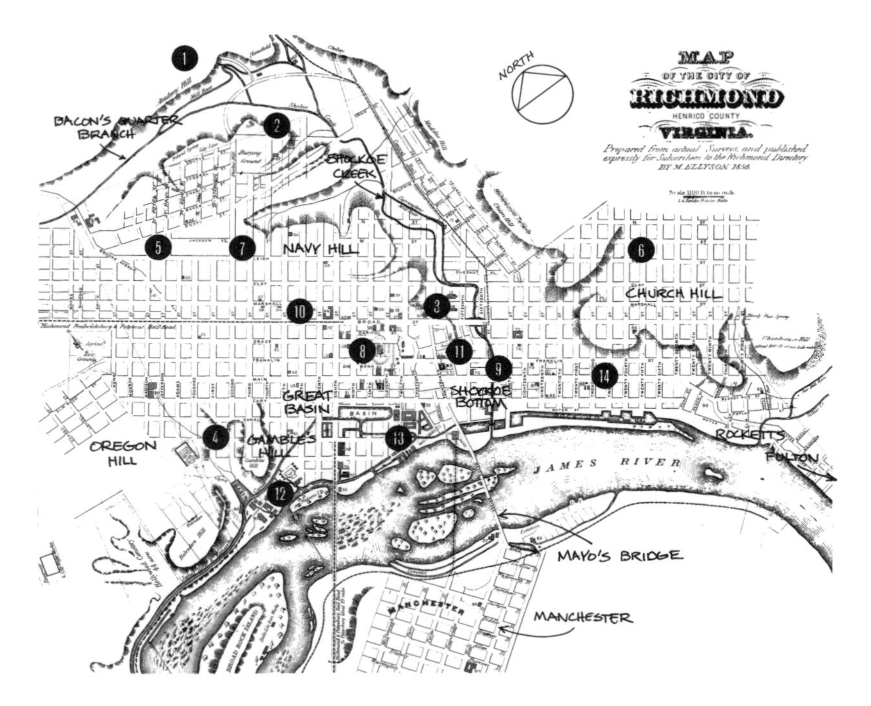

Figure 3.1. Richmond, c. 1860
SOURCE: M. Ellyson, *Map of the City of Richmond*, 1856; maps of various cemetery properties (City of Richmond, Department of Public Works, Bureau of Survey and Design, 1934).

List of sites
1. Free black subscription cemeteries
2. Public free black and slave cemeteries
3. First African Baptist Church
4. Second African Baptist Church
5. Third African (Ebenezer) Baptist Church
6. Fourth African Baptist Church (basement of white-owned Leigh Street Baptist Church)
7. Third Street Bethel A.M.E. Church
8. Capitol Square
9. First Market
10. Second Market
11. Center of the Slave Trade
12. Tredegar Iron Works
13. Warehouse and Flour Mill District
14. Tobacco Factory District

commerce and people mixed in Shockoe Valley, where Shockoe Creek originally meandered to the city warehouses and docks around 17th Street. Shockoe Creek passed within a few blocks of the old city market, of some of Richmond's largest hotels, domestic spaces, and auction houses, and of Wall Street, an extension of 15th Street between Franklin and Main, the center of Richmond's burgeoning domestic slave trade.

Moving west, visitors standing in Capitol Square could look down on the financial institutions of Main Street, liberally interspersed with dry goods stores and the shops of artisans, who often lived above their establishments. Just south stood the "great basin," a man-made lake in the middle of the city where canal boats turned around after their journeys from western Virginia on the James River and Kanawha Canal. The smokestacks of several industries could be seen to the west between the canal and the river, including the massive Tredegar Iron Works.

In antebellum Richmond, the differentiation both of public space by function and of neighborhoods by race was less evident than it would be in the early years of the next century. Although there were distinct clusters of both white immigrants and African Americans in certain areas of Richmond by 1860, nothing approximating a "segregated" neighborhood existed. A rare listing of "Free Colored Housekeepers" in the 1852 city directory suggests the distribution of free black men and women (see Figure 3.2).[5] Clusters of free black residents along Broad and Main Street most likely represent the shops of artisans who jostled for business with white native and immigrant shopkeepers. The other major concentrations of free black men and women were in the low-lying areas of Shockoe Valley and Bottom and in the northwestern region of the city, an area later known as Jackson Ward.

The hiring-out and living-out systems of antebellum Richmond meant that numerous slave men and some slave women boarded out in a variety of arrangements including "in boarding houses owned by white or free Black proprietors, rent[ing] small, shack-like houses behind white residents' homes, and stay[ing] with family members who were employed as domestic servants." Some lived with, and sometimes were related to, free black men and women. Other slaves lived in the outbuildings of businesses and factories, although few manufacturers provided housing. The residences in the wealthier sections of Richmond were equipped with numerous outbuildings, and slaves working as domestics, even those hired-out rather than owned, most likely lived in white households or in such outbuildings.[6] Within these shared living spaces, white owners and black slaves lived separate lives, developing their own distinct social worlds. But they also interacted in many public areas such as thea-

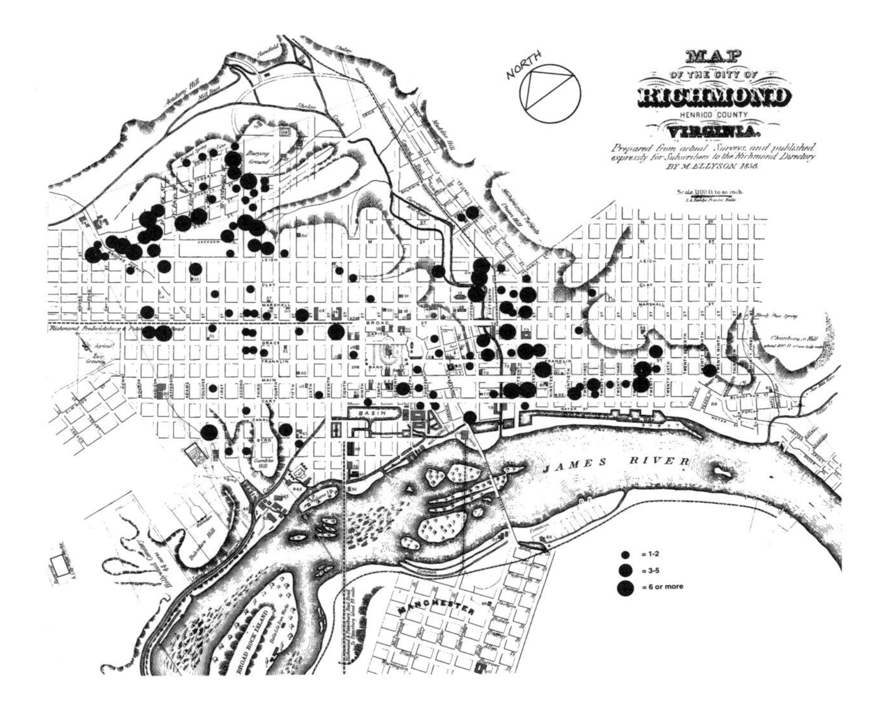

Figure 3.2. Free Black Population, 1852: Location of Approximately One Fifth of the Total Free Black Population in 1852

SOURCE: M. Ellyson, *Map of the City of Richmond*, 1856. Addresses listed in William L. Montague, *The Richmond Directory and Business Advertiser for 1852* (Richmond, 1852), section titled "Free Colored Housekeepers."

NOTE: smallest black dot = 1-2; middle black dot = 3-5; largest black dot = 6 or more.

ters, where black and white audiences were segregated, and churches, especially before the beginning of separate black religious institutions in the 1840s.[7]

By the late nineteenth and early twentieth centuries, Richmond developed a suburban periphery (Figure 3.3). Richmond tripled its land size, and at the same time segregation of commercial, financial, industrial, and residential areas increased, especially among the white middle class.[8] The 1888 introduction of the electric streetcar facilitated the development of white middle-class enclaves north and west of the city, among the more prestigious of which were Highland Park, Ginter Park, and Windsor Farms. These, as well as Manchester

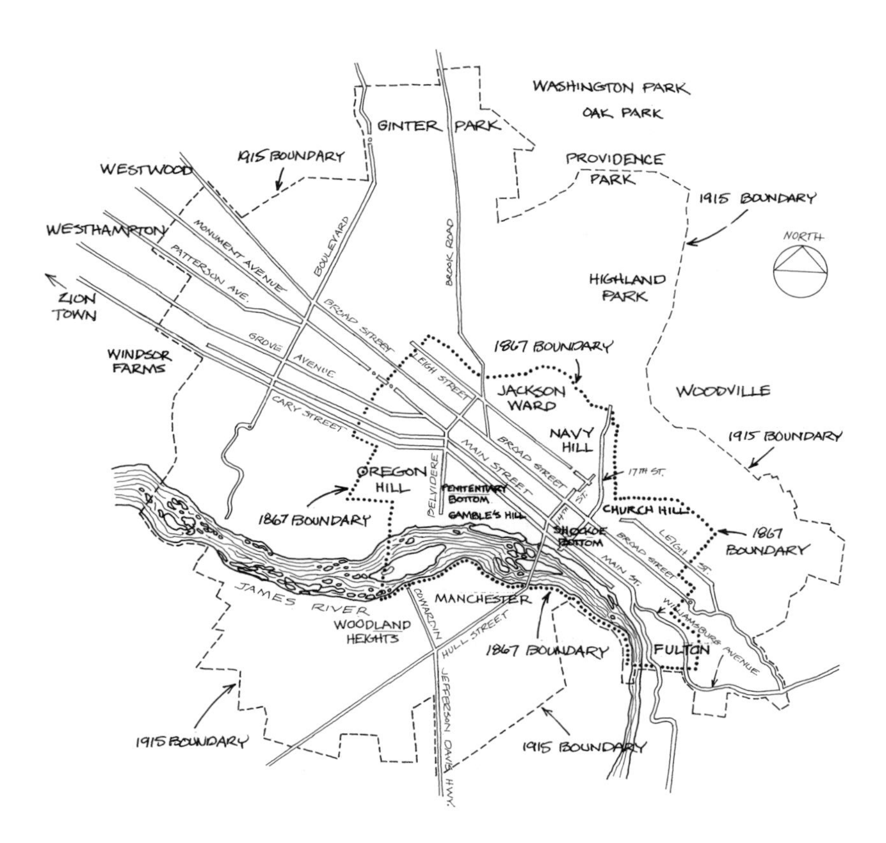

Figure 3.3. Richmond Neighborhoods and Territorial Expansion, 1967-1915
SOURCE: *Clarke's New Guide Map of Richmond and Suburbs* (E. C. Clarke, 1920); *Map Showing Territorial Growth of Richmond* (City of Richmond, Department of Public Works, Bureau of Survey and Design, July 1929, issue of 1931); Howard H. Hartan, *Zion Town—A Study in Human Ecology*, Publications of the University of Virginia Phelps-Stokes Fellowship Papers No. 13 (Charlottesville, 1935).

and adjacent suburbs south of the James River, such as Woodland Heights, were annexed into the city in the first two decades of the twentieth century. Although few black Richmonders could hope to share in suburban life, some black settlements outside the urban core did develop. One, Washington Park, was home to domestic workers employed in suburban white homes; another, Zion Town, was an area of land owned by black men and women emancipated after the Civil War that survived the encroachment of white suburbs. Dairy farm workers from King William and Hanover Counties developed Providence Park in the 1870s, continually moving between the employment and educational opportunities of the city and work in their native counties.[9] The outlines of the physical development of the black urban core followed a familiar storyline—the dispersed residential pattern, the continuing concentration in the central city of not only working-class and artisan black Richmonders but also business and professional people; the lack of basic city services—water, paved streets, street lights, adequate police protection, refuse collection—or of public recreational facilities—parks, playgrounds; and the dilapidated and overcrowded school buildings, one located directly across the street from the city jail.[10]

Civic Space and Public Ritual

Symbolic acts and public ceremony had deep meaning for North Americans in the nineteenth century. Social historians point to the reliance on public discourse, rhetoric, and ceremony in an age when literacy and mass communication were limited. Perhaps even more important, historians have begun to accept the idea that common people understood the complex meanings of political and artistic performances and events. Lawrence Levine, for example, has demonstrated that the denizens of the Bowery did not attend Shakespearean plays simply for their more bawdy or violent aspects; rather, they understood the subtle human drama drawn out in the tales of treachery, kinship, and flawed character. Susan G. Davis has focused on parades and other public festivities as symbolic presentations of the orders of society that in turn spawned "counter-parades" where those of different class, race, sex, and/or ethnicity revealed the inequities embedded in official ceremonies and publicly set forth their own ideas about history, politics, and social hierarchy. These challenges to authority also could become contested in intracommunity struggles over history, politics, and social order.[11] Tracing the history of celebrations and parades and one of their constitutive elements—militias—in the Richmond streets provides a venue for looking at black rights, citizenship rites, and ritualistic negotiations of manhood and womanhood.

John O'Brien has noted that in the immediate aftermath of emancipation, black Richmonders developed their own political calendar, celebrating four civic holidays: January 1, George Washington's birthday, April 3 (Emancipation Day), and July 4.[12] They thus inserted themselves in preexisting national political traditions and at the same time expanded those traditions. White Richmonders watched in horror as former slaves claimed civic holidays white residents believed to be their own historic possession, and as black residents occupied spaces, like Capitol Square, that formerly had been reserved for white citizens. Following black residents' July 4 parade and celebration in 1866, the *Richmond Dispatch* complained that Afro-Richmonders took "complete possession of the day and of the city. The highways, byways, Capitol Square, were black with moving masses of darkeys." Following Washington's birthday earlier that year, white Richmonders had announced they would prefer "the Twenty-second of February and the Fourth of July . . . be abolished in this part of the country hereafter" than to have such desecrated by the "disorderly, disgraceful, indecent, and contemptible set of beings" who had taken over the national holidays and even dared decorate the Capitol Square statues of Thomas Jefferson and George Mason with wreaths and small flags.[13]

The complicated nature of such contests for civic space is particularly evident in the practices of black militias. The Freedmen and Southern Society project editors have observed that "more than any other post-bellum figure, the black soldier represented the world turned upside down: the subversion of slavery, the destruction of the Confederacy, and the coming of a new social order that promised to differ profoundly from the old."[14] Throughout the late nineteenth century, first as self-defense units, then as official parts of state militia, and finally as ceremonial traditions, black men (and for a time women) took to the city streets in military style not only to claim civic space but also to challenge gendered exclusions within this arena of civic space.

By the summer of 1866, black men in Richmond had organized at least three voluntary militia units. These men marched, sabers drawn, in the April 3 emancipation celebration parade and the July 4 parade that year. The militias' nightly drills on city streets in preparation for these parades disturbed white Richmonders. This was especially true in the first decades of emancipation when such companies also served as self-defense units for Afro-Richmonders, guarding black residential areas and black schools against attacks, leading protests against segregated streetcars, rallying black voters and warning "would-be white aggressors against intimidating them." White Richmonders expressed grave concern about the black militia units, and white newspapers regularly questioned the ability of black men to maintain military discipline (and even to wear appropriate attire), doubting such units would ever be called into active service. These companies nevertheless received official recognition both in December 1866, from Reconstruction Governor Pierpont, and in 1876, when a conservative government approved the organization of the First Colored Battalion of the Virginia State Militia.[15] These governmental actions affirmed in part that the black militia fit into an existing ethnic tradition of pre-Civil War Irish and German units, which were as much ritualistic as they were militaristic. This tradition allowed participation in the black units to be officially recognized as part of men's political liberties.[16]

Yet even as they claimed that masculine tradition, black militias in Richmond, at least for a time, also challenged the notion that this part of the civic domain was an exclusively male preserve. By the late 1870s, black women had also organized a militia company. This women's militia was apparently for ceremonial purposes only because, reportedly, it was active only before and during emancipation celebrations. Its members conducted preparatory drills on Broad Street. Frank Anthony, the man who prepared and drilled the women's company, demanded military precision and observance of regular military commands.[17] Unlike militia men, who came from working-class, artisan, business, and professional backgrounds, the women were, no doubt, working class. Although they served no self-defense role, their drilling in Richmond streets and marching in parades challenged ideas and assumptions about appropriate public behavior held by both white southerners and white Unionists. Although the men's militias may have been acceptable in part because they fit into a masculine tradition, the women's unit not only challenged the idea of black subservience but also suggested wholly new forms and meanings of respectable female behavior. We do not know how long this women's unit survived or the causes of its demise. But we can speculate that in addition to horrifying white Richmonders, such a unit may have become unacceptable to a number of black Richmonders; increasingly, concerns about respectable behavior were connected to the public behavior of the working class and of women. This black women's militia, however, suggests the fluidity of gender in the early years of emancipation. For a brief time, these women declared that no area of political participation or public ceremony was strictly a male domain.

The effort to claim civic space by participating in Richmond's militia tradition had more than gendered problematics. In asserting their rights in the public domain, black militias demanded acceptance in the larger culture through biracial participation in public ceremony. However, many of the honorific occasions in Richmond that called for militia were events tied to the commemoration of the Confederacy. On one hand, to demand or accept inclusion in such was incongruent with many black Richmonders' own political traditions. On the other hand, the lack of recognition and exclusion from civic rites were also problematic to some militiamen. Thus, in October 1875, when Confederate General George E. Pickett was buried, black militia units joined white units in the procession, although the black men marched "without Arms." Later that month, when Stonewall Jackson's statue was unveiled in Capitol Square, black militias asked to participate and were assigned a position "in the rear of the whole, distinct from the white procession," despite objections from some ex-Confederates. However, newspaper reports following the event suggest that the black units did not participate. In October 1887, black Richmonders debated the initial decision of black militias to participate in the cornerstone-laying ceremony for the Robert E. Lee monument. The militia eventually declined to attend, although the reason is unclear—perhaps it was the members' own political opposition, community pressure, or, as reported, "the tardy invitation was an insult that did not allow them to practice their drills or clean their uniforms."[18]

By the end of the century, in an era of increasing disfranchisement and segregation justified in gendered as well as racial terms, Virginia's black militias lost government approval. During the Spanish-American War, after controversy over the appointment of white officers and the resistance of the Virginia Sixth (composed of the former militia units) to racist practices in Tennessee and Georgia camps, the governor disbanded Virginia's black military units without sending them to active duty.[19] Many black Richmonders perceived this as a denial not merely of black political rights but of manhood. Yet this official action did not remove uniformed black men from the city streets. The reuniforming of black men in military style in the city streets became a central concern of men such as John Mitchell, Jr., who promised that the January 1, 1899, emancipation celebration would "show up the largest quota of uniform men ever seen in Richmond on such an occasion." Minus a state-sponsored militia, the vehicle for such a display was now the Uniformed Rank of the Knights of Pythias, which divided into battalions, wore full dress regalia, rode as cavalry or marched with military precision, engaged in military-style parades and mock battles that were community-wide entertainments, and even developed a cadet program to train young boys.[20] All of these activities were intended to suggest the defensive preparation of black Richmonders and, by equating uniforms and military precision with respectability, to use the city streets to parade black manhood, thus reasserting black men's rights in the political arena. As African American men were denied what they considered their citizenship rights to military participation, these ritualistic signifiers—once only one component of a wider definition of manhood—became crucial. In the process, reuniforming through ceremonial drills, parades, and mock battles took on an intracommunity meaning that made manhood more a matter of status, of one's ability to purchase a uniform, and of one's claim to be of the "best" class. Yet these drills publicly proclaimed that the dissolution of the black militia could not be accomplished by the government only; their authority existed just as much within the black community.

We can also trace the contests among black Richmonders for physical, civic, and historical space through their celebrations, parades, and other public rituals. Parades—big and small, for funerals, society mass meetings, fraternal conventions, holidays, or other occasions—were a central feature of black life in the city throughout the late nineteenth and early twentieth centuries. The Emancipation Day parades of April 3, 1866, and April 3, 1868, both went through the main streets of the city (although exact routes are not known) and terminated at Capitol Square, a particularly symbolic space because antebellum law had defined this area off-limits to African Americans (see Figure 3.4).[21] Through this restriction, white Richmonders had asserted not only physical control but also psychological control of the symbols and mythology of the state, constructing African Americans as outside the polity. Through these parades, black Richmonders claimed the city as a whole, pronounced their rights to civic space, and also seized the power to define public memory, insisting that their version of the day's history become public history:

> NOTICE! THE COLOURED PEOPLE of the City of Richmond WOULD MOST RESPECTFULLY INFORM THE PUBLIC, THAT THEY DO NOT INTEND TO CELEBRATE THE FAILURE OF THE SOUTHERN CONFEDERACY, as it has been stated in the papers of this City, but simply as the day on which GOD was pleased to Liberate their long-oppressed race.[22]

Later emancipation celebrations traveled shorter routes, often more confined to black neighborhoods. More important, they became arenas for black Richmonders' struggles with each other. For example, when residents debated whether to celebrate January 1 or April 3, or in 1884, when some black Richmonders paraded on April 3 and others on April 20 to commemorate the ratification of the Fifteenth Amendment, the streets became the site of contests over differing conceptions of emancipation and freedom and over who would hold the power to define their history.[23]

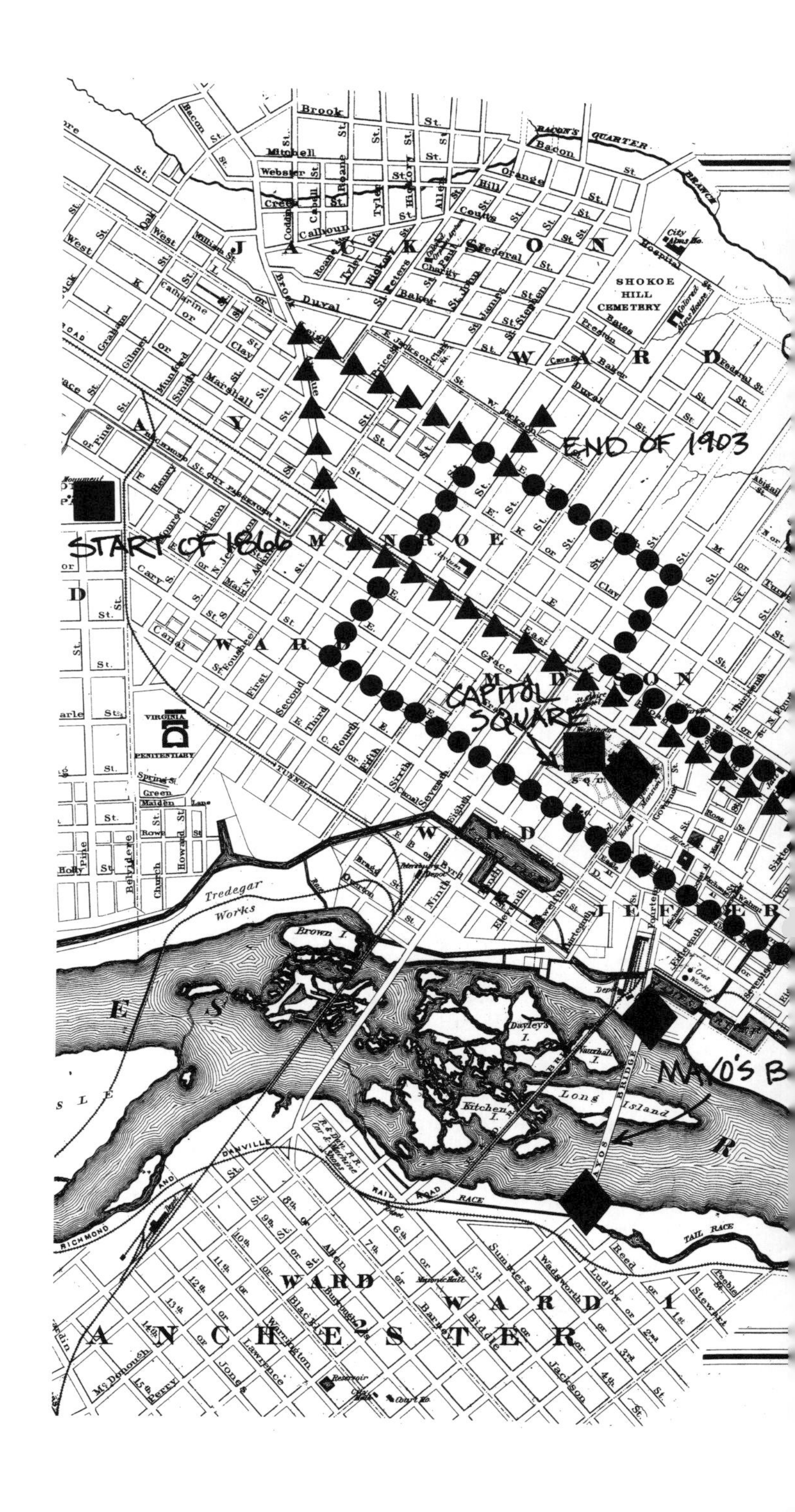
START OF 1866
END OF 1903
CAPITOL SQUARE
SHOKOE HILL CEMETERY
VIRGINIA PENITENTIARY
Tredegar Works
Brown I.
Long Island
Kitchen I.
Vauxhall I.
Dayley's I.
MANCHESTER

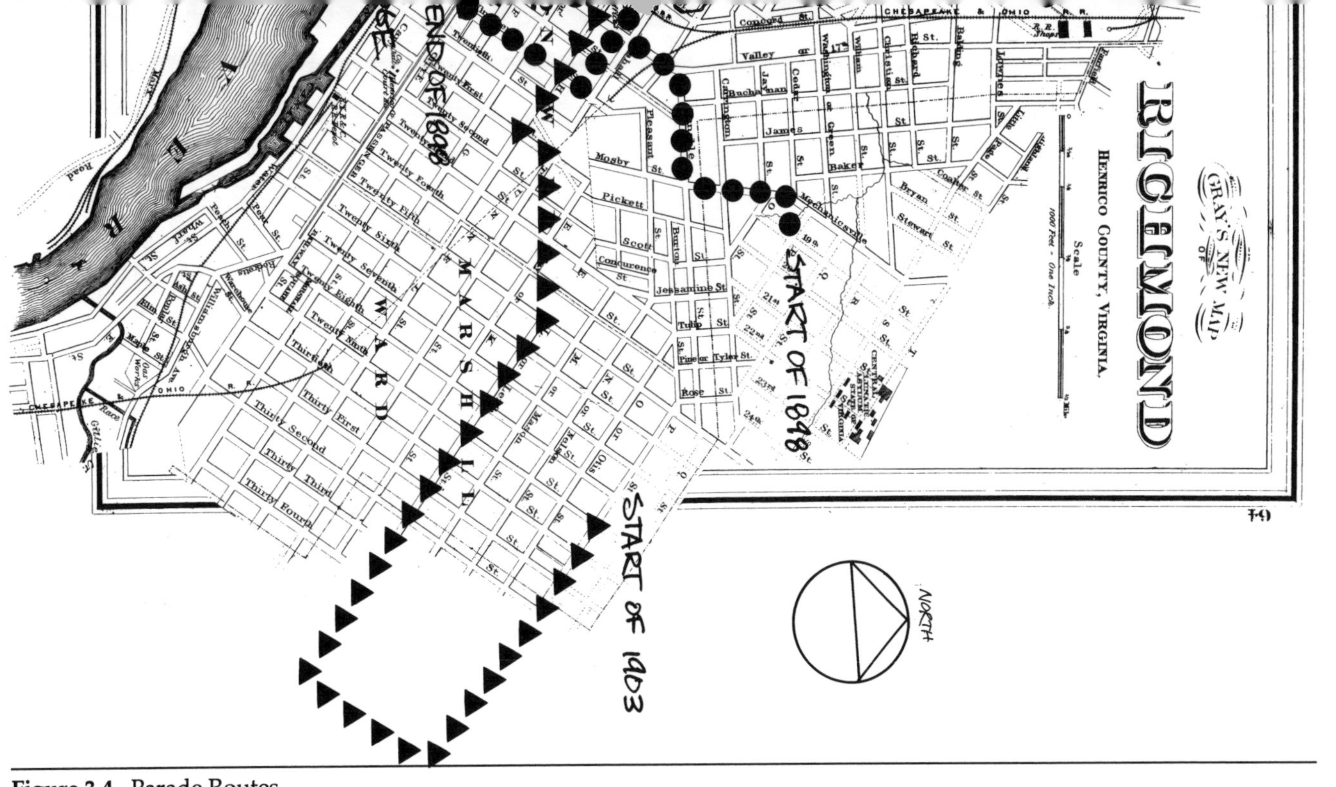

Figure 3.4. Parade Routes

SOURCE: *Gray's New Map of Richmond*, 1882; Richmond *Citizen*, April 4, 1866; *Richmond Republic*, April 9, 1866; *Richmond Dispatch*, April 9, 1866, April 7, 1868; *Richmond Planet*, September 17, 1898, September 19, 1903.

■ = Start and end point of Emancipation Day parade, 1866

▲▲▲ = Route of Knights of Pythias parade, 1903

◆ = Known points of Emancipation Day parade, 1868

●●● = Route of Lilies of the Valley parade, 1898

Although these explicitly civic rituals provide insight into intra- as well as interracial political discourse, much can also be gained from mapping more routine public rituals as well. Lorenzo Jones, who grew up in Church Hill in the early twentieth century, recalled at least one society parade a week, usually held on Sunday.[24] What might a mapping of these parades illuminate about black life in the city? Both an 1898 Lilies of the Valley mutual benefit society parade and a 1903 Uniformed Rank, Knights of Pythias parade left from and terminated at halls belonging to the organizations, traveling routes that connected African American neighborhoods throughout the city and also marching down Broad Street through the central business district (Figure 3.4). The Lilies of the Valley parade, for example, began in Church Hill, traveled to the outskirts of Penitentiary Bottom, and through Jackson Ward.[25] These public rituals suggest ways in which black Richmonders worked to create a sense of community among a widespread and disparate people with competing needs and interests. At the same time, they provide a window on class and gender relations. Large numbers of black Richmonders in the late nineteenth and early twentieth centuries were chronically unemployed and underemployed and may, therefore, have experienced high degrees of geographic mobility. Ethel Thompson (Overby), for example, noted that her family moved frequently: "If you did not have the money to pay the rent, you moved about often."[26] By constantly "reinventing" through ritual their organizations and neighborhoods, might the society parades have served to diminish the degree of alienation historians sometimes attach to very mobile populations? Might the members of the Independent Order of Saint Luke have achieved a similar effect by performing the same rituals at the same time in different neighborhoods? In April 1900, they held simultaneous mass meetings in Jackson Ward and Church Hill, members assembling and marching short distances to the meeting sites. There were sex-based differences in each march route: The women and men gathered at different places, and the women marched shorter distances (in Jackson Ward the women did not march through the streets at all but started from the basement of the church and marched upstairs to the mass meeting).[27] How might we explain this ritual pattern in the mutual benefit society that came as close as any Richmond institution to establishing female equality within the total organization? We have yet to fully discover the social, political, and cultural significance of such rituals or to fully examine the basis of participation. What might be learned from the dress, the banners in the parades and along the routes, the ceremonies at the beginning and end, the occasions for parading, and the participation of working- and middle-class, men, women, and children? Full exploration of ceremonial and ritualistic uses of the streets is an aspect of African American urban history that promises rich rewards.[28]

Struggles for civic space were fought not only in the streets but also within community institutions, such as the church. Attention to questions of space and the spatial construction of political, social, and even economic discourse suggests increasing class and gender differentiation among African Americans in the postemancipation South and opens up new ways to think about the negotiation of community. In the immediate post-Civil War era, black Richmonders enacted their understandings of democratic political discourse through intracommunity and Republican Party mass meetings at which men, women, and children fully participated (including voting). They carried these notions of political participation into the state capitol, engaging from the gallery in the debates on the constitutional convention floor. The church as a central foundation of the black public sphere was central to African Americans' realization of a fully democratic political discourse. In the immediate postslavery era, church buildings doubled as meeting halls and auditoriums. As a political space occupied by men, women, and children, ex-slave and formerly free, literate and nonliterate, the availability and use of the church for mass meetings enabled the development of political concerns in democratic space.[29]

By the 1870s, white Republicans seeking to bolster their hold on party politics tried to remove political meetings from the black church. They relegated meetings to smaller spaces under their control, which precluded mass participation, and they closed the galleries, allowing only official delegates to attend and participate. Despite these efforts, black Richmonders continued to hold mass meetings, often when dissatisfied with the official Republican deliberations. Throughout the late nineteenth century, however, the use of church space became contested among black Richmonders, and in the process political participation itself became contested and increasingly class and gender based. By the 1880s, a series of debates within black Richmond over the use of church facilities led to a prohibition against mass political meetings at First African Baptist Church. Because no other facility within the community could accommodate a true mass meeting and those outside the community were closed to black Richmonders, the closing off of First African meant that indoor mass gatherings of Afro-Richmonders were no longer possible. At the same time, debates over appropriate behavior, preferred forms of worship, and the nature of rational discourse combined with the limited ability to hold true mass meetings to produce a discourse of entitlement in which some persons—those of education, those of learned speaking styles, and eventually those who were male and middle class—had greater authority and rights in this political arena. Increasingly, the more regular forums for political discussions were literary societies, mutual benefit societies and fraternal order meetings, women's clubs, labor organizations, street corners, kitchens, washtubs, and saloons—all of which served to retain mass involvement in politics while transplanting political discussions to specialized places where working class and middle class, male and female, were often set apart from each other. As this happened, political rhetoric and ideology became more class and gender stratified. The complicated issues of how southern African Americans moved to a more class- and gender-based politics while also seeking to maintain a democratic agenda can partially be explored through the changing geography of their political discourse.[30]

The Boundaries and Meanings of Black Richmond

In considering how black Richmonders conceptualized their urban environment, we interrogate the cultural meanings they gave to the spaces they shared and the rhetoric and ideologies of urban space they developed. We suggest not only the street maps but also something of the mental maps that black Richmonders may have laid out and traveled.[31] Our investigation treats city space as more than merely fixed residential and work patterns mapped on linear blocks; we see city space as an amalgam of fluid public spaces and institutions culturally defined by the inhabitants. Elizabeth Blackmar has noted how "the crafted landscape functions symbolically; it is the physical incarnation of social priorities."[32] Similarly, we attend to the built environment as a means of exploring social, political, and economic ideology.

In the immediate post-Civil War era, black Richmonders erected buildings that tangibly testified to their emancipation. In December 1865, former slaves, who had been secretly worshipping together since 1860, transformed a stable on Main Street into the Fifth African Baptist Church. In the spring of 1866, members of Second African Baptist decided not only to rebuild their church, burned down by white residents angered by the church's political and educational activities, but to construct "a substantial edifice" built "entirely of brick" to replace the former wooden structure. In the summer of 1866, black residents constructed the Navy Hill school while black militia stood guard to prevent its destruction by white Richmonders opposed to black education. In these and countless other ways, black Richmonders imprinted their freedom in the urban landscape. Imagine, for example, what particular meanings it must have given to black Richmonders' images of freedom when, in 1867, black men, former slaves, "knocked out the cells, removed the iron bars from the windows, and refashioned" Lumpkin's Jail, the old slavetrading pen on 15th Street between Broad and Franklin, into a school for freedmen.[33]

By the late 1880s, black Richmonders began emphasizing "race progress" as a way of giving African Americans a history and status through which they could claim their rights. The construction of a black urban environment of larger and more elaborate businesses, churches, and homes was used to signify this historical progress. Some of this, as Walter Weare has noted in his discussion of late-nineteenth- and early-twentieth-century black expositions, had the purpose of "testify[ing] before skeptical whites . . . and placing the proof of 'race progress' on elaborate display."[34] Black-owned banks, of which there were four in Richmond by 1903, took on an especially symbolic role, standing, according to Richmond schoolteacher, minister, and poet Daniel Webster Davis, as "conclusive evidence of a high degree of civilization."[35]

The establishment of churches, banks, and businesses as the proof of "race progress" was not principally directed at white Richmonders, however. When newspapers and speakers heralded each new brick residence; each church organ, beaded ceiling, set of pews or stained glass windows; each business "erected by an Afro-American builder, assisted by Afro-American laborers and for Afro-Americans," they proclaimed more than black Richmonders' material worth.[36] Each visible evidence of progress—the construction of these buildings as well as the telling of their tales of achievement in newspapers, books, speeches, poetry—was part of a ritual of memory, struggle, and hope.[37] Black men and women used major buildings as mnemonic devices. For example, the Grand Fountain United Order of True Reformers opened its bank on April 3, 1889, and its store on April 3, 1900. The members waved the banner "In 1860 slaves, in 1890 bankers" at emancipation parades and produced literature with titles like *1619-1907 From Slavery to Bankers.*[38] It would be a mistake to read the celebrations of economic mobility as merely part of the late-nineteenth-century Horatio Alger-type emphasis on individualism and progress; black Richmonders often represented individual achievement as collective prosperity, not only removing the logic behind their social subordination in the larger society but also providing for each other "a new visual landscape of possibility."[39]

Exploring the cultural meanings of black urban space has opened up new understandings for us and presented us with new problems of interpretation and presentation. Our struggles to integrate the material, conceptual, and representational spaces all embodied in an entity called Jackson Ward are suggestive. We have often been unconscious of the contradictions within our own placements. Thus, Elsa has been able to insist on Jackson Ward as the creation of black Richmonders, arguing that one evidence for such is its continued political, social, and economic importance even after white city councilmen gerrymandered the district out of existence in 1903. At the same time, she has insisted that we map the ward by the boundaries created by the white city council. Gregg has insisted on Jackson Ward's origins in the 1871 gerrymandering by the white city council to insure black political impotence. Yet he has been emphatic that we could not literally map the ward's boundaries because Jackson Ward as a place has so many different meanings and boundaries based in black Richmonders' own images and landscaping. In reality, Jackson Ward is, simultaneously, all of these literal and symbolic spaces. Should we now proceed to produce a visual representation of the ward, we could do so only by mapping many different spaces existing in legislative books, in autobiographies and interviews, in newspapers and elsewhere, seeing where these maps converge, overlap, diverge, and seeking meanings in the interstices.[40] Jackson Ward is a function of history, collective memory, mythology, and power; it is also a function of legislation, politics, and inequality. In confronting black residents' conceptions of the boundaries and the nature of the ward, we have found it important not merely to understand how stories of this space became inflated but also to realize how some of these stories became their own history as black Richmonders created a "countermemory," reremembering the creation of the ward as an act undertaken by black people, a distinction that turned it into a place of congregation as well as segregation.[41] We have learned how necessary it is for students of African American urban history to recognize mythology and public memory not merely as useful historical sources but also as important historical forces.

Memory is, of course, contested. An examination of Maggie Lena Walker's neighborhood suggests the complexities of weaving together the threads of history and memory in Jackson Ward (Figure 3.5). Walker, founder and president of the St. Luke Penny Savings Bank (now the Consolidated Bank and Trust Company, the oldest continuously existing black-owned and black-run bank in the United States), is one of the iconic figures in this now national historic district; her home on Leigh Street is a National Park Service site. We began by plotting her neighborhood, getting a sense of her daily

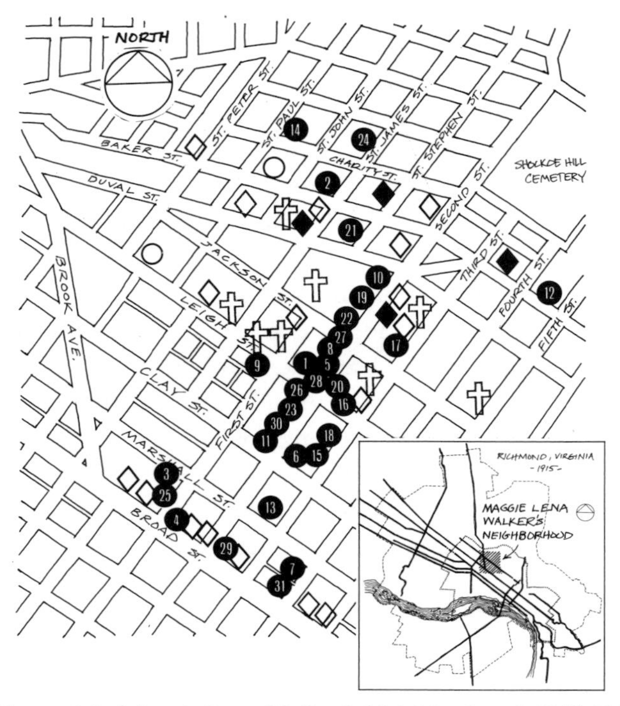

Figure 3.5. Maggie Lena Walker's Neighborhood, 1905-1915
SOURCE: Sites listed in one or more Richmond city directories, 1905-1915.

✞ = Churches, 1907

◆ = Saloons, black owned, 1907

○ = Public schools, 1907

◇ = Saloons, white owned, 1907

List of sites, 1905-1915

1. The Walker home
2. St. Luke Hall
3. St. Luke Penny Savings Bank
4. St. Luke Emporium
5. Grand Fountain United Order of the Reformers bank, hall, and office building
6. Mechanics Savings Bank
7. Nickel Savings Bank, Jackson Ward branch
8. American Beneficial Insurance Company
9. National Ideal Benefit Society
10. Richmond Beneficial Insurance Company
11. Southern Aid Insurance Company
12. Richmond Hospital
13. Women's Central League Training School and Hospital
14. Friends Asylum for Colored Orphans
15. Colored Workingwomen's Industrial Home and Nursery
16. YMCA
17. Knights of Pythias Castle
18. Knights of Pythias Castle Hall
19. Hayes Hall
20. Price's Hall
21. Sixth Virginia Social Club
22. Richmond Athletic and Social Club
23. Hippopdrome Theatre
24. Globe Theatre
25. Jamestown Pool and Billiard Parlor
26. Miller's Hotel
27. Mrs. P. C. Easley Confectionary
28. George Brown Photographic Studio
29. Negro Development Corporation
30. Negro Historical and Industrial Association
31. Richmond Planet building

surroundings. We found that her block from First to Second Street on Leigh was an enclave of black professionals (including three professional couples) in the midst of working-class blocks sprinkled with other professionals. She lived just steps from "Two Street," the main business and entertainment thoroughfare of Jackson Ward, and a block from the Hippodrome, which opened in 1915 for vaudeville acts and later featured the new popular entertainment of movies along with a variety of shows. When the Rayo Theatre opened in 1924 on Second Street between Marshall and Clay, bringing Gippy Smith and his Six Kings of Jazz, Boisy DeLegge and his Bandana Girls, and " 'Some Wild Oats'—The Picture They Fight to See," the Jazz Age was in Maggie Lena Walker's front yard.[42] Saloons were as prolific as churches in her neighborhood, and Leigh Street was a major thoroughfare to and from the nightlife of Jackson Ward. Those who could not afford a Hippodrome ticket might hang out on the corner of Second and Leigh, some of them singing or dancing till they procured the price of a ticket or practicing their steps before they went off to one of the dance halls. The St. Luke Bank was close by, and Walker's short walk or ride to St. Luke Hall, the headquarters of the 100,000-member insurance company she headed, took her through some of the poorest streets in the city—unlike most white bankers and insurance executives who lived far removed from the downtown commercial district in a white world increasingly segregated by class. The St. Luke Emporium, a department store the Order opened on Broad Street, traditionally a center of white business, was still only three blocks from her home.[43] Equally close were other black banks and insurance companies.

The question we would pose is how Walker's political and economic vision was shaped by the spaces she inhabited. We find a woman of privilege determined to maintain her privilege while at the same time working to eradicate the injustices that came from others' lack of privilege. We would suggest that a very distinct conception of race and class was bred in her daily geography. She talked and acted on a politics and economics that came very much out of thinking about how females and males of all ages and classes live with each other on a daily basis. Under her leadership, St. Luke

Hall developed as a physical space that reproduced class, gender, age, and other distinctions within black Richmond; at the same time, the Order's regulation that meetings could not be held in private homes meant that the mutual space of St. Luke Hall brought all members together.[44] We do not mean to reinforce the romantic contemporary discussions of some golden age of black life when the close proximity of the middle class and working class translated into a homogeneous, nonconflictual community, but to rather emphasize the need to place Walker's (or any other person's) economic and political visions within her daily geography, mapping her daily contact with other residents of similar and differing statuses and with the broad range of street life in Jackson Ward.

We also consider the space Maggie Lena Walker has come to occupy in public memory. We are intrigued by the "picture" of Walker that has emerged as a mainstay of public memory—that of a plain, even austere, businesswoman. We compare that picture with another of Walker that suggests a woman of great style and some flair, a high-spirited, fun-loving, as well as domineering and insecure, person (Figures 3.6, 3.7).[45] The "official" public memory of Walker has removed much of the flair and dynamism of her life and has required us to see her only as a businessperson, churchgoer, and social activist. Invisible is the woman who frequented the Hippodrome and who built a second-story porch overhanging Leigh Street so she could still engage in street life even while maintaining a respectable distance from the street.[46] That street life included both legal and illegal business ventures, those considered "uplifting" and those considered "sinful." Just as the memory of Maggie Walker has been sanitized, so too the "official" public memory of Jackson Ward rarely recognizes the entrepreneurship of the street corner vendors, numbers runners, and prostitutes in its veneration of this black business district.[47] Of course, the official and vernacular images of Walker are in part products of her own self-fashioning. This suggests that one avenue for exploring the "perils and prospects" of black business development, especially for a woman, may be attention to how she creates and re-creates herself in different spaces.

Figure 3.6. This stylish image of Maggie Lena Walker is a rare glimpse of a woman usually pictured as plain and austere.
SOURCE: Reprinted by permission from the Valentine Museum, Richmond, Virginia.

Figure 3.7. Styling herself the "St. Luke Grandmother," Maggie Lena Walker often invoked a maternal, authoritarian posture.
SOURCE: Reprinted by permission from the Valentine Museum, Richmond, Virginia.

The question of daily geography and the creation of oneself and one's community through spatial meanings raises questions of urban spectatorship, social identity, and definitions of the city and of "black Richmond." Except for the relatively small number of men engaged in trades like huckstering, the black Richmonders who may have had the widest gaze on the city were women, those thousands who worked as laundresses or domestic servants, and who, by virtue of their employment, had to traverse and were seen as "belonging in" the widest range of spaces. For example, one woman living in Fulton did laundry for white families on Fulton Hill and in Highland Park and Ginter Park; she also made meals and carried them to the factories to sell every day at lunchtime (see Figures 3.1 and 3.3). One domestic servant, living in Jackson Ward, traveled by streetcar to the west end to work in the home of a white family but also traveled daily, again by streetcar, to the downtown business district to carry lunch to her employer. She may have also been required to market and perform other duties that would take her through other parts of the city.[48]

This is not to suggest any privileged positioning of these women in the urban landscape; in fact, their wide gaze on the city, their use of the streets, and their need to travel to their homes by night after a day's work were viewed by many white Richmonders as evidence of their immorality.[49] One white newspaper urged city officials to adopt a "curfew on disreputable women" and arrest "all unescorted black women on the streets after ten in the evening," thus implying that most black women were disreputable.[50] Given the long hours of domestic servants and the fact that those who did not live in were likely to return home late at night, such views classified a substantial number of working-class black women as immoral merely because their employment required late hours. At the same time, the use of the streets and the familiarity of large numbers of black women with large parts of the city suggests that the notion of the city as male terrain does not adequately convey the landscape of black Richmond. Here we are also suggesting a relationship between geography and knowledge. Their large gaze on the city may have given working-class black women broader social and political knowledge and allowed them greater participation in politics than white middle-class women who were more confined in the late nineteenth century to prescribed spaces of the city.

If work provided black men less mobility through the urban terrain than it did women, leisure time may have provided men more mobility and thus may have been been central to their perception of the city. The diary of Edward McC. Drummond provides some insight into an extended black metropolis. Drummond was a seventeen-year-old porter at the Regal Shoe Store when he began his diary in 1910, recording his daily travels and activities. Drummond's everyday life in Richmond spanned a wide geographic area, bending neat concepts of "neighborhood" and "community." Drummond made his home on Brook Road just outside the city limits, and his daily afternoon and evening walks took him to Jackson Ward, Church Hill, Navy Hill, Manchester, Fulton, Gamble's Hill, and many other areas of the city (Figure 3.3). Mutual benefit society meetings, lectures, movies, funerals, as well as parties, teas, and visits to a variety of churches took him far beyond the confines of his home neighborhood.[51]

Just as Drummond in his daily life laid claim to many parts of the city, numerous black Richmonders mapped their homes and neighborhoods in ways that ignored the corporate boundaries of the city or the social and ideological divisions between urban and rural. Even before Manchester (the city south of the James) was annexed by Richmond in the early twentieth century, African Americans residing within Richmond's corporate boundaries had made Manchester part of their extended community. A June 1865 petition to President Andrew Johnson considered Richmond and Manchester as an entity. Likewise, participants in an 1868 Emancipation Day parade crossed over the James, refusing to pay the toll on Mayo's Bridge, drawing residents of the two areas together and objecting to this arbitrary boundary (Figure 3.4).[52] A variety of social and economic conditions reinforced the continued migration of Richmonders from countryside to city and back. Some white operators of highly diversified farms in the Richmond area turned to day labor in the post-emancipation era and employed African Americans living in the city. Likewise, African Americans in surrounding counties were drawn to construction jobs and other seasonal employment in the city. The city markets daily drew men and women into town to sell their produce, small game, and flowers. The residents of suburban and nearby rural communities traveled to downtown Richmond as the streetcar system expanded; for many, Jackson Ward became the site of weekly shopping trips or afternoons out at the movies. Family and church connections as well as excursion trips regularly brought people in and out of the city.[53] The result was a constant discourse between rural, urban, and even suburban elements of black culture in the streets of Richmond.[54]

One of the primary achievements of large organizations such as the Independent Order of Saint Luke was the maintenance of regular networks of contact between black Virginians in cities and countryside. Local St. Luke councils established in towns and villages across Virginia connected officials in Richmond with local affairs in places as remote as Covington and as small as Christian, Virginia, on a regular basis and brought significant numbers of rural folk into the city to mix with their urban brothers and sisters for special celebrations and especially conventions. Chapters established in states along the eastern seaboard from North Carolina to Massachusetts served in part to help those who had migrated—seasonally or permanently—retain a connection to Richmond.[55]

All of this is to suggest that the boundaries of black Richmond cannot easily be drawn on a static map. Rather, we want to understand the actual space in which people lived by focusing on their daily lives and activities and on how they understood the city. When we do so, we may also begin to reexamine the nature of race relations in the late nineteenth and early twentieth centuries. For example, the spatial construction—at home and at work—of the nineteenth-century Richmond workforce facilitated interaction and potential political alliances among black and white working-class males. Yet it encouraged not only a physical but also an ideological distance between black and white working-class women and provided no real living or working spaces that black and white middle-class Richmonders had in common. Because of the close proximity in which working-class black and white residents still lived in the late nineteenth century, working-class neighborhood saloons and groceries were often frequented by and became sites for interaction between black and white patrons. In 1891, for example, African Americans lived in 25.6 percent of the households on Williamsburg Avenue, a main street of the working-class neighborhood of Fulton, and resided in seven of the nine blocks.[56] John O'Grady, Jr., recalling his experiences growing up in Fulton in the early twentieth century, remembered his father's saloon, which served both black and white patrons. As segregation became more prevalent, the senior O'Grady was forced to find a solution to mixing at the bar: "The bar being long, there was a screen on rollers separating the two races, all served by the same bartender, and when there were more of one race at a time than the other, the screen would be moved up or down."[57] This screen could serve to mask the friendly interaction, discussions, recreation, and even political alliances that may have continued in the saloon and on the sidewalks surrounding it; at the same time, the screen may have been a point of contention and open conflict if, for example, in a crowded saloon black and white patrons disagreed over moving the dividing line, providing more space on one side and lessening the space on the other. Close proximity could lead to political alliances; it could just as well lead to conflict. No doubt most of the time it did both.

In the late nineteenth century, thousands of black and white laboring and artisan men encountered each other daily in the iron and tobacco factories, on the docks, and in building trades such as plastering and bricklaying.[58] Working-class women were less likely to have opportunities to socialize around common workplace experiences. Although the number of black and white women in the city's tobacco factories expanded in the late nineteenth century, the industry was rigidly stratified by race as well as gender; thus, black and white women never worked in the same places. In various ways, white male employers and white female reformers portrayed white women as pure, innocent, moral, and endangered, in need of protection from the city and the factory. Black women were perceived as dirty and dangerous—to the city, to the factory, and to moral white women. White women needed to be segregated not only from men—white and black—but also from the dirty and dangerous black women.[59] Black and white women's positions in the workplace itself were, therefore, separated not only by physical rifts but by large ideological canyons as well.

The degree to which black and white women socialized in their neighborhoods merits investigation. Certainly, some did meet in dance halls and saloons, even though the police court judge in Richmond ruled it illegal for women to frequent barrooms.[60] Late-nineteenth-century social reformers—black and white—often saw the mixing of black and white men and women in dance halls as a sign of the unrespectability of such places and their patrons. For the most part, however, women—black and white—had fewer institutionalized public leisure spaces than did men. A larger portion of women's time away from places of paid employment was connected to their own domestic space;[61] issues of social equality therefore may have been more pronounced in interactions between black and white women than in those between men which occurred in public and semipublic leisure spaces. Nevertheless, domestic chores in working-class neighborhoods provided many opportunities for interaction, even after segregation was more firmly established. It is likely, for example, that black and white families without indoor plumbing or backyard wells drew their water from the same springs or water troughs. Similarly, in working-class neighborhoods, black and white women may have patronized the same grocery stores or vegetable and fish trucks. They may have met as they supervised their children's play, picked rags or scavenged, frequented secondhand stores, or did their wash. No doubt many black and white working-class women frequented the city markets at the end of the day near closing time when the prices of meat, especially, were often lowered. Numerous sites of informal interaction may have highlighted working-class women's similar economic conditions and domestic chores; they could also be venues for working-class women's recreation—perhaps an interracial recreation.[62] Just as easily, these sites could be places for conflict.

How easily working-class women's interactions translated into political alliances is less clear. Available sources shed little light on the interaction of women in Richmond's Knights of Labor. Black women, many of them employed as domestic workers, and white women organized a number of local assemblies, but on a segregated basis.[63] In Richmond, as elsewhere in the South, one significant factor in black and white working-class women's domestic arrangements may have worked against political alliances. Scholars have noted the ability of white working-class women to hire black working-class women to do their household work or laundry. Those they hired often lived in their neighborhoods. When Maggie Alease Taylor (Jackson Howard), for example, who lived on the black side of 27th Street, had to leave school before graduation, she went to work cleaning houses and washing dishes for women on the white side of 27th Street.[64] These relationships of unequal power, and possible exploitation, shaped the interactions of black and white working-class women.

Yet the continued attempts in the late nineteenth century at biracial political and labor coalitions such as the Republican Party, the Readjuster Party, and the Knights of Labor were made possible not merely by the continued interaction of black and white men at worksites but also by the reinforcement of these workplace connections in neighborhood meeting places even in an overall context of distrust and violence. The Knights of Labor movement drew on the common ground of religious imagery and the language of fraternalism among black and white working-class families, sustaining biracial consumer boycotts and strikes in the 1880s.[65]

By the early twentieth century, working-class neighborhoods underwent changes that decreased the opportunities for interaction between black and white working-class men and women. As skilled white workers followed the new white middle class of clerks, salesmen, and professionals to the suburbs, the composition of older working-class neighborhoods such as Oregon Hill and Penitentiary Bottom, on the western and northern boundaries of the Tredegar Iron Works site, profoundly changed. By 1920, Oregon Hill remained white, but the skill level of residents had declined significantly. Penitentiary Bottom, which in the nineteenth century was home to both black and white residents, had evolved into an overwhelmingly black neighborhood of laborers, tobacco hands, and domestic workers.[66]

Even so, common interests and conditions could still draw people together. As Earl Lewis has noted in his study of Norfolk, Virginia, new forms of mass culture could create common spaces. The recollections of Lorenzo Jones and Bessie Bailey (Baldwin), both of whom grew up in Church Hill in the early twentieth century, are suggestive. Jones, who details which blocks were occupied by white families and which by black, still considered his neighborhood integrated in comparison to what would come later, and he remembered the crowds—black and white—who gathered on Fourth Street to watch wire service reports of the World Series games. Baldwin recalled that one of the few radios in her neighborhood was owned by a "couple who lived on 34th Street and people would come on a Sunday morning, people would come from Oakwood Ave., white people, to listen to this couple's radio."[67] Richmonders could simultaneously conceive of their neighborhoods in terms of interaction and within a system of segregation. The fact that working-class people, black and white, lived much of their life outdoors meant that the possibilities for interaction were frequent. Whether those interactions were friendly or hostile, they reinforced a vision of the late-nineteenth- and early-twentieth-century city as shared space.

The use of the sidewalks, streets, parks, and other public spaces by the working class sometimes led to condemnation of their activities as immoral and unrespectable by an increasingly privatized middle class. Debates over outdoor life, dancing, and dress suggest some of the nuances of class, gender, respectability, and leisure in African American urban life.[68]

Moral Geography: Class, Gender, and Respectability

The landscape of urban leisure provides a venue for examining the everyday rituals of urban life and the "moral geography" of southern urbanization.[69] Spatial inequalities made many activities engaged in by the working-class more visible because, lacking private facilities, their work and leisure were more public. Elite African Americans, as elite European Americans, often naturalized ideologically and rhetorically their class and status privileges. Thus, middle-class black Richmonders often spoke of working-class men's and women's inability (as well as unwillingness) to observe the more privatized and restrictive conventions as evidence of work ethics and morality, ignoring factors of time, space, and money. It was the public visibility of working-class activities that often made them threatening to a middle-class increasingly worried about image as a sign of progress and a means of obtaining rights. The public behavior of the working class was considered an affront to propriety and decorum by some African Americans and a menace to public order and private property by white Richmonders. It was precisely white men's and women's perceptions of these activities that often concerned black middle-class reformers.

Working-class women, for example, often did their laundry or other chores outdoors because of the small space of their homes, the lack of indoor plumbing, and because they could oversee their children's play and socialize with their neighbors as they worked. Having the responsibility of caring for their own, their relatives', or their neighbors' children, working-class women also often conducted much of their leisure life in the streets or from their windows and doorways. White observers portrayed these "women of the negro laboring classes" as having "the . . . quality of careless, aimless happiness. They possess all the time there is. They hang out of windows, stand in doorways, on the street, careless of home, children, appearance, bubbling with laughter." Black reformers, such as a columnist of the *Richmond Planet,* railed against leisure time spent in the streets, admonishing young women, for example, to entertain their male companions at home rather than on street corners, advice that made courting on the street a sign of lax morality rather than insufficient domestic space. Although working-class children most often played in the streets, some middle-class parents required their children to play inside fenced yards, not merely for safety purposes but to keep their children away from what they perceived as the bad influence of those children who used the streets as their playground.[70] Questions of space were thus projected as questions of respectability and morality.

Dancing, for example, was a common leisure activity of both working-class and middle-class men and women. The spaces in which dancing occurred could help to determine its relative respectability. In the immediate post-Civil War period, when church buildings were the only public spaces owned by African Americans, churches stood as democratic political space and fairly democratic leisure space as well. Every week, churches hosted grand festivals, lawn parties, juvenile operettas, musicales, military drills, cake walks, or debates and other literary entertainments. In the arena of the church, cake walks were a respectable activity. As the leisure landscape changed in the late nineteenth century and as more facilities—saloons, billiards halls, dance halls, society halls, parks, and picnic grounds—became available to African Americans, church authorities were able to limit the activities held in their buildings. They also believed these limits necessary to clearly distinguish the church from these secular places. By the 1880s, many activities, such as dancing, were barred from many church facilities, and the issue of whether it was immoral for church members to dance became a matter of serious debate.[71]

Church records, ministers' sermons, newspaper reports, and personal accounts reveal the dynamics of these debates and suggest something of the class and gender dimensions of the geography of leisure.[72] In April 1884, when the Invincible Social Club held a party at Evan's Hall, the attendees who "danced to the entrancing strains of sweet music" included three of Richmond's black physicians and two of its newspaper editors.[73] Elite men and women believed respectability could be maintained when dancing was done at private parties where one's partners would all be acquaintances there by invitation (even when a fee was charged those who could attend were still limited to those approved by the hosting organization). They assumed respectability was easily compromised at public dance halls where one might dance with any stranger who laid out the price of the ticket. The Reverend Anthony Binga, Jr., pastor of First African, Manchester, for example, cited as one of the signs of the immorality of dancing the "unholy passions" it could excite in strangers:

> Look at the young girl or some one's wife borne around the room in the arms of a man; his arms are drawn around her waist; her swelling bosom rests against his; her limbs are tangled with his; her head rests against his face; her bare neck reflecting the soft mellow light of the chandelier, while the passions are raging like a furnace of fire. But who is the individual with whom she is brought into such close contact? She does not know; neither does she care. The most she cares to know just now is, that he is a graceful dancer.[74]

Binga's sermon also highlights the importance of gender in the discussion of public morality. Although church prohibitions on dancing applied to all members, women were more likely to be excluded for engaging in it.[75] As Ellen Ross has observed in her work on the London working class, it was women who "indeed *embodied* respectability or the lack of it, in their dress, public conduct, language, housekeeping, childrearing methods, spending habits, and, of course, sexual behavior."[76] Thus, prohibitions against dancing and drinking were greater for women. The clearest evidence of the ways in which attitudes toward dancing often reflected different expectations for men and women may be found in one report regarding late-nineteenth-century militia-sponsored events. As churches began to prohibit dancing, one militia, the Petersburg Blues, decided "that for the sake of the many married and unmarried females, whose presence was desired to grace the occasion," dancing would be eliminated from their socials. Soon, however, the command agreed that this was too big a sacrifice to require of the militiamen, and the Blues adopted a new plan:

> Our socials were to last till 10:30 p.m. Announcement would be made that the entertainment by the command was at end. Those morally obligated to their churches or having personal conscientious scruples against dancing would leave. Men with friends who desired to remain for an hour and dance could do so upon their personal responsibility.[77]

Given the clear assumption that the morality of dancing was an issue for women only, one can only wonder how those female friends who did remain after 10:30 were perceived by the women who left and the men who stayed. Yet the Blues affairs were private, by invitation, and sponsored by the "best" people and thus faced less opposition than did the public dance halls.

Of course, as Brian Harrison has pointed out, "Respectability was always a process, a dialogue with oneself and with one's fellows, never a fixed position."[78] A middle class that heralded public displays of material consumption, as in the construction of larger homes "suitable for race advancement," could at the same time denounce as wasteful working-class displays, such as popular excursion trips. Many working-class families may have seen the ability to travel together on these one-day trips as important signs of their standing rather than as extravagance.[79] For many working-class women in neighborhoods where survival depended on mutuality, how a woman kept her house and children, managed her finances, and participated in the sharing network that paid rent, watched children, cared for sick, and fed each other, were far more important indicators of respectability than whether or not she "wiggled" when she danced. In 1896, when Mrs. Harriet Beverly and other women of her neighborhood, seeing a white man carrying a nine-year-old girl away and believing he would sexually assault her, followed and rescued her, the importance of work and sociability in public space was reinforced. They and other women like them no doubt rejected not merely the practicality but also the desirability of ensconcing themselves and their families in private space.[80]

White Richmonders used public space to construct black women in masculine or at least unfeminine and immoral ways—as in public whippings for crimes, refusing to allow them to ride in parts of the streetcars reserved for white "ladies," outrage over elegant feminine dress, and relegation to sex-integrated workplaces in factories. Many African Americans sought to use public space and behavior to reconstruct black women in "feminine" and "respectable" ways. The emphasis on dress was part of this concern. White observers remarked on black women in the streets and factories with their "brilliantly colored turbans," or on working-class women's Sunday clothes, which were "a riot of color." But black women perceived as well-dressed could also become a point of antagonism for white Richmonders. In 1892, for example, the *Richmond Dispatch* reported displeasure at "white ladies" being "crowded out of the seats in the public parks by colored teachers clothed in 'purple and fine linen' bought with money which might more properly [have] gone in the pockets of the white young ladies." The *Dispatch* saw black women teachers as "wearing this apparel at the expense of white tax-payers," evidence that white citizens' money was being used to finance an extravagant, wasteful, and most important, nonsubservient lifestyle of black women who more appropriately belonged in domestic service. Among black Richmonders, however, especially the elite, these women's elegant dress was a sign of status and respectability, not of extravagance. Some observers questioned the Reverend John Jasper's refusal to adopt a dress code for his Sixth Mount Zion Baptist Church congregation that would prevent women from wearing loudly colored hats. Jasper answered that women so dressed should be welcomed in church as the ones in the greatest need of religion, and in so doing defended the women's presence by pronouncing them sinners rather than believers. For some observers, the "unrespectability" of the women's apparel overrode the "respectable" behavior of attending church.[81]

These examples should alert us to the fluidity of definitions of respectability and suggest that people—men and women, working class and middle class—defined their own standards partly from dominant codes but largely from their own experiences. In much African American history, "respectability" has often been spoken of as a middle-class value. But as Roy Rosenzweig reminds us in his study of white workers in Massachusetts, "respectability [was] perfectly compatible with the maintenance of ethnic and working-class values . . . respectability was not necessarily a strategy for individual advancement into the middle class. It could be a part of a worldview that emphasized church, home, family, ethnic community, group solidarity, and a stable working-class identity."[82] Some working-class Afro-Richmonders did seek social and economic mobility and acceptance within the middle class and the larger society through the adoption of a set of personal behaviors, but others sought to adopt the same set of personal behaviors in the maintenance of their working-class family and neighborhood. Who, if not a domestic worker subject to constant sexual harassment, might have more reason to proclaim a Victorian desexualized persona in hopes of bringing protection to herself and stability to her job and family? Who would have more reason to advocate temperance than those working-class people whose families might be devastated by the loss of valuable financial resources, strained family relations, and even violence that sometimes resulted from drink? That many working-class people supported more restricted sexual behavior or temperance does not suggest their aspiration to or imitation of the middle class, but rather the degree to which such reforms could speak to realities and values that could be as deeply embedded in working-class culture as in middle-class culture.

Similarly, many in the middle class defied the standards of respectability often ascribed to those in their position. In the early twentieth century, both Mrs. Johnson and Mrs. Price supervised the dance floors above their husbands' funeral parlors. Johnson ejected any dancers who engaged in that "new kind of dancing that called for shaking and clutching and stepping high." Price, on the other hand, not only allowed that "funny walking dance" but actually invited it by playing the "wigglin'" music that Johnson and others no doubt thought unrespectable, even immoral, indicative of "joy boys" and "easy ladies." Price did, however, pull apart any couples who danced too closely together.[83] Among those who most vigorously questioned, opposed, or defied the church's prohibitions against dancing, theater going, and other entertainments were middle-class women and Virginia Union University students, many of whom were preparing for the ministry.[84] Helen Jackson (Lee)'s parents prevented her from sitting on the front porch of their Third Street home "listening to the syncopated sounds coming through the open doors and windows" of the corner saloon where that "loose woman," Queenie Ross, played ragtime piano. They also perceived the foot-stomping, tambourine-playing "Holy Rollers who had a store-front church in the next block" as uncultured. Yet they could not keep these sounds from reaching their daughter, "who many nights . . . fell asleep to the combined sounds of the Holy Rollers' hollers and Queenie's piano." It was not just from the streets that Helen and her brother absorbed this music, however. In their home they heard not only classical music but also race records—jazz and blues—and her parents enjoyed it as much as Helen when they could get her talented but shy brother to "rock" the piano. Charles and Nanny Jackson's principal concern was with Helen enjoying such music in the neighborhood rather than in their home; they worried about the publicness, the indiscriminate mixing, and the lifestyle that they assumed went with the enjoyment—as audience or entertainer—of these rhythms in public spaces. Some condemned the music itself as primitive or immoral, but the Jacksons and many other well-to-do black Richmonders taught their children to enjoy the music while condemning the public venues in which it was often presented.[85]

The idea that certain spaces were safe and respectable—that is, moral—and that therefore the people and activities within them were more moral carried over to entire areas of town. White Richmonders, for example, used a geographic terminology for rowdy, boisterous, violent behavior, referring to such as the "Jackson Ward yell."[86] In Richmond, as elsewhere, prostitution was carefully contained by the police to areas of town away from elite, white residential areas. By the late 1870s, a red-light district had emerged in the city, bounded by 14th, 15th, Main, and Broad Streets. Even within these four blocks, "streets, lanes, and alleys devoted to Negro prostitutes for white men . . . Negro prostitutes for colored men," and "white prostitutes for white men, some of whom secretly sold their favors to colored men" were clearly delineated. Surrounded by the old, decaying hotel district, by factories, and by the remains of Richmond's slave markets, this area was also home to many factory workers, day laborers, and domestic workers. Other low-lying, poor areas such as "Chinch Bottom" below Jail Alley or "Cash Corner" near the Cary Street tobacco factories also housed prostitutes along with other working-class people. The designation of these districts as immoral allowed police to make "wholesale arrests" on "no other charge than that" the arrested "lived in a certain street." One such incident occurred in August 1910, when numerous black men and women in one of these districts were "herded together like cattle" and "hurried away to jail after being hustled out of their beds at midnight, absolutely without any warrant or right to search their premises."[87]

How people who lived in these areas were or were not incorporated into black community institutions is a question that merits study. We do know that some black people shared a view of certain areas as immoral. When insurance executive B. L. Jordan asked the police to arrest a white man for attempting to molest his fourteen-year-old sister-in-law, his primary evidence was a letter from the man requesting the girl meet him. Jordan successfully argued that merely the neighborhood the man chose for the meeting was evidence of his bad intentions, as no respectable person would go there or even know of it. Similar protection from the law was not forthcoming for the ten-year-old girl sexually molested by a white insurance collector; the fact that she was home alone caring for her baby brother in a poor neighborhood allowed her to be categorized as unrespectable, and her charges against the respectable white man were dismissed. Although B. L. Jordan was no doubt right about the man's intentions toward his sister-in-law, this incident along with indiscriminate arrests by the police and the lack of protection afforded to many working-class people by the judicial system suggests the degree to which a significant number of black Richmonders may have been viewed as unrespectable solely on the basis of their address.[88]

Conclusion

Our aim has been to suggest the ways in which a broad focus on the historical, cultural, and social mappings of the city may enrich our work in African American history. Further investigation of street rituals, architectural interventions, and public memory will enable us to understand more fully the varied conceptualizations of history that have shaped African American culture. Additional explorations will illuminate the ways in which the city appeared as text in African American rhetoric and political ideology, for example, investigating the changing perceptions of the city as a place of opportunity and hope, or a place of danger and despair for differing groups of people. The city—its spaces, its forbidden and inviting areas, its pleasures and dangers, even its boundaries—existed in people's minds as much as on street maps. We have tried here to limn the form and dimensions of the " 'invisible landscapes' that people carr[ied] in their heads" as well as the physical landscapes they had to negotiate daily. Black Richmonders inhabited a landscape "dense with historical imagery," but they also used the urban landscape to articulate their own stories of emancipation, freedom, progress, and success.[89] Like all histories, these too, were contested. Black Richmonders not only manufactured a built environment that could generate new meanings of possibilities, they also struggled for control of those meanings and symbols. Among the principal places in which they did so were the contested arenas of leisure space and public behavior. We also have tried to recognize that Jane and Jim Crow were not only "city slickers,"[90] they were also natty dressers, appearing in a variety of sophisticated attire. It is, therefore, necessary for historians to pay close attention to the actual spaces in which black and white residents carried out their daily lives, seeing the possible simultaneity of relationships of hierarchy and relationships of camaraderie. We hope in these ways to more fully and centrally situate African American urban history in the city.

Notes

1. See, for example, James C. Scott, *Weapons of the Weak: Everyday Forms of Peasant Resistance* (New Haven, 1985); James C. Scott, *Domination and the Arts of Resistance: Hidden Transcripts* (New Haven, 1990); Rosalind O'Hanlon, "Recovering the Subject: *Subaltern Studies* and Histories of Resistance in Colonial South Asia," *Modern Asia Studies* 22 (February 1988), 189-224; Robin D. G. Kelley, "An Archaeology of Resistance," *American Quarterly* 44 (June 1992), 292-98; Lila Abu-Lughod, "The Romance of Resistance: Tracing Transformations of Power Through Bedouin Women," *American Ethnologist* 17, 1 (1990), 41-55; Stuart Hall, "Notes on Deconstructing the Popular," in Raphael Samuel, ed., *People's History and Socialist Theory* (London, 1981), 227-41.

2. See, for example, the essays in Lynn Hunt, ed., *The New Cultural History* (Berkeley, 1989). For an insightful analysis of the debates about poststructuralist theory and historical research, see Kathleen Canning, "Feminist History After the Linguistic Turn: Historicizing Discourse and Experience," *Signs: Journal of Women in Culture and Society* 19 (Winter 1994), 368-404.

3. "Roundtable: The Recentering of Twentieth Century African American History," Social Science History Association annual meeting, Minneapolis, Minnesota, October 1990; see especially Thomas C. Holt, "Comment" (cited with permission of Holt); also, Henry Louis Taylor, Jr. and Vicky Dula, "The Black Residential Experience and Community Formation in Antebellum Cincinnati," in Henry Louis Taylor, Jr., ed., *Race and the City: Work, Community, and Protest in Cincinnati, 1820-1970* (Urbana, 1993), 96-125.

4. Particularly helpful to our thinking about urban space has been Elizabeth Blackmar, "The Urban Landscape," *Journal of Architectural Education* 30, 1 (September 1976), 12-14; Robert Rotenberg and Gary McDonogh, eds., *The Cultural Meaning of Urban Space* (Westport, 1993); and D. W. Meinig, ed., *The Interpretation of Ordinary Landscapes: Geographical Essays* (New York, 1979). We focus herein on issues of public and semipublic space but we recognize the need to more fully interrogate issues of private space as well.

5. The addresses of approximately one fifth of the total free black population were listed in William L. Montague, *The Richmond Directory and Business Advertiser for 1852* (Richmond, 1852), section titled "Free Colored Housekeepers." See also Marie Tyler-McGraw and Gregg D. Kimball, *In Bondage and Freedom: Antebellum Black Life in Richmond, Virginia, 1790-1860* (Richmond, 1988), 48-52.

6. Midori Takagi, "Female Slave Participation in the Urban Market Economy: Richmond, Virginia, 1780-1860," *Southern Women: The Intersection of Race, Class and Gender*, Working Paper No. 8 (Memphis, 1994), 6; Gregg D. Kimball, "African-Virginians and the Vernacular Building Tradition in Richmond City, 1790-1860," *Perspectives in Vernacular Architecture*, vol. 4, Thomas Carter and Bernard L. Herman, eds. (Columbia, Mo., 1991), 121-29; Tyler-McGraw and Kimball, *In Bondage and Freedom*, 48-52.

7. On black and white audiences at antebellum Richmond theaters, see Patricia C. Click, *The Spirit of the Times: Amusements in Nineteenth-Century Baltimore, Norfolk,*

and Richmond (Charlottesville, 1989), 34-38. On churches, see John T. O'Brien, "Factory, Church and Community: Blacks in Antebellum Richmond," *Journal of Southern History* 44, 4 (November 1978), 509-36; and Tyler-McGraw and Kimball, *In Bondage and Freedom.* The tour of the Valentine Museum's Wickham House is organized to emphasize the related but separate lives of all the inhabitants—white and black, free and enslaved—of this home and neighborhood. See Gregg D. Kimball and Barbara C. Batson, "Shared Spaces, Separate Lives," Gallery Guide (Richmond, 1993).

8. For the development of Richmond, see Christopher Silver, *Twentieth-Century Richmond: Planning, Politics, and Race* (Knoxville, 1984).

9. The other black settlements were Oak Park, Woodville, and Westwood. The suburban development and later annexation is traced in Silver, *Twentieth-Century Richmond.* Howard H. Harlan, *Zion Town—A Study in Human Ecology,* Publications of the University of Virginia Phelps-Stokes Fellowship Papers No. 13 (Charlottesville, 1935); Miles B. Jones, "The Providence Park Baptist Church, 1876-1976," *Journal of the Richmond Oral History Association* 1 (Winter 1977), 23-28.

10. Thomas J. Woofter, Jr., *Negro Problems in Cities* (Garden City, 1928); Gustavus A. Weber, *Report on Housing and Living Conditions in the Neglected Sections of Richmond, Virginia* (Richmond, 1913); Charles L. Knight, *Negro Housing in Certain Virginia Cities* (Richmond, 1927); Silver, *Twentieth-Century Richmond;* W.T.B. Williams, "Colored Public Schools in Southern Cities," *Ninth Annual Report of the Hampton Negro Conference* (Hampton, Va., 1905), 36.

11. Lawrence Levine, *High Brow/Low Brow: The Creation of Social Hierarchy in America* (Cambridge, Mass., 1988); Susan G. Davis, *Parades and Power: Street Theatre in Nineteenth-Century Philadelphia* (Philadelphia, 1986). See also Mary Ryan, "The American Parade: Representations of the Nineteenth-Century Social Order," in Hunt, ed., *The New Cultural History,* 131-53; David Glassberg, *American Historical Pageantry: The Uses of Tradition in the Early Twentieth Century* (Chapel Hill, 1990).

12. John Thomas O'Brien, Jr., "From Bondage to Citizenship: The Richmond Black Community, 1865-1867" (Ph.D. diss., University of Rochester, 1974), 326.

13. *Richmond Dispatch,* July 6, 1866; *Richmond Enquirer,* February 23, 1866.

14. Ira Berlin, Joseph Reidy, and Leslie S. Rowland, eds., *Freedom: A Documentary History of Emancipation, 1861-1867,* vol. 2, *The Black Military Experience* (New York, 1982), 735.

15. O'Brien, "From Bondage to Citizenship," 320-23, 363-64, 366-69, 379, 478n; Peter Rachleff, *Black Labor in the South: Richmond, Virginia, 1865-1890* (Philadelphia, 1984), 40, 53; Michael B. Chesson, *Richmond After the War 1865-1890* (Richmond, 1981), 193; Michael B. Chesson, "Richmond's Black Councilmen, 1871-96," in Howard N. Rabinowitz, ed., *Southern Black Leaders of the Reconstruction Era* (Urbana, 1982), 213-14; D. B. Williams, *A Sketch of the Life and Times of Capt. R. A. Paul* (Richmond, 1885), 44-45; *Richmond Dispatch,* January 24, February 26, 1878; Virginia, *Report of the Adjutant General of the State of Virginia for the Year 1897* (Richmond, 1897), 38-39; Virginia, *Report of the Adjutant General of the State of Virginia for 1898-1899* (Richmond, 1899), 6; *Richmond Planet,* June 21, July 5, 1890, December 21, 1895. For a history of the black militia in the state, see William Henry Johnson, *History of the Colored Volunteer Infantry in Virginia, 1871-1899* (Richmond, 1923).

16. In antebellum Richmond, all white men were liable for militia muster every year, although they could obviate this duty by paying a fine. Some of Richmond's more prosperous citizens formed elite units such as the Richmond Light Infantry Blues, which performed ceremonial duties and were mustered to suppress insurrectionary activity such as the John Brown raid in 1859. Other antebellum units rallied around ethnic identities, such as the Richmond German Rifles, later known as

the Virginia Rifles, and the Montgomery Guards, an Irish unit. The German Rifles took part in celebrations and ceremonies with both German and American themes and also served a protective function in the German community, especially during the Know-Nothing agitation in the 1850s. Both the Montgomery Guards and the Virginia Rifles served in the Confederate Army during the Civil War. See Louis H. Manarin and Lee A. Wallace, Jr., *Richmond Volunteers: The Volunteer Companies of the City of Richmond and Henrico County, Virginia, 1861-1865*, Official Publication No. 26, Richmond Civil War Centennial Committee (Richmond, 1969); John A. Cutchings, *A Famous Command, the Richmond Light Infantry Blues* (Richmond, 1934); Klaus G. Wust, "German Immigrants and Nativism in Virginia, 1840-1860" in *Twenty-Ninth Report, Society for the History of the Germans in Maryland* (Baltimore, 1956), 31-50; Herman Schuricht, *History of the German Element in Virginia* 2 vols. (Baltimore, 1898); James Henry Bailey II, *A History of the Diocese of Richmond: The Formative Years* (Richmond, 1956), 145-48.

17. Wendell P. Dabney, "Rough Autobiographical Sketch of His Boyhood Years," (typescript, n.d.), 17-18, microfilm copy in Wendell P. Dabney Papers, Cincinnati Historical Society, Cincinnati, Ohio.

18. *Richmond Daily Dispatch,* October 22, October 25, October 26, October 27, 1875; Virginius Dabney, *Richmond: The Story of a City* (Garden City, 1976), 232-33; Michael B. Chesson, "Richmond After the War 1865-1890" (Ph.D. diss., Harvard University, 1978), 350-51.

19. The story of the Sixth Virginia Volunteers can be followed in the *Richmond Planet,* April 1898-May 1899.

20. See, for example, *Richmond Planet,* January 11, September 19, 1902. Quote is from *Richmond Planet,* September 17, 1898.

21. Richmond *Citizen,* April 4, 1866; *Richmond Republic,* April 9, 1866; *Richmond Dispatch,* April 9, 1866, April 7, 1868; O'Brien, "From Bondage to Citizenship," 327-28, 334-42; "An Ordinance Concerning Negroes," in *The Charters and Ordinances of the City of Richmond* (Richmond, 1859).

22. Broadside No. 1866:13, dated April 2, 1866, Virginia Historical Society, Richmond, Virginia.

23. The changing story of emancipation, reflected in black Richmonders' parades, is examined in Elsa Barkley Brown, "Telling Stories: The Invention of Black Richmond" (paper presented at Colloquium Series, W. E. B. Du Bois Institute for Afro-American Research, Harvard University, March 1994). Parades were also used to connect African Americans residing within the corporate limits of the city with those outside and even in other cities. See, for example, the 1890 Knights of Pythias parade, which began in Petersburg, marched to and through Manchester, across the bridge to Richmond, through Church Hill, and ended in Jackson Ward. *Richmond Planet,* August 23, 1890.

24. Interview with Lorenzo Jones by Barbara Roane, July 17, 1979, Richmond Independence Bicentennial Commission Oral History Program, Series 4, Special Collections and Archives, James Branch Cabell Library, Virginia Commonwealth University, Richmond, Virginia (hereafter VCU).

25. *Richmond Planet,* September 17, 1898, September 19, 1903.

26. Ethel Thompson Overby, *"It's Better to Light a Candle Than Curse the Darkness": The Autobiographical Notes of Ethel Thompson Overby,* edited by Emma Thompson Richardson (1975), 1.

27. *Richmond Planet,* April 15, 1900.

28. On antebellum rituals and celebrations, see Geneviève Fabre, "African-American Commemorative Celebrations in the Nineteenth Century," in Geneviève

Fabre and Robert O'Meally *History and Memory in African-American Culture* (New York, 1994), 72-91; Shane White, " 'It Was a Proud Day': African Americans, Festivals, and Parades in the North, 1741-1834," *Journal of American History* 81 (June 1994), 13-50. Robert A. Hill contends that the 1920s Universal Negro Improvement Association allowed women a public leadership role in political rituals held inside Liberty Hall but cast them in a subordinate and supportive role in those political rituals, such as parades, which took place on New York City streets. Robert A. Hill, "Making Noise: Marcus Garvey in Procession, August 1922," in Deborah Willis, ed., *Picturing Us: African American Identity in Photography* (New York, 1994), 181-205. Elsa Barkley Brown explores the importance of ritual in the Independent Order of Saint Luke in " 'Not Alone to Build This Pile of Bricks': Institution Building and Community" (paper presented at the Age of Booker T. Washington: Conference in Honor of Louis Harlan, University of Maryland, College Park, May 1990).

29. Elsa Barkley Brown, "Negotiating and Transforming the Public Sphere: African-American Political Life in the Transition From Slavery to Freedom," *Public Culture* 7, 1 (Fall 1994), 109-25.

30. Barkley Brown, "Negotiating and Transforming the Public Sphere," 126-41.

31. For the concept of mental maps, see Roger M. Downs and David Stea, *Maps in Minds: Reflections on Cognitive Mapping* (New York, 1977) and Peter R. Gould and Rodney White, *Mental Maps*, 2nd ed. (Boston, 1986). For two different but equally informative analyses of the relationship between imagining and mapping, see Florence Ladd, "A Note on the 'World Across the Street,' " *Harvard Graduate School of Education Association Bulletin* 12 (1967), 47-48, and Larry Wolff, *Inventing Eastern Europe: The Map of Civilization on the Mind of the Enlightenment* (Stanford, 1994).

32. Blackmar, "Urban Landscape," 13. See also Jane Jacobs's exploration of the conflicts between city planners' layout of streets, parks, neighborhoods, shopping districts, and cultural centers, and urban residents lived mapping of arenas for work and play in *The Death and Life of Great American Cities* (New York, 1961).

33. *Inventory of Church Archives of Virginia. Negro Baptist Churches in Richmond* (Richmond, 1940), 18; O'Brien, "From Bondage to Citizenship," 354-55, 364; Leon S. Litwack, *Been in the Storm so Long: The Aftermath of Slavery* (New York, 1979), 172.

34. Walter B. Weare, "New Negroes for a New South: Adaptability on Display," in Elizabeth Jacoway, Dan T. Carter, Lester C. Lehman, and Robert C. McMath, Jr., eds., *The Adaptable South: Essays in Honor of George Brown Tindall* (Baton Rouge, 1991), 90-103. Black Richmonders led the construction of the "Negro Building" and organization of its exhibits for the 1907 Jamestown Tercentennial Exposition. When it was closed, they attempted to have the exhibits permanently reinstalled in Richmond as a Negro National Museum. The space they hoped it would occupy is telling; to get white Richmonders' support, the organizers agreed not to attempt to place it on Grace or Franklin Streets but also made it clear that they "did not want it set way back in Jackson Ward." See Daniel Webster Davis and Giles Jackson, *The Industrial History of the Negro Race of the United States* (Richmond, 1908); Peabody Clipping File No. 93, Collis P. Huntington Library, Hampton University, Hampton, Virginia; quote is from vol. 1, p. 169.

35. The Savings Bank of the Grand Fountain United Order of True Reformers opened in 1889; the Nickel Savings Bank in 1898; Mechanics Savings Bank in 1902; and the St. Luke Penny Savings Bank in 1903. Daniel Webster Davis, *The Life and Public Services of Rev. Wm. Washington Browne* (Philadelphia, 1910), 108; Browne was the leader of the True Reformers in the late nineteenth century.

36. *Richmond Planet*, January 4, February 15, March 15, March 22, March 29, May 10, May 13, June 7, June 28, July 12, November 5, November 15, December 6, 1890;

January 10, May 16, 23, 1891; January 5, 1895; September 10, November 12, 1898; August 26, 1899.

37. See, for example, the *Richmond Planet's* front-page coverage of the True Reformers' new bank building in 1891; the newspaper's regular front-page picture biographies in 1895; biographies of black Richmonders written by other black Richmonders; the writings of popular poets; or the speeches of businesswomen. *Richmond Planet,* May 16, May 23, 1891; 1895, passim; Williams, *A Sketch of the Life and Times of Capt. R. A. Paul;* Davis, *Life and Public Services of Rev. Wm. Washington Browne;* Wendell P. Dabney, *Maggie L. Walker and the I.O. of St. Luke: The Woman and Her Work* (Cincinnati, 1927); R. J. Chiles, "He Saw the Point," *Richmond Planet,* May 16, 1891; speeches in Maggie Lena Walker Papers, Maggie L. Walker National Historic Site, Richmond, Virginia (hereafter MLW Papers).

38. W. P. Burrell and D. E. Johnson, Sr., *Twenty-Five Years History of the Grand Fountain of the United Order of True Reformers, 1881-1905* (Richmond, 1909); *Richmond Dispatch,* October 17, 1890; Grand Fountain United Order of True Reformers, *1619-1907 From Slavery to Bankers* (Richmond, 1907).

39. This is Elizabeth Ewen's phrase for an entirely different phenomenon, the symbols of mass consumer culture, such as billboards and electric lights. "City Lights: Immigrant Women and the Rise of the Movies," in Catharine R. Stimpson, Elsa Dixler, Martha J. Nelson, and Kathryn B. Yatrakis, eds., *Women and the American City* (Chicago, 1981), 43.

40. Janice L. Reiff suggests computerized graphics as one means of exploring the various geographic boundaries of community articulated in oral histories, diaries, and other sources. Overlaying these images one could, for example, compare men's and women's descriptions of community, or see "if people who have jobs inside it describe the community the same way as people who don't." *Structuring the Past: The Use of Computers in History* (Washington, D.C., 1991), 35-37.

41. We draw here on Earl Lewis's discussion of how African Americans in Norfolk, Virginia, used the language of space to translate segregation into congregation; Earl Lewis, *In Their Own Interests: Race, Class, and Power in Twentieth-Century Norfolk, Virginia* (Berkeley, 1991), chap. 4. On Richmonders' conceptions of Jackson Ward, see video created for the exhibition "Jackson Ward: A Century of Community" (Richmond, 1987), and "Two Street," video created for the exhibition "Second Street" (Richmond, 1989).

42. Bernadine Simmons, "A Street Second to None," *Richmond Celebrates,* Special Arts Issue, 1987, 60-61.

43. The Emporium was intended as a space where women could participate in the growing consumer culture as customers, salesclerks, and managers without suffering the indignities of second-class service and menial jobs in white establishments. One strategy of white businessmen who opposed the store was to turn the space into a "rough" rather than a "respectable" area by opening up barrooms on each side. For a discussion of the Emporium, see Elsa Barkley Brown, "Womanist Consciousness: Maggie Lena Walker and the Independent Order of St. Luke," *Signs: Journal of Women in Culture and Society* 14 (Spring 1989), 610-33.

44. Barkley Brown, " 'Not Alone to Build.' "

45. We speak here not only of the most frequently reproduced visual images of Walker but also of the ways in which she is envisioned in historical studies and contemporary lore. Our alternative image is based in part on a reading of her diaries as well as in our mapping of her daily life.

46. Diaries, MLW Papers; Bruce Potter, "Maggie Walker House to Open," *Richmond News Leader,* July 12, 1985.

47. Public comment on the "Second Street" exhibition, which was sponsored by the Black History Museum of Virginia, Richmond Renaissance, and the Valentine Museum, testifies to the wide range of memories about Jackson Ward and Second Street. Visitor comment books posted in the exhibition and videotapes of public comment sessions contain many discussions of the major business and entertainment enterprises of Second Street, but also a significant number revealed concerns about the moral atmosphere on the street. One person noted, "When I was growing up in Richmond 1934-53 my parents didn't allow me to go on 2nd St. We used to sneak down there to the Globe and Hipp. Theatres and to the Skating Rink." Visitor comment books for "Second Street" exhibition, Archives, Valentine Museum, Richmond, Virginia. For a discussion of prostitution, see "Murder in a Jackson Ward Assignation House," *The Idea* 5 (October 7, 1911), 3-4.

48. Scott C. Davis, *The World of Patience Gromes: Making and Unmaking a Black Community* (Lexington, Ky., 1988), 31; *Richmond Planet*, April 30, May 14, 1904. Daphne Spain has argued that domestic work "kept women safely within the private sphere." We are suggesting that it took women fully into the public sphere not only because of their travels through the city but also because these work spaces became public space. See Daphne Spain, *Gendered Spaces* (Chapel Hill, 1992), 172.

49. For an insightful discussion of the assumption that the city was male terrain and that those women who "gazed back" were immoral, see Judith R. Walkowitz, *City of Dreadful Delight: Narratives of Sexual Danger in Late-Victorian London* (Chicago, 1992). See also Mary P. Ryan, *Women in Public: Between Banners and Ballots, 1825-1880* (Baltimore, 1990).

50. *Richmond Republic,* August 29, 1865, cited in O'Brien, "From Bondage to Citizenship," 199.

51. Diary of Edward McC. Drummond, 1910-1912, Drummond Papers, Valentine Museum, Richmond, Virginia.

52. See also the Baptist Ministers Association and the Mothers League, late-nineteenth-century organizations that incorporated residents within the boundaries of both Richmond and Manchester. *New York Tribune,* June 17, 1865; *Richmond Dispatch,* April 7, 1868; *Richmond Planet,* 1890s, passim.

53. Gregg Michel, "From Slavery to Freedom: Hickory Hill, 1850-1880," in Edward L. Ayers and John C. Willis, eds., *The Edge of the South: Life in Nineteenth-Century Virginia* (Charlottesville, 1991), 123; W. P. Burrell, "Report of the Committee on Business and Labor Conditions in Richmond Va.," *Proceedings of the Hampton Negro Conference,* No. 6, July 1902 (Hampton, Va., 1902), 43; *Richmond Times-Dispatch,* May 8, 1949; Elizabeth Dabney Coleman, "Richmond's Flowering Second Market," *Virginia Cavalcade* 4, 4 (Spring 1955), 8-12; interview with Mrs. Estelle F. Carter and Mrs. Sadie C. Sears, April 5, 1980, Henrico County, Virginia, VCU; *Richmond Planet,* September 13, 1898.

54. For one example of this discourse, see Earl Lewis's discussion of food patterns in Norfolk, *In Their Own Interests,* 92-93.

55. On the St. Lukes, see *Fiftieth Anniversary—Golden Jubilee Historical Report of the R.W.G. Council of I.O. St. Luke, 1867-1917* (Richmond, 1917); MLW Papers; Lillian Payne Papers, Valentine Museum, Richmond, Virginia; *Richmond Planet,* 1890-1930, passim.

56. Gregg D. Kimball, "Race and Class in a Southern City: Richmond, 1865-1920" (paper presented at Organization of American Historians annual meeting, Louisville, Kentucky, April 1991).

57. Typescript reminiscences of John O'Grady, Jr., 6 (copy in possession of authors, courtesy of Nora Witt).

58. For a description of black men in the building trades, see Kimball, "African-Virginians and the Vernacular Building Tradition." In 1890, black workmen still made up 29 percent of the brick- and stonemasons and 89 percent of the plasterers in Richmond. Department of the Interior Census Office. *Compendium of the Eleventh Census: 1890. Part III* (Washington, D.C., 1897), 718-19.

59. The late-nineteenth-century opposition of endangered versus dangerous women is developed by Mary Ryan, *Women in Public;* the late-nineteenth- and early-twentieth-century opposition of morality and dirt by Mary Douglas, *Purity and Danger: An Analysis of Concepts of Pollution and Taboo* (London, 1966). Dolores E. Janiewski, drawing on Douglas, notes the symbolic roles that occupational hierarchies play, stratifying workers by "varying degrees of purity." *Sisterhood Denied: Race, Gender, and Class in a New South Community* (Philadelphia, 1985). See also Dolores Janiewski, "Subversive Sisterhood: Black Women and Unions in the Southern Tobacco Industry," "Southern Women: The Intersection of Race, Class and Gender," Working Paper No. 1 (Memphis, 1984), especially 1-6. Tera Hunter applies these concepts to understanding the context of black women's labor in white households throughout the New South. Tera W. Hunter, "Household Workers in the Making: Afro-American Women in Atlanta and the New South, 1861 to 1920" (Ph.D. diss., Yale University, 1990). See also Phyllis Palmer, *Domesticity and Dirt: Housewives and Domestic Servants in the United States, 1920-1945* (Philadelphia, 1989).

60. The evidence of black and white women's contact in dance halls and saloons is contained primarily in newspaper accounts throughout the late nineteenth century of police arrests for disorderly conduct. For Judge John Crutchfield's view on women in barrooms, see, for example, *Richmond Times*, September 2, 1899.

61. For an insightful commentary on the difficulty of defining leisure in studies of working-class women given the degree to which their domestic work and leisure were intertwined, see Elizabeth Roberts's review essay in *Gender and History 6*, 2 (August 1994), 303-5.

62. On leisure and household shopping, see Melanie Tebbutt, "Women's Talk? Gossip and 'Women's Words' in Working-Class Communities, 1880-1939," in Andrew Davies and Steven Fielding, eds., *Workers' Worlds: Cultures and Communities in Manchester and Salford, 1880-1939* (Manchester, England, 1992), 49-73; Andrew Davies, *Leisure, Gender and Poverty: Working-Class Culture in Salford and Manchester, 1900-1930* (Buckingham, England, 1992), 130-38.

63. Jonathan Garlock, *Guide to the Local Assemblies of the Knights of Labor* (Westport, 1982).

64. Dolores Janiewski, "Seeking 'a New Day and a New Way': Black Women and Unions in the Southern Tobacco Industry," in Carol Groneman and Mary Beth Norton, eds., *'To Toil the Livelong Day': American Women at Work* (Ithaca, 1987), 168; Hunter, "Household Workers in the Making," 142-47; interview with Mrs. Maggie Alease Taylor Jackson Howard, n.d., History of Church Hill Project, Series 12, VCU.

65. See Peter J. Rachleff, *Black Labor in the South: Richmond, Virginia, 1865-1890* (Philadelphia, 1984), chaps. 7-11.

66. For the changing face of Richmond neighborhoods, see Silver, *Twentieth-Century Richmond.* The source for the neighborhoods around Tredegar is Richard Love, "From the Hilltop to the Bottom: Trajectories in Urban Development" (unpublished paper, cited by permission of Love).

67. Lewis, *In Their Own Interests*, 96; interview with Lorenzo Jones; interview with Mrs. Bessie Bailey Baldwin by Akida T. Mensah, October 9, 1982, History of Church Hill Project, Series 12, VCU.

68. Our discussion of respectability is most directly informed by Ellen Ross, " 'Not the Sort That Would Sit in the Doorstep': Respectability in Pre-World War I London Neighborhoods," *International Labor and Working Class History* 27 (Spring 1985), 39-59; Ellen Ross, "Survival Networks: Women's Neighborhood Sharing in London Before World War I," *History Workshop Journal* 15 (Spring 1983), 4-27; Peter Bailey, " 'Will the Real Bill Banks Please Stand Up?': Towards a Role Analysis of Mid-Victorian Working-Class Respectability," *Journal of Social History* 12 (Spring 1979), 336-53; Christine Stansell, *City of Women: Sex and Class in New York, 1789-1860* (New York, 1986); Roy Rosenzweig, *Eight Hours for What We Will: Workers and Leisure in an Industrial City, 1870-1920* (Cambridge, 1983); and discussions with Victoria Wolcott, Tera W. Hunter, and Jerma Jackson.

69. The scholarly emphasis on moral regions is longstanding; see, for example, Robert Park, "The City: Suggestions for the Investigation of Human Behavior in Urban Environment," *American Journal of Sociology* 20 (1916), reprinted in Richard Sennett, ed., *Classic Essays on the Culture of Cities* (New York, 1969), especially 128-30. More recently, see, for example, Neil Larry Shumsky, "Tacit Acceptance: Respectable Americans and Segregated Prostitution, 1870-1910," *Journal of Social History* 19 (Summer 1986), 665-79; Joanne Meyerowitz, "Sexual Geography and Gender Economy: The Furnished-Room Districts of Chicago, 1890-1930," *Gender and History* 2 (Autumn 1990), 274-96; Timothy J. Gilfoyle, *City of Eros: New York City, Prostitution, and the Commercialization of Sex, 1790-1920* (New York, 1992). We have borrowed the term *moral geography* from Perry R. Duis, *The Saloon: Public Drinking in Chicago and Boston 1880-1920* (Urbana, 1983); our definition, however, differs somewhat from Duis's. By moral geography we mean both the coding of some behaviors as immoral and the confinement of these activities to certain regions of the city, and the adoption of a moral mapping which codes neighborhoods and their residents as moral and immoral, respectable and not, by virtue of their location rather than their behavior.

70. Lillian W. Betts, "The Richmond of To-Day," *The Outlook* 65 (August 25, 1900), 977; *Richmond Planet*, September 17, 1898, July 15, 1899; Dabney, "Rough Autobiographical Sketch," 10. Christine Stansell, in a study of antebellum New York, examines the way white middle-class women's prescription against white working-class women's and children's use of the streets turned "a particular geography of sociability" into "evidence of a pervasive urban pathology." *City of Women;* quote is on 203.

71. In 1883, for example, a dispute arose in Ebenezer Baptist when some members allowed dancing at an entertainment the proceeds of which were to benefit the church. *New York Globe*, September 29, 1883.

72. Focusing on Atlanta, Tera W. Hunter analyzes the geography of urban leisure, exploring "the relationship between work and play; class conflict among African Americans; and, competing concepts of gender, godliness, and sexuality" in " 'Sexual Pantomimes,' 'Hurtful Amusements,' and the Blues Aesthetic" (unpublished paper, cited with permission of Hunter).

73. *New York Globe*, April 26, 1884. Throughout the late nineteenth century, elite black Richmonders put their quiet, demure social life on public display in the pages of the *Richmond Planet*. These elaborate details of the dress, food, entertainment, and guest lists at their private affairs were intended as evidence of respectability and race progress.

74. A. Binga, Jr., "Dancing and Evil Fruit," in A. Binga, Jr., *Sermons on Several Occasions*, vol. 1 (Richmond, 1889), 195.

75. This is based on our reading of Minutes, First African Baptist Church, Books 2 and 3, 1875-1930, microfilm copy in Archives, Virginia State Library, Richmond, Virginia.

76. Ross, " 'Not the Sort That Would Sit in the Doorstep,' " 39.

77. Johnson, *History of the Colored Volunteer Infantry*, 41-42. Although this refers specifically to the policy of the Petersburg militia, the units in Richmond and Petersburg shared many of the same rules; additionally, because militia units shared many social activities it is likely that a number of Richmonders were present at these functions.

78. Brian Harrison, "Traditions of Respectability in British Labour History," in Brian Harrison, *Peaceable Kingdom: Stability and Change in Modern Britain* (Oxford, 1982), 16.

79. Although excursions remained a popular activity throughout the late nineteenth century, often sponsored by churches, clubs, and mutual benefit societies, they also came under attack by those who thought working-class people should save the money spent on such and should save their energies so as to be more productive laborers. See, for example, the heated debate over excursions at the Acme Literary Society meeting reported in *New York Globe*, August 18, 1883. For an example of these discussions on the national level, with an emphasis on the moral as well as financial dangers of excursions, especially for "our girls," see Margaret Murray Washington, "Club Work as a Factor in the Advance of Colored Women," *Colored American Magazine* 11 (August 1906), 85.

80. *Richmond Planet*, October 31, 1896.

81. Betts, "Richmond of Today," 977; "A Ramble in Virginia: From Bristol to the Sea," *Scribner's Monthly* 7 (April 1874), 664; *Richmond Dispatch*, September 1, 1982. See also Orra Langhorne, a southern white woman's address to the 1899 Hampton Negro Conference wherein she advocated "thick-soled shoes and simple dresses, and hats . . . as necessary factors in the improvement of the race." *Hampton Negro Conference Number III. July 1899* (Hampton, Va., 1899), 45. A useful discussion of dress and morality is Mariana Valverde, "The Love of Finery: Fashion and the Fallen Woman in Nineteenth-Century Social Discourse," *Victorian Studies* 32 (Winter 1989), 169-88.

82. Rosenzweig, *Eight Hours for What We Will*, 78-79.

83. Simmons, "A Street Second to None," 19, 60-61; *Richmond Afro-American*, March 7, 1981. For a social history of African American vernacular dance and the locations in which it developed, see Katrina Hazzard-Gordon, *Jookin': The Rise of Social Dance Formation in African American Culture* (Philadelphia, 1990).

84. Minutes, Book 3, First African Baptist Church; *The University Journal* 5, 5 (March 1905), 74.

85. Helen Jackson Lee, *Nigger in the Window* (Garden City, 1978), 26-29. Similarly, Elizabeth and John Dabney's adherence to their church's prohibition had kept them from dancing as their children were growing up in the 1870s and 1880s, but they watched with keen interest every new step performed by their children's friends at entertainments in their home. Dabney, "Rough Autobiographical Sketch," 60-62.

86. *Richmond Times*, April 4, 1899. Similarly, Victoria Wolcott identifies a dichotomous "language of geographic identity" in the twentieth century "discourse of urban migration and resettlement": southern and rural life were equated with "disorder, dirt, and licentiousness" and northern, urban life with "self-restraint . . . cleanliness," and "respectability." "Mediums, Messages, and Lucky Numbers: African-American Female Leisure Workers in Inter-War Detroit" in Patricia Yaeger, ed., *The Geography of Identity* (Ann Arbor, forthcoming).

87. Dabney, "Rough Autobiographical Sketch," 69-70; "Still Another Red-Light District," *The Idea* 4 (April 2, 1910), 8-9; "Wholesale Arrests: Negroes Treated Like Cattle," *The Idea* 4 (August 13, 1910), 4.

88. *Richmond Planet,* October 25, November 1, December 20, 1902. See also "Resolutions," *Ninth Annual Report of the Hampton Negro Conference,* 13, for the assumption that alley homes were inherently immoral and unfit places to properly rear and train children.

89. Gould and White, *Mental Maps,* preface; Glassberg, *American Historical Pageantry,* 1.

90. Hunter, "Household Workers in the Making," 140; John Cell, *The Highest Stage of White Supremacy: The Origins of Segregation in South Africa and the American South* (Cambridge, 1982), 134.

4

Connecting Memory, Self, and the Power of Place in African American Urban History

Earl Lewis

hen asked to talk about the 1930s, Susan J. somewhat anxiously and cautiously volunteered: "The Depression came on . . . and the taxes . . . and I lost it. That got to me more than I ever let anything get to me, leaving out death." The "it" was her brother Luke's (Bud Luke, she said) house.[1] Susan had paid the mortgage on the dwelling once, but her father needed money and refinanced so she was asked to pay the mortgage a second time. This experience, she maintained, she would remember forever.

When asked about Marcus Garvey, her memory clouded considerably. Although she was a young adult living in Norfolk when the

Author's Note: An earlier version of this essay was presented at a meeting of the Organization of American Historians, April 15-18, 1993, Anaheim, California. I would like to thank Katherine Corbett, Linda Shopes, Kenneth Goings, Raymond Mohl, and Elsa Barkley Brown for comments on earlier version.

famed leader made several trips to the city and adjacent municipalities, she had a vague memory of him at best. When I, in somewhat exasperated tone, proclaimed Garvey was one of the most important figures in African American history, she calmly conceded the point. From my questions, she knew he was important and that she was expected to know him; yet she did not. Are we to question Susan's memory of Norfolk? No, because in many respects her sense of place was unimpeachable. She knew Norfolk, its sights, sounds, and people; moreover, it was her home for more than fifty years. But she also knew the pain of losing a house that had twice belonged to her. Such memories loomed larger than others because they remained in the foreground of her experience and helped define who she was.

This example illustrates the importance of examining the connections among place, memory, and urban history. It suggests that the memories that African Americans have of a particular place are defined from the inside out. Race, gender, class, ethnicity, age, and other parts of one's identity help situate those memories. As a result, larger historical processes, although important, are not how most people write their own individual narratives of place. Instead, they remember the pain and joy, triumph and despair, conflict and resolution that marked their daily lives.

Ironically, in our attempt to write social history from the ground up, urban historians, like most historians, have paid more attention to the accuracy of particular recollections and far less attention to interpreting those recollections. Yet phrases like race relations, ghettoization, and even proletarianization are not how people remember their lives in the urban setting; rather, such framing concepts reflect the intervention of the historian and highlight the interface of our imagination, understanding, and memory with that of the subjects of our study. Underscoring this point, David Thelen observed, "The struggle for possession and interpretation of memory is rooted in the conflict and interplay among social, political, and cultural interests and values in the present."[2] Because of this inherent struggle and conflict, memory functions as a contested area of deeply held but at times highly idiosyncratic beliefs. At the same time, a society's cohesiveness hinges on its ability to create national or group memories, which enlist the support of large segments of the population. Few have studied the nexus between individual memories and group behavior. More important, as this essay will attempt to highlight, we know less than we should about the conjunction between memory, identity, and the importance of place. Such an investigative pursuit should lead us to rethink key categories in urban and African American history and to reexamine the relationship between race and place.

The Importance of Place

For some time, "place" has been the imagined linkage between the present and the what-had-been for African Americans. After all, place connected a diverse creolized population to an ancestral homeland that few had seen or would see, and that ultimately never existed. Long before African Americans were colored, or Negro, or black, or any of the names they were to be called, they were Ewe, Yoruba, Asante, Ibo, and Wangara. It was the imaginings of place that molded those disparate experiences into a corporate identity in the United States. Through names, music, poems, stories, dreams, and nightmares, Africa came to mean something. In time, we would call this movement a diaspora, an African diaspora, which had profound social, spiritual, and political consequences for a people who had imagined ties to a special place.[3]

The themes of movement and place took a new and important turn in the twentieth century, leading to a new interest in Africa American urban history. For the past twenty-five years, historians of African Americans who examined and analyzed the meaning of place in the twentieth century started with the mass migrations of blacks from the rural to urban setting. Between 1900 and 1920, roughly 1.5 million African Americans left the rural South; between 1940 and 1970, more than triple that number left. The staggering demographic transition had profound implications because the shift precipitated the general movement of blacks cityward. In 1900, the least urbanized segment of the population, African Americans, by 1960 were the most urbanized. As a result, beginning in the late 1960s and early 1970s, many inner cities became remarkably blacker, especially after the cathartic rebellions and eruptions of the mid- and late 1960s. By the mid-1970s, therefore, many major cities featured black majorities.[4]

The dramatic redistribution of the twentieth-century African American population had the unintended consequence of blurring the meaning of place once again. In a number of urban communities, the new arrivals stood out and were singled out. They clearly hailed from a different place. Sometimes their dress or speech gave them away; other times they were distinguished by their associations, housing arrangements, or employment. Migrants, although in a new place, wore the imprimatur of another place.[5]

Concerned as they were with migration to the city or settlement within the urban context, few historians stopped to ask what "place" meant to African Americans—migrants or long-time residents. Scholars did not ignore the centrality of place, but as Hershberg remarked more than a decade ago, many viewed the city as a site or place to execute life's pedestrian affairs.[6] Seldom did scholars stop to ask how memory configured an understanding of place.

Proponents of the ghetto-formation approach sharply analyzed the social impediments that circumscribed opportunities for blacks. Despite some textual variation, they all concluded that occupational, geographic, and residential mobility limited blacks to major urban ghettos.[7] Ghettos in this context functioned as places marred by limited opportunity, privations, and social unease; moreover, in an ironic twist, because the ghetto came to embody all that is negative, sociologists and journalists identified it as a place to flee from or escape—very much like the rural South decades before.[8]

Neither the ghetto formation literature that grew out of the social concerns of the 1960s nor the proletarianization critique that followed satisfactorily questioned how African Americans thought about or mapped place. As Trotter, architect of the proletarian critique, mentions, few people self-consciously moved to the urban North to live in a ghetto—that was not the framing experience or motivation for their actions.[9] Most left to secure a better future for themselves and their families.[10] In the process, they made the journey from migrants to residents of the new places they called home.

At the same time, neither did most migrants think of themselves as proletarians nor view their life course through the lens of proletarianization. Most undoubtedly sought well-paying, secure employment, and at a certain level they understood the relationship between individual choice and structural change. As scholars ranging from Joe Trotter to Roger Lane have noted, from the late 1800s through the 1960s, African Americans throughout the nation were all too aware of the barriers to full occupational mobility and material comfort. But, as scores of residents in northeastern cities told anthropologist John Gwaltney in the 1970s, foremost they wanted to live a decent life. Edith Baker, one of Gwaltney's collaborators, retorted after a caseworker criticized her exacting housekeeping rules, "Lady, if I had a woman to clean my house while I'm out cleaning somebody else's, I could do what you say, but I'm not living in filth for nobody, least of all a child that I have birthed and taught right from wrong." Gwaltney called these learned "truths" core black culture. Because of the current interest in social construction, some might find such a conceptual orientation uncomfortably close to racial essentialism. But it is clear, as other examples in his and other studies reveal, that black urban dwellers understood all too well the two-sidedness of urban living. Another caseworker could have as easily reprimanded the woman for keeping an untidy house, had this been her style. For folks like Baker, place became inextricably tied to a profound sense of decency.[11]

The Role of Memory

Meanwhile, the Susan J. example given at the outset suggests there is no single memory of a given place. The fact that Susan J. had no memory of Garvey should not lead us to question Garvey's importance to a larger Norfolk. Scores knew him, followed him, and proclaimed him their Moses. Instead, the example forces us to ponder how memory rewrites the meaning of place. Place, after all, is a location on a map, an imagined belonging, the scene of a bitter memory or beautiful happening; place, though often fixed, was always transportable. Scores moved away from a place only to reclaim it in their new locations, forming clubs and associations that marked their ties. Most important, for many, *place* was home.

It is the centrality of home in understanding the power of place that urban historians must more effectively explore. Homeboys and homegirls are not mere creations of a hip-hop culture. For generations, African Americans, even those who despised the place of their birth, proclaimed their allegiance to that place. Richard Wright captured a larger sentiment when he wrote:

> I was leaving the South to fling myself into the unknown, to meet situations that would perhaps elicit from me other responses. . . . Yet, deep down, I knew that I could never really leave the South, for my feelings had already been formed by the South. . . . So, in leaving, I was taking a part of the South to transplant in alien soil.[12]

In the process, the memories of a distant place fused with current conditions, forcing Wright and thousands of other migrants to expand the meaning of home.

As the social columns in black newspapers richly illustrate, thousands journeyed to and from communities in valiant efforts to stay connected. Surely, some allowed unwanted memories to retreat from public awareness, including links to places no longer considered home. Others, however, celebrated rather than denied their earlier existences. Susan and Clifton J. regularly returned to Norfolk County to see family friends decades after they moved away. Norfolkians who called New York home advertised themselves as members of the Sons of Norfolk association. State clubs in Chicago, the Bay Area, and elsewhere, organizations of affinity that advertised one place of origin, bridged the social and psychological distance produced by migration.[13]

In the urban context, especially in Norfolk, home meant both the household and the community. This is an important consideration because it forces us to probe the interior of historical construction. Mr. and Mrs. J. W. Jones moved to New York, often returning to Norfolk to see her sister Perlie Hall, as they did in 1925. Clearly, affective attachments brought them south. But we can also speculate that a part of them remained southerners—perhaps, even Norfolkians.[14]

We need at this juncture to further probe the constitution of the self because as psychotherapist Bruce Ross reminds us, in certain ways, all memory is autobiographical. And like the psychotherapist, he contends, the historian must consider the feeble ability of humans to recall the past. Ross writes,

> History and personal memory have in common that they try to determine which facts and events are true. We can . . . label history as "objective" and memory as "subjective," but such labels are relative at best; there are many parallels and much remains only probable in validating the contents of either.[15]

Ross's entreaty is useful as a point of departure. Literary scholars, meanwhile, have noted the remarkably structured nature of autobiographies.[16] Most detail a story of life-long progress. More than that, early slave narratives were characterized by the inordinate concern with telling a recognizable story and authenticity. Frequently, prominent whites introduced the authors, attesting to their veracity and claims of authenticity. In so doing, they inserted the memory of their exalted social status as a marker. Remember, they seemed to say, you recognize me, I am a noted citizen and neighbor, one who has accumulated the proper social credits, and one you can trust.[17]

Memory is too structured in certain ways, although we have yet to understand the intricacies of those structures. As a result, some cognitive psychologists have attempted to distinguish between *semantic memory* and *episodic memory*. The former refers to the ability to recall things that have little to do with us personally; the latter pertains to the recollection of highly individual experiences. But as critics have noted, such a conceptualization is time and culture determined. In premodern, nonwestern societies, where time is neither linear nor cyclical but spherical, or in those societies organized in other ways, the universal and the individual are indistinguishable. We know, however, that no one locus in the brain stores all memory; instead, the brain functions as a superhighway, transporting the discrete and global to access points for retrieval. Along the way, the information deposits itself at rest stops, encouraged to go on when associations in the conscious realm trigger a recollection. Those recollections cohere to form a broad memory, a trace of a memory, or no memory at all.[18]

As Fentress and Wickham remind us, "Memory is a complex process, not a simple mental act."[19] Few people can or choose to reveal all of themselves—even to themselves. In fact, some events remain instantly retrievable, whereas others fade quickly and often permanently. Still, recalling the past, which begins as an individual endeavor, has tremendous social importance. Nations build monuments, create festivals and holidays, consecrate buildings, and name battlefields for the purpose of securing a societal memory, a way of tying the most isolated resident to the larger fabric of society. And in many respects, the veracity of those built memories are not as important as the process and purpose of memory construction. "For our purposes," writes Thelen, "the social dimensions of memory are more important than the need to verify accuracy. . . . People develop a shared identity by identifying, exploring, and agreeing on memories." Therefore, as we reformulate our critical understanding of the relationship between place and memory, we must reconsider the factors that shape memory.[20]

Moreover, to call history objective and memory subjective inflates the value of the former and unduly deflates the latter. History represents the coming together of many memories, including those of the scholar. In his recent book on public memory, commemoration, and patriotism, John Bodnar positions class and ethnic difference as salient features of the construction of public memory in the United States. He details, for instance, how European ethnic groups or the Knights of Labor proclaimed their own versions of a public memory.[21] In other areas, we have been asked to look more closely at the conjunction of race, class, and gender as the constitutive elements of the self.[22] Yet as we think more critically about memory, we are required to go further and reassess our understanding of the overall construction of the self.

For example, throughout the twentieth century, black people in northern, western, and southern cities overcame the differences that separated them to fight in their own interests. At a certain level, they recognized race as a construction, a way of organizing their world because of real differences in power based on skin color and social characteristics. Race, therefore, functioned as both a social fiction and a social reality. It was a social fiction because African Americans constantly delineated the factors that divided them, even challenging the efficacy of a biological explanation of race. It was a social reality because few could deny that racial membership influenced one's access to jobs, income, housing, education, and social services. Thus, when historical actors tell us that class, color, gender, and religious differences are not merely artifacts of the historian's imagination, we should listen. After all, these other factors represent central aspects of the lives of city residents. Yet it is also worth remembering that for most of the twentieth century, race became the vehicle for the public presentations of African American needs and demands in the urban setting.[23]

As important, for far too long historians have ignored the relationship between race and the other aspects of the self. Often, scholars have spoken as if people had a choice, or that the choice was simply between race and gender, or class and race. Indicative is the tendency to still use words like *privileging* to describe how people organized their social worlds and access parts of their identities at discrete historical moments. Part of the problem is how we interpret events, and part of the problem is the current discourse on what motivates historical actors. For example, more than six decades passed before Charles Grandy shared his memories of slavery and attempted to explain why blacks migrated to Norfolk. He told a WPA interviewer, "Nobody owned the niggers; so dey all come to Norfolk, look lak to me." In the concluding months of World War II, Dolly Jones told an interviewer she had taken steps to vote because "five thousand qualified voters would make a lot of difference in the attitude of other people toward us, and [in] many of our desires for a better Norfolk."[24]

In broad outline, both comments draw our attention to the enabling actions of African Americans. They contextualize the struggles that began in the 1860s and concluded in certain ways in the 1960s. They even explain why most African Americans refused to flee the South during the age of Jim Crow, and instead stayed put to fight and improve local conditions. For Grandy, the memory of slavery functioned as an important counterreference: Blacks fled to the city because it defined the meaning of freedom. On the other hand, coming of age in a Jim Crow city redirected the strategies Jones and her neighbors adopted. The franchise was no panacea but it could be added to the arsenal of the city's black communities.

Place, Memory, and the Self

Race, however, is only one element of the self that accounts for the comments offered. How people foreground and background other elements of the self must become part of our critical repertoire. Age, marital status, sexual orientation, and other social markers provide critical cues for us and for those we study. Grandy noted the number of blacks who came to Norfolk during the Civil War because of when it happened. He was a relatively young man who saw the war as an end to a painful existence; for him, movement to Norfolk represented a new beginning. As Litwack and others have observed, others felt more at ease in the place they called home and refused to migrate.[25] Grandy's age, life cycle, gender, and work experiences are as important as his race in explaining his actions and possible motivations. In fact, to understand Grandy's racial self we have to keep in view subtle shifts in alignment of other elements of his person.

In thinking about memory, it might help to think of the self as multipositional. Several critical theorists have advanced the notion of *multiple subjectivities.*[26] Accordingly, each aspect of the self constitutes another subject. Such a formulation has the danger of obfuscating the ways in which we foreground or background aspects of the self. As a result, scholars sometimes come dangerously close to explaining the inevitable contradictions as irreconcilable and crazy. Few, however, are clinically schizophrenic. Most of us are engaged in the awkward process of simultaneously negotiating the past, present, and future.

As an alternative, the subject or self should be considered singular and positions of the self multiple—or multipositional. It is important, however, to remember that we can never see the total self. At best, we glimpse the totalizing self. It is a self that refuses to surrender to a simple mathematics. Race plus class plus gender does not approximate the complexities of the self because no one is simply additive. The notion of multipositionality takes into consideration a complex social calculus, a calculus that allows us to add, subtract, multiply, and divide parts of our identities at the same time. Such a perspective allows us to examine how race is shaped by other aspects of the self and, in turn, how race shapes those aspects.[27]

But we must also remember that the process of identity formation is neither linear nor always intuitive. At various times, one part of our identity is struggling to displace another. Memories of losing a house during the depression pained Susan J. because it rekindled memories of her subordinate status within her own family, even as it revealed her firm commitment to the family's needs. Her parents had two sets of children; the oldest group could easily have been the younger ones' parents. In her family of eighteen, Susan came from the younger group and her brother, Luke, the older. Birth order combined with race, gender, and age to situate her place within her family and to frame her memories.[28]

In the sharp rejoinders to those who have highlighted the disintegration of African American family life, too many of us have failed to discuss the honest pain of Susan J. and others. Women of a variety of backgrounds sought refuge in the city, often as a way of altering familial responsibilities. Susan's sense of familial obligation is clear, however; so, too, is the power of the men in her life. After all, as an unmarried, working daughter she was forced to pay for a family house that was in her older brother's name. Luke did not lack resources; he worked as a teacher in the Titustown section of Norfolk, near his house. Yet her income was tied to the family's economy more than was her brother's. Looking back on this episode from the distance of fifty years meant filtering through the anger, rage, and feeling of loss all over again. For her, it meant remembering an earlier episode about an important place marked by both conflict and resolution. Race, family, class, and gender became part of the mix of emotions, experiences, and memories that defined that moment.[29]

For those of us concerned with African American urban history, the need to problematize identity construction is clear and important. We know that in every city studied there was one community and several communities. When the black-owned Norfolk *Journal and Guide* complained about poor city services between 1920 and 1945, often it dramatized its concerns by highlighting the troubles of the black middle class, explicitly ignoring the plight of working-class blacks. Many black residents felt equity certainly escaped them when even the "best" among them were denied key rights or privileges. As one resident bemoaned,

> We pay equal taxes, but because colored people live in these streets the city won't repair the roads. They are rich people living in these houses, all Negroes. Several of them own cars. . . . Now look on the other hand at this street. It's a white street, all smoothly repaired. What a beautiful surface; see the difference![30]

Read through the prism of race alone, important examples of intraracial class cleavages fade from view.

Equally dangerous, however, is reading this example literally. Some, such as Michael Katz, have drawn a sharp distinction between residential proximity and class-integrated neighborhoods in black communities. He insists that "most urban African Americans . . . always were poor, and the small middle class that did exist distanced itself from its less-fortunate neighbors."[31] Such a claim is as incorrect as it is correct. As urbanists have noted for more than a century, in cities such as Philadelphia, Chicago, New York, Cleveland, Milwaukee, and Norfolk, real differences divided African Americans, among them class and status.[32] Nonetheless, internal conversations frequently blunted the recognition of sharply drawn class distinctions. After all, when teaching became a primarily female occupation after 1910, who did the women marry? From the skimpy evidence that is available, many married men with working-class jobs and better paying salaries. Moreover, if the autobiographical literature is any indicator, many African American households—or at the very least families—were not only intergenerational but housed members of an array of social classes.[33] Class in the black community must be viewed as part of an intraracial discourse. Oftentimes, a middle-class existence hinged on the community's agreement; as a consequence, most middle-class blacks lacked the luxury of removing themselves from their working-class relatives and neighbors, despite rhetoric to the contrary. During much of this century, but especially after 1920 and before 1970, they depended on working-class blacks for their livelihoods and acknowledged social status. Too much physical or social distance threatened the implicit racial covenant as well as their slightly elevated place in a race conscious society.[34]

Yet how do we strike the proper balance? In a recent discussion of the actions of the powerful and powerless, James Scott coined the phrase *infrapolitics*. This neologism pointed to the offstage or hidden transcripts that social actors used to critique the actions of their adversaries. Nonetheless, this thoroughly useful way of assessing power relations misses one important point: Between the hidden and public transcript one can find the semipublic transcript—that is, a song, story, poem, folktale, or toast.[35] This coded public message is audible but indecipherable to those outside the community of reference because most outsiders do not understand the importance of certain symbols, cues, and events. Of course, the ability to interpret these symbols is learned behavior, passed from adult to child through ritual, folktale, and social practice. Typical were the heroic tales of masculine figures such as Shine and John Henry, or the urban tales that paired African, Jewish, Irish, and Italian Americans.[36] As a result, certain public complaints not only set the middle class apart from the working class, but knitted them together through a language of shared suffering. To see the former and miss the latter means that we too quickly inflate the significance of public and private transcripts and minimize the importance of semipublic transcripts.

In large part, semipublic transcripts worked because of the way in which the accumulation of isolated incidents coalesced into a collective memory. First, residential patterns reinforced memory formation. In a city like Norfolk, municipal ordinances for racial zones controlled where one lived by the 1910s. Meantime, in northern and certain western communities steering by real estate agents, restrictive covenants, and the threat of racial violence worked as well as legal sanctions. As a result, regardless of location, through the 1960s, most upwardly mobile residents lacked the luxury of removing themselves—despite several court challenges. Consequently, through 1945, although a classic ghetto failed to materialize in most southern locations, legal and extralegal actions effectively defined the geographic character of urban black communities. And in both the North and the South, notwithstanding notable internal divisions, the urban layout increased the likelihood of residential proximity and communal action among African Americans.[37]

Second, the social construction of race led many blacks to define whites as the "other." At home, for the period 1910-1945, many black residents lived in an almost all-black world, one shaped but not totally defined by limited interactions with whites. In such a world, notions about power changed. In churches, on windowsills, street corners, or other places of congregation, African Americans sermonized, joked, sang, and for moments at a time altered power relations. More important, in these settings concepts like *minority, difference,* and *other* meant something other than what we have come to accept. When read from inside the black community outward, place helped re-situate the colored "other."[38]

In these secured zones, for instance, black urban dwellers created mythic characters like the urban folk hero Shine or writer Langston Hughes's everyday man, Jesse B. Simple (aka Semple). The former originated as an expression of the frustrated desires and hopes of the century's first two generations of black migrants. Fueled by dreams of beating the odds and securing a permanent foothold in the political economy, African Americans created Shine, who invariably beat the odds to realize personal fulfillment. Originally recorded in Mississippi and Louisiana, the tale underwent subtle changes in wording and meaning as blacks migrated to urban centers.[39] Through humor, exaggeration, and misdirection, the storytellers in the process critiqued the pretense of white omnipotence. From his place in Harlem, meanwhile, Simple surveyed conditions in the United States and commented—often wryly and satirically—on the foibles of whites and blacks. Both characters dealt with the urban environment as a utopia and a dystopia, paralleling the efforts of prominent black authors like Richard Wright, Ralph Ellison, James Baldwin, and Toni Morrison.[40]

One favorite Shine story features him as a stoker on the luxury oceanliner Titanic, just before it retires to the bottom of the ocean. With death imminent, Shine makes plans to flee the doomed vessel. Before he can depart, he is approached by the captain's wife and daughter, symbols of all that is within sight but out of reach. In an obvious commentary on the presumption of white male power and ability, the two decide that Shine is the only one capable of saving them; in exchange for their lives, they offer him sex. Shine teases that there is sex on land and on the sea, but at this moment he prefers sex on land to the sea, leaving the two to save themselves. In a final scene, the captain, the symbol of white male power, begs for Shine's help. The captain promises all the money he has if Shine saves him. Again, Shine prefers the comfort of land to the hazards of the sea. He leaves the captain to his own devices, jumps in the water, and races the sharks to the shore. In most versions of the story, Shine is on 125th Street in Harlem, in the company of several beautiful women, when the world receives word of the fate of the Titanic and its passengers.[41] To the thousands of African Americans who heard and retold this tale, Shine beat the white man and ignored the pleas of white women by finding and seeking comfort in his own surroundings. Through these stories, scores of black city dwellers critiqued their worlds and defrocked whites at the same time. For moments at a time, they also changed the vectors of power by seeing themselves as other than subordinates.

As important, these stories and the location of their transmission encourage an examination of the sites of relaxation and leisure in the urban environment. Robin Kelley has suggested that

> for members of a class whose long workdays were spent in backbreaking, low-paid wage work in settings pervaded by racism, the places where they played were more than relatively free spaces in which to articulate grievances and dreams. They were places that enabled African Americans to take back their bodies, to recuperate, to be

> together. . . . Despite opposition . . . black working people of both sexes shook, twisted, and flaunted their overworked bodies, drank, talked, flirted, and . . . reinforced their sense of community.[42]

We might extend his commentary to the stoops of Philadelphia row-houses in the 1950s, where the Philadelphia sound originated, and the buses, parks, and public places in the 1970s, where urban youth brought their boomboxes and transformed the space and place, commandeering it for their own purposes. Kelley is no doubt correct in insisting that some of this behavior stemmed from a learned oppositional practice, but the learning and passing on of those practices brings us back to the importance of memory for understanding the salience of place. For on those stoops and in those public spaces, mythic heroes linked together several generations of urban dwellers. Anchored in time and connected in space, those generations learned and relearned what it meant to be black in America; in the process, they manufactured memories of the commonalities of their histories. Thus, the Shine and Simple tales were both symbols of those memories and instruments of memory creation.

Third, racist comments, brutal police action, public embarrassments, or racial harassment became part of a social script. As Lawrence Levine, among others, has argued, the reading of this social script keyed African American survival and empowerment strategies before and after slavery.[43] The critical reading of this script did something else: It mediated the distance between individual actions and group imperatives. At its core, the concept of multipositionality works best at the individual level, explaining how individual actors foreground or background aspects of the self. But it also marks the intersection between individual experience and group relations.

In the political economy of twentieth-century Norfolk, for example, most African Americans occupied a similar social, political, and economic space. They were politically disfranchised citizens of working-class backgrounds who lived in one of four or five neighborhoods. Working-class and middle-class blacks shared a problem of inadequate city services. Therefore, the earlier quotation detailing inadequate city services, although class specific in character, was a general lament. Regardless of their station, most could verify the veracity of the complaint; this process of verification was influenced by the memory of other wrongs inflicted on the group. In such instances, the racial self moved to the foreground as other aspects of the self realigned. As a result, Jerry O. Gilliam, a self-proclaimed member of Norfolk's black working class, told one interviewer in the late 1930s:

> The whites deprive the Negro of privileges here like the mythical rat bites. I once slept in a back room in Washington, D.C., and big rats would run over me every night. A friend told me to get out of there or the rats would eat me to death. . . . "The rat," he said, "will take a small bite off your toe, and then blow on it so you won't feel, then take another nibble and blow some more, until the blood starts flowing and you bleed to death in your sleep." That's the way white people lull Negroes to sleep in Norfolk and then bite them till they're bled to dry.[44]

Gilliam could in one moment be a biting critic of the "college-bred" boys he disdained and at another moment a judge of racial practices that affected all blacks. For him and others, such varied stances seemed natural. Each turned on the importance of improving conditions at home because home was a place worth improving.

At the same time, attentiveness to certain memories and not others blinds us to other interpretative possibilities. As Elsa Brown has noted, women were as active in the political affairs of reconstruction Richmond as men. If we rely solely on who held elective office, whose voices were recorded in the newspapers, and whose counsel was sought by whites, we miss the ways black women used church, labor, and women's auxiliaries to direct the men in their communities and the affairs of those communities.[45] It is essential, therefore, to consider what we do not hear as well as what we are told.

Moreover, because the historian divides the world into work and home with a more perfect demarcation than most historical actors, the relationship between work and home must become part of any discussion of place. After all, the majority of black women who worked for wages in the urban South labored as domestic or personal servants. Going to work meant laboring in someone else's home. For the few who lived in, home and work meant the same thing at a certain level. Yet as the numerous studies of black domestics reveal, few were allowed to forget that they were hired help. Long and irregular hours, sexual harassment, and dismissal robbed all but the most fortunate of sentimental memories.[46]

In most urban settings, African Americans sought employment, pursued labor activism, and agitated for change because they thought better working conditions, improved wages, and wider opportunities redounded to the entire community. From the 1870s through the present, from the tobacco factories of nineteenth-century Richmond through the docks of twentieth-century Norfolk and New Orleans, the automobile plants of Detroit, and the steel mills of Birmingham and Pittsburgh, black workers joined the cause of organized labor. Doing so was not always easy or advisable. Racism plagued the nation's labor associations just as it infected the body politic. Through the 1950s, some union brotherhoods barred blacks; in other cases, capital promised and delivered more than labor ever could. Still, where the benefit was clear, African Americans openly and earnestly pursued the cause of organized labor. Born in 1909, New Orleans native Sylvia Woods, later a union stalwart in Chicago, remarked of her father, "He was a union man."[47] This simple declaration defined what it meant to be working class. Even when some eschewed labor, they could never comfortably separate what they did at work from their lives at home.

Of course, union and labor organizations were not the only ways that African Americans asserted control over their work lives. Politics spanned the spectrum from exercising the franchise to "goofing off" at work. Without question, African Americans, endowed with memories of cruel employers and racist vengeance, found various ways of empowering themselves at work. In some cases, building on the distinction between "stealing" and "taking" that had been part of African American survival tactics since slavery, they simply "took" what they needed to survive. On other occasions, they absented themselves from work, feigned illnesses and injuries, and sabotaged the work environment.[48] How we understand and interpret these actions hinges on our ability to analyze memories and their connections to place, to assess the politics of memory.[49]

By the time of his death in 1974, David Daniel Alston had earned the title "a giant in labor relations." The North Carolina native migrated first to Richmond, where he married and started his career as a member of the urban working class, arriving in Tidewater just before World War I. After working for a dredging company for twenty-five cents per day, he moved to Baltimore, where he landed a better-paying job as a longshoreman. His wife opted to stay on the south side of the Chesapeake Bay, so they still called Norfolk home. As he recalled, "It was hard coming home every weekend and going back to Baltimore for work, but I was able to get a job on the Norfolk and Western coal piers in 1918 and stayed there 28 years."[50] During this period, Alston replaced George Millner as the leading black labor figure in the Hampton Roads area. Within two years of leaving the docks, he had become president of the District Council of the International Longshoremen's Association (1946), and International Vice President of the ILA (1947). He would also serve as vice president at-large of the Virginia State Labor Federation (1939) and the first full-time black organizer for the state body (1945). Alston, who said, "I have always felt that if a man has the right to tell me how much a pair of shoes cost, I have the right to tell him how much a day's work will cost," ended his career as Senior Vice President of the ILA.[51]

The evidence is growing that David Alston represented the bulk of black laborers in key respects. Certainly, few shared his level of accomplishment. He negotiated some of the most difficult contracts in the history of Hampton Roads longshoremen; he had a building named for him; and he was the recipient of four testimonial dinners. But like the majority of his contemporaries, he tied the improvement of his community, his home, to improving the status of black workers.[52] Because of this, he readily took an active role in the affairs of his community. Alston belonged to several fraternal organizations including the Elks and the Masons; he regularly offered a tithe at First Baptist of Lamberts Point; and he offered his services to both the NAACP and the Boy Scouts.[53] To study what Alston accomplished at work in isolation from his view of the home distorts more than it clarifies. The result is a partial portrait of the confluence of memory, self, and place in African American urban history.

The Historian's Memory

It is important to remember, moreover, that memory is the joint possession of the historian and the historical actor. Although some of us may be troubled by the current practice of self-reflection, we should not shy away from a full discussion of our roles as partial architects of the histories we write. "Objectivity" and "subjectivity" are not polar extremes along a continuum.

Pinpointing the historian's place in the project of recovery and analysis means examining our relationship to the power of place as well. Presumably, personal experience influences the subjects we choose to study and the histories we opt to write. This is not to say we abandon the training learned in graduate school; rather, we must differentiate between objective and subjective and biased and unbiased. After all, none of those examined by Peter Novick who questioned the worth of objectivity recommended the substitution of bias; in part, this offends the process of socialization so key to becoming a professional historian.[54]

Still, local memory and urban history are not the exclusive properties of historical subjects. As someone born and raised in the Norfolk area, I brought a certain memory to my research and writing as well. The city's sights, sounds, and smells were not only revealed in the documents but observed over a short life. Even though I tried to carefully separate my memories from those discussed in my book on Norfolk, there were moments of intersection. As a child, I witnessed streets overwhelmed by rain-choked drains. I can never forget the humbling act of rolling up my pant legs and wading through dirty, thigh-high waters to reach my grandparents' house in the heart of black Norfolk. Even if it were not true, I (and many like me) believed our group, our community, endured such penalties of life far more than other groups, other communities. Evidence of earlier complaints simply reinforced the notion of inequities. At no time was I overwhelmed by my memories, but neither could I erase them.

Ironically, African American residents of the city trusted me because I had my own memories. Memories established me as a member and not an outsider, in much the same way that participant-observation researchers must gain the trust of those they study. Of course, the meaning here is of far greater significance. Vincent Harding observed a decade ago, "The responsibility of the black scholar is constantly to be alive to the movement of history and to recognize that we ourselves are constantly being remade and revisioned."[55] Residents, who valued examples of prodigious research, sound judgments, and keen analytical skills, embraced me because I was one of their own coming home to tell their story. As someone's child, grandchild, or friend, they trusted that I too had an understanding of *home.*

Susan J. understood this as well. She recognized the value of my personal and scholarly memory. She tolerated my intrusions and encouraged my efforts. The final product was important to her, I believe. No one had written a history of the successes and failures of African Americans in Norfolk. Far better than some academics, she knew my memory was as important to them as their memory was essential for me.[56]

In sum, as historians reexamine the importance of place, closer attention must be paid to how memories facilitate the writing and rewriting of utopian and dystopian meanings of home. As part of this process, we must display a greater interest in how people construct their identities. This means that we must view historical subjects as multipositional actors, who foreground and background aspects of themselves depending on the social context and historical period. Moreover, because most memories start as individual recollections, then through a process of socialized learning they become attached to group or national perspectives, it is imperative that we interrogate those memories with a fresh attention to what they say as well as what they do not say. As we do so, we must take care to analyze how our memories rewrite the experiences we document, describe, and analyze. Then, and only then, will we be in a position to fully appreciate the memory, the self, and the *power* of place in African American urban history.

Notes

1. As quoted in Earl Lewis, *In Their Own Interests: Race, Class, and Power in Twentieth Century Norfolk, Virginia* (Berkeley, 1991), 120; interview, January 2, 1981.

2. David Thelen, "Memory and American History," *Journal of American History* 75 (March 1989), 1127.

3. Work has been done on the making of an African diaspora or Africa as an imagined place in the cosmology of nationalist ideology. These studies have explored cultural transformation, continuity, and discontinuity, but few have puzzled over the intergenerational process of group creation and identification. For an example of the former, see Vincent Bakpetu Thompson, *The Making of the African Diaspora in the Americas, 1441-1900* (New York, 1987).

4. For an effective review of this literature and the particulars, see Kenneth L. Kusmer, "The Black Urban Experience" in *The State of Afro-American History*, Darlene Clark Hine, ed. (Baton Rouge, 1986), 91-122; and Joe W. Trotter, Jr., ed., *The Great Migration in Historical Perspective* (Bloomington, 1991), 1-21.

5. This literature is immense. A sample of those works include Emmett J. Scott, *Negro Migration During the War* (New York, 1920; reprinted Arno Press, 1969); Charles S. Johnson, "How Much Is Migration a Flight From Persecution?" *Opportunity* 1 (September 1923), 272-74; Louise V. Kennedy, *The Peasant Turns Cityward: Effects of Recent Migration to Northern Cities* (New York, 1930); H. G. Hamilton, "The Negro Leaves the South," *Demography* 1 (1964), 273-95; and Karl and Alma Taueber, "Changing Characteristics of Negro Migration," *American Journal of Sociology* 70 (1964), 429-41. Some of the recent works include Florette Henri, *Black Migration* (Garden City, 1976); Neil Fligstein, *Going North: Migration of Blacks and Whites From the South, 1900-1950* (New York, 1981); Peter Gottlieb, *Making Their Own Way: Southern Blacks' Migration to Pittsburgh, 1916-1930* (Urbana, 1987); and James R. Grossman, *Land of Hope: Chicago, Black Southerners, and the Great Migration* (Chicago, 1989).

6. Theodore Hershberg, ed., *Philadelphia* (New York, 1981), 4-7.

7. Representative works include Gilbert Osofsky, *Harlem: The Making of a Ghetto* (New York, 1963); Allan H. Spear, *Black Chicago: The Making of a Negro Ghetto, 1880-1920* (Chicago, 1967); and Kenneth L. Kusmer, *A Ghetto Takes Shape: Black Cleveland, 1870-1930* (Urbana, 1976).

8. For an example of this new trend, see Nicholas Lemann, *The Promised Land: The Great Black Migration and How It Changed America* (New York, 1991). Recently, sociologists Douglas Massey and Nancy Denton extended this argument. According to Massey and Denton, the fortunes of the underclass turn on the abilities of inner-city residents to escape the malevolent conditions of the ghetto. *American Apartheid: Segregation and the Making of the Underclass* (Cambridge, 1993).

9. Joe W. Trotter, Jr., *Black Milwaukee: The Making of an Industrial Proletariat, 1915-45* (Urbana, 1985), especially appendix 7.

10. Trotter, ed., *The Great Migration in Historical Perspective* provides a nice overview of the motivations and variations in experience and expectation.

11. Trotter, *Black Milwaukee*, chaps. 1, 3, 4; Roger Lane, *The Roots of Violence in Black Philadelphia* (Cambridge, 1986), argues that exclusion from industrial employment forced Philadelphia blacks to turn to petty crime and prostitution. As Gavin Wright has noted, at times the structure of the labor market created employment opportunities for blacks, but most often in the poorer-paying sectors. Gavin Wright, *Old South, New South: Revolutions in the Southern Economy Since the Civil War* (New York, 1986), chap. 6; John Langston Gwaltney, *Drylongso: A Self-Portrait of Black*

America (New York, 1993), 188. Similar perspectives are suggested in Bob Blauner, *Black Lives, White Lives: Three Decades of Race Relations in America* (Berkeley, 1989); Elijah Anderson, *Streetwise: Race, Class, and Change in an Urban Community* (Chicago, 1990); and Mitchell Duneier, *Slim's Table: Race, Respectability, and Masculinity* (Chicago, 1992), although each also discusses the ever-changing meaning of decency and ensuing conflicts between the "d heads" (Anderson's phrase) and the streetwise youth who have staked a claim to many inner-city communities.

12. Richard Wright, *Black Boy* (New York, 1966), 284.

13. Lewis, *In Their Own Interests*, 102-9; Grossman, *Land of Hope*, 156; and Shirley Ann Moore, "Blacks In Richmond, California, 1930-1945" (Ph.D. diss., University of California at Berkeley, 1989).

14. Lewis, *In Their Own Interests*, 104.

15. Bruce M. Ross, *Remembering the Personal Past* (New York, 1991), 151.

16. See, for example, William L. Andrews, *To Tell a Free Story: The First Century of Afro-American Autobiography, 1760-1865* (Urbana, 1988); and William L. Andrews and Nellie McKay, guest eds., *Black American Literature Forum* 24 (Summer 1990).

17. Andrews, *To Tell a Free Story*, 1-31.

18. James Fentress and Chris Wickham elaborate on the general themes in *Social Memory* (Cambridge, 1992), especially 20-23. Thelen offers a nice description of the physiology of memory processing in "Memory and American History," 1120-21.

19. Fentress and Wickham, *Social Memory*, x.

20. See, for example, John Bodnar, *Remaking America: Public Memory, Commemoration and Patriotism in the Twentieth Century* (Princeton, 1992); Gaines M. Foster, *Ghosts of the Confederacy: Defeat, the Lost Cause, and the Emergence of the New South* (New York, 1987); George Lipsitz, *Time Passages: Collective Memory and American Popular Culture* (Minneapolis, 1990); and Michael Kammen, *Mystic Chords of Memory: The Transformation of Tradition in American Culture* (New York, 1991). Thelen, "Memory and American History," 1122.

21. Bodnar, *Remaking America*, chaps. 1-5 and conclusion.

22. The call for a race, class, and gender approach has been signaled so often that it has the character of a mantra. Although an important intervention when first issued, it now has the effects of stunting other equally rewarding developments. Most useful is the developing literature on identity construction emanating from the scholarship in critical theory, postmodernism, and poststructuralism. For examples of useful discussions, see Gerald Early, ed., *Lure and Loathing* (New York, 1993); Lawrence Grossberg, Cary Nelson, and Paula Treichler, eds., *Cultural Studies* (London, 1992), especially chaps. 6, 12, 19, 24, 27, and 38; and Linda Alcoff, "Cultural Feminism Versus Post-Structuralism: The Identity Crisis in Feminist Theory," *Signs* 13 (1988), 405-36.

23. Differences and solidarity have been twin themes in the historiography from the very beginning. See, for example, W. E. B. Du Bois, *The Philadelphia Negro* (New York, 1967; originally published 1899), which discusses class, criminal, status, and social behavioral differences among blacks as well as the search for community cohesion. Building on the pioneering study of differences among black families by E. Franklin Frazier (*The Negro Family in the United States*, Chicago, 1967 edition; originally published 1939), St. Clair Drake and Horace Cayton continued this line of investigation in *Black Metropolis* (New York, 1945), detailing the considerable differences among black Chicagoans. Although somewhat muted in the early years, historians in the ghetto school explored internal differences as well. See, for example, Osofsky, *Harlem*, particularly chap. 9; Spear, *Black Chicago*, chaps. 3, 4, 8-10; and Kusmer, *A Ghetto Takes Shape*, chaps. 5, 6, and 11. Trotter, Lewis, and others have both expanded the discussion of differences among blacks and explored the tension be-

tween group solidarity and group unity. For evidence of race as a nonbiological, social fiction, see Alain LeRoy Locke, *Race Contacts and Interracial Relations,* ed., Jeffrey C. Stewart (Washington, 1992), 1-14.

24. Lewis, *In Their Own Interests,* 9-10 and 196.

25. Leon Litwack, *Been in the Storm So Long* (New York, 1979), chap. 1.

26. See Ben Agger, "Critical Theory, Poststructuralism, Postmodernism: The Sociological Relevance," *Annual Review of Sociology* (1991), 112. Alcoff, "Cultural Feminism Versus Poststructuralism," 420-21; see also Nicola Gavey, "Feminist Poststructuralism and Discourse Analysis," *Psychology of Women Quarterly* 13 (1989), 459-75, especially 464-66; and bell hooks, *Yearning: Race, gender, and cultural politics* (Boston, 1990), chap. 2, for a small but representative sample of approaches.

27. A fuller discussion of the implications and mechanics of this approach are outlined in Earl Lewis, "Invoking Concepts, Problematizing Identities: The Life of Charles N. Hunter and the Implications for the Study of Gender and Labor," *Labor History* 34 (Spring-Summer 1993), 292-308.

28. Interview, January 2, 1981.

29. Joanne Meyerowitz, *Women Adrift: Independent Wage Earners in Chicago, 1880-1930* (Chicago, 1988); interview, January 2, 1981.

30. Lewis, *In Their Own Interests,* 80.

31. Michael Katz, ed., *The Underclass Debate: Views From History* (Princeton, 1993), 446. Katz may have simply meant physical distancing, but a degree of social distancing is an assumed result of physical separation.

32. See Du Bois, *The Philadelphia Negro;* E. Franklin Frazier, *The Negro Family in the United States;* St. Clair Drake and Horace Cayton, *Black Metropolis;* Osofsky, *Harlem,* particularly chap. 9; Spear, *Black Chicago,* chaps. 3, 4, 8-10; and Kusmer, *A Ghetto Takes Shape,* chaps. 5, 6, and 11; Trotter, *Black Milwaukee,* chaps. 2-6; and Lewis, *In Their Own Interests,* chaps. 2-3, 5-7.

33. The best source on household construction, marriage among teachers, and autobiographical data is Russell C. Brignano, *Black Americans in Autobiography* (Durham, 1984). He catalogued and abstracted more than 600 autobiographies, memoirs, and personal narratives. My assessment of the marrying patterns of teachers stems from a close analysis of more than three dozen autobiographies and reviews of pertinent oral histories.

34. The literature on intraracial class negotiations is voluminous and growing. For example, see Du Bois, *The Philadelphia Negro;* "The Talented Tenth," in *The Negro Problem,* Booker T. Washington, ed. (New York, 1903); Allison Davis, Burleigh B. Gardner, and Mary R. Gardner, *Deep South: A Social Anthropological Study of Caste and Race* (Chicago, 1941); Drake and Cayton, *Black Metropolis;* E. Franklin Frazier, *Black Bourgeoisie* (Glencoe, 1957); Nathan Hare, *The Black Anglo Saxons* (New York, 1965); Oliver C. Cox, *Caste, Class and Race: A Study in Social Dynamics* (Garden City, 1948); Sidney Kronus, *The Black Middle Class* (Columbus, 1971); John H. Bracey, August Meier, Elliott Rudwick, eds., *The Black Sociologist: The First Half Century* (Belmont, 1971); James E. Blackwell, *The Black Community: Diversity and Unity* (New York, 1975); Loretta J. Williams, *Black Freemasonry and Middle-Class Realities* (Columbia, Mo., 1980); Michael J. Bell, *The World From Brown's Lounge: An Ethnography of Black Middle Class Play* (Urbana, 1983); Annie S. Barnes, *The Black Middle Class Family: A Study of Black Subsociety, Neighborhood and Home in Interaction* (Bristol, 1985); Bart Landry, *The New Black Middle Class* (Berkeley, 1987); Alice F. Coner-Edwards and Jeanne Spurlock, eds., *Black Families in Crisis: The Middle Class* (New York, 1988); Sharon M. Collins, "The Making of the Black Middle Class," *Social Problems* 30 (April 1983), 369-82; Noel A. Cazenave, " 'A Woman's Place': The Attitudes of Middle-Class Black Men," *Phylon*

(March 1983), 12-32; Lynn Weber Cannon, "Trends in Class Identification Among Black Americans From 1952 to 1978," *Social Science Quarterly* (March 1984), 112-26; Thomas J. Durant, Jr. and Joyce S. Louden, "The Black Middle Class in America: Historical and Contemporary Perspectives," *Phylon* 47 (December 1986), 253-63; Shelby Steele, "On Being Black and Middle Class," *Commentary* 85 (January 1988), 42-47.

35. James Scott, *Domination and the Arts of Resistance* (New Haven, 1991), 1-27.

36. Lawrence Levine, *Black Culture and Black Consciousness* (New York, 1977), 370-440. Daryl Cumber Dance, *Shuckin' and Jivin'* (Bloomington, 1978), chaps. 10 and 12.

37. Massey and Denton, *American Apartheid*, 47, 48, and 76.

38. The writings of Edward Said, Homi Bhabha, Gaytri Spivak, and others need to be reconsidered, as does the notion of minority discourses so essential to the work of Abdul JanMohamed and a generation of cultural and literary scholars. Locality, bound by intense temporal specificities, has greater meaning then currently noted.

39. Levine, *Black Culture and Black Consciousness*, 427-29.

40. Early Simple tales appeared in the *Chicago Defender*. They were subsequently collected into five books of verse: *Simple Speaks His Mind, Simple Takes a Wife, Simple Stakes a Claim, The Best of Simple,* and *Simple's Uncle Sam.* Arnold Rampersad, *The Life of Langston Hughes: I Dream A World, vol. II, 1941-1967* (New York, 1988), 62-67. Charles Scruggs, *Sweet Home: Invisible Cities in the Afro-American Novel* (Baltimore, 1993), 1-12.

41. Levine, *Black Culture and Black Consciousness*, 427-29.

42. Robin D. G. Kelley, " 'We Are Not What We Seem': Rethinking Black Working-Class Opposition in the Jim Crow South," *Journal of American History* 80 (June 1993), 84.

43. Levine, *Black Culture and Black Consciousness.*

44. Lewis, *In Their Own Interests*, 167.

45. Elsa Barkley Brown, "Constructing a Life and a Community: A Story of the Life of Maggie Lena Walker," *OAH Magazine of History* 7 (Summer 1993), 28-29. Evelyn Brooks Higginbotham, *Righteous Discontent: The Women's Movement in the Black Baptist Church, 1880-1920* (Cambridge, 1993).

46. David Katzman, *Seven Days a Week: Women and Domestic Service in Industrializing America* (New York, 1978); Susan Tucker, "A Complex Bond—Southern Black Domestic Workers and Their White Employers," *Frontiers* 9 (1987), 6-13; Elizabeth Clark-Lewis, "This Work Had a End: African American Domestic Workers in Washington, D.C., 1910-1940" in *"To Toil the Livelong Day": America's Women at Work, 1780-1980,* Carol Groneman and Mary Beth Norton, eds. (Ithaca, 1987), 207-8.

47. Eric Arnesen, "Following the Color Line of Labor: Black Workers and the Labor Movement Before 1930," *Radical History Review* 55 (1993), 53-87; Lewis, *In Their Own Interests*, 46-65, 137-148, and 173-187; Peter Rachleff, *Black Labor in the South* (Philadelphia, 1984); August Meier and Elliott Rudwick, *Black Detroit and the Rise of the UAW* (New York, 1979); Eric Foner, *Nothing but Freedom: Emancipation and Its Legacy* (Baton Rouge, 1983); Joe W. Trotter, Jr., *Coal, Class, and Color: Blacks in Southern West Virginia* (Urbana, 1990); Eric Arnesen, *Waterfront Workers of New Orleans* (New York, 1991); and Alice and Staughton Lynd, eds., *Rank and File: Personal Histories by Working-Class Organizers* (Boston, 1973), 113.

48. Kelley, " 'We Are Not What We Seem,' " 75-112.

49. The politics of memory is discussed in Scott A. Sandage, "A Marble House Divided: The Lincoln Memorial, the Civil Rights Movement, and the Politics of Memory, 1939-1963," *Journal of American History* 80 (June 1993), 135-67.

50. *Ledger Star*, November 8, 1974; *The Virginian-Pilot*, November 9, 1974.

51. *The Virginian-Pilot,* June 19, 1967, November 9, 1974; *Norfolk Journal and Guide,* April 25, 1959, November 16, 1974. *Norfolk Labor Journal,* September 6, 1945.

52. The most recent evidence of this is Bruce Nelson, "Organized Labor and the Struggle for Black Equality in Mobile During World War II," *Journal of American History* 80 (December 1993), 962-88; Michael Honey, *Southern Labor and Black Civil Rights: Organizing Memphis Workers* (Urbana, 1993); Lewis, *In Their Own Interests;* James R. Grossman, "The White Man's Union," in Trotter, ed., *The Great Migration in Historical Perspective,* 83-105.

53. *The Virginian-Pilot,* June 19, 1967, November 9, 1974; *Norfolk Journal and Guide,* April 25, 1959, November 16, 1974; *Norfolk Labor Journal,* September 6, 1945.

54. Peter Novick, *That Noble Dream: The "Objectivity Question" and the American Historical Profession* (Cambridge, 1989), chaps. 15 and 16.

55. Vincent Harding, "Responsibilities of the Black Scholar to the Community," in *The State of Afro-American History,* 280.

56. This point was reiterated on March 5, 1993, when I delivered a public lecture in Norfolk on black life during World War II. Several area residents reclaimed me as one of them and applauded my decision to write the book.

"Unhidden" Transcripts

Memphis and African American Agency, 1862-1920

Kenneth W. Goings
Gerald L. Smith

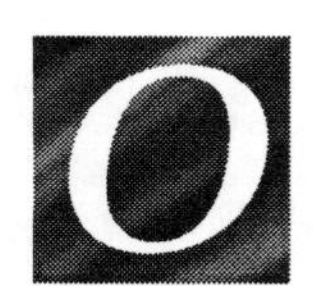

n May 22, 1917, the Memphis *Commercial Appeal* (hereafter *CA*) printed on its front page two items referring to the African American community. The headline of the column-one story read, "Mob Captures Slayer of the Rappel Girl, Ell Persons *to Be* [emphasis added] Lynched Near Scene of Murder" (Figure 5.1). The article informed Memphians that the lynching would take place between 9:00 and 9:30 that Tuesday

Authors' Note: An earlier version of this essay was written for David Katzman's NEH Summer Seminar for College Teachers at the University of Kansas, 1992, titled "The Origins of African-American Urban Communities." Kenneth W. Goings gratefully acknowledges the insights and comments of Professor Katzman and the other members of the seminar. Another version of this essay was presented by Goings and Smith at a session titled "Racial Violence in the Twentieth-Century Urban South" at the American Historical Association annual meeting, Washington, D.C., 1992. The authors would like also to thank Raymond A. Mohl and Stephanie Shaw, who provided helpful criticisms and comments.

MOB CAPTURES SLAYER OF THE RAPPEL GIRL

Ell Persons to Be Lynched Near Scene of Murder.

MAY RESORT TO BURNING

Negro Taken From Officers at Potts Camp, Miss.

ARMED FORCE IS GROWING

Figure 5.1. Front Page, Column 1 Headline, Memphis *Commercial Appeal*, May 22, 1917

morning near the Wolf River Bridge. It also reported that armed men from all over the tristate area were descending on Memphis for the event.[1]

The second item was the daily cartoon "Hambone's Meditations." It featured a small African American male with a misshapen head (in phrenologist circles the bump on the back of his head denoted sexual proclivities and could be also be found in animals) who would be dressed up in a variety of worker/servant costumes that directly related to his "humorous" little speech for that particular day (Figure 5.2).[2] Docile, illiterate, and deferential toward whites, Hambone projected established stereotypes of African Americans. "Hambone's Meditations" epitomized the kind of behavior Memphians (white) expected African Americans to display.[3] The announced lynching of Ell Persons indicated the consequences facing those who chose to behave otherwise.

Antoinette Rappel, a sixteen-year-old white student at Treadwell School in Memphis, had been raped and murdered. She had also been decapitated and her eyes left in a "frozen expression of horror." Although a white man was seen coming from the wooded area where the crime had been committed, deputy sheriffs focused their investigation on the arrest of a black assailant. After arresting several black men, authorities became convinced that Persons, a woodcutter who lived less than a mile from the murder scene, was the guilty party. After undergoing the "second-degree" for twenty-four hours, Persons supposedly confessed his guilt.[4] James Weldon Johnson of the NAACP investigated these allegations after the lynching and found no evidence at all to support the case against Persons.[5]

Sheriff Mike Tate wanted to make absolutely sure that Persons was the murderer. He requested that Rappel's body be exhumed so her pupils could be photographed by the police department's Bertillon expert, Paul Waggener. A French scientist (Bertillon) had claimed that photography was useful in reconstructing the last image seen by a victim who had died a terrifying death. Although Rappel's right eye had decomposed, the results from a high-powered camera lens convinced authorities that Persons's forehead and hair were still visible in Rappel's eyes. There was no need to continue the investigation. Persons was now clearly guilty. In anticipation of a lynch mob, Persons was taken to Nashville to await arraignment and trial. But while in transit back to Memphis, a mob seized him from deputy sheriffs.[6]

Figure 5.2. Front Page, Hambone Cartoon, Memphis *Commercial Appeal,* May 22, 1917

The subsequent carnival-like lynching of Ell Persons was a major attraction for Memphians (white). Vendors were on hand to sell pop, sandwiches, and chewing gum. Women wore their best clothes to the event. Parents wrote notes to schoolteachers requesting their children be excused to witness the lynching. An estimated five thousand spectators gathered as Persons was tied to a stake in the ground, then drenched with ten gallons of gasoline and burned alive. Men, women, and children snatched pieces of his clothing and shreds of the rope used to capture him. Later that afternoon, Persons's head and one of his feet were thrown from a passing car into the midst of a group of African Americans standing near Beale Avenue and Rayburn Boulevard, the center of the African American community in Memphis.[7]

The lynching of Ell Persons was one of several major physical attacks against the African American community in Memphis before the 1920s. These attacks were linked by four common threads. First, each was an overt attempt to control an African American community that seemed not to know its "place." Second, African American migrants to the city played a central role in each of these events by providing a large and ever-increasing population base that helped bolster the assertiveness of Memphis African Americans. The third thread running through these attacks was the role played by the civil authorities. City officials as well as police officers were involved overtly or covertly in each of these attacks, demonstrating that indeed they were not simply "random" fragments of spontaneous racial violence but calculated attempts by the white community and/or its official representatives to suppress African Americans. The fourth thread connecting all these attempts was the fact that none of the lynchings was successful in the attempt to crush the African American community, which continued to grow and to assert itself.

Robin Kelley, in his important essay " 'We Are Not What We Seem': Rethinking Black Working-Class Opposition in the Jim Crow South," rightly calls on historians of the African American urban experience to reject the implicit assumption that working-class African Americans were voiceless and actionless in the face of Jim Crow.[8] Kelley notes that "beneath the veil of consent lies a hidden history of unorganized, everyday conflict waged by African-American working people." Kelley then outlines these "hidden transcripts" of political culture—activities ranging from jokes and songs to destruction of property. Quoting the work of anthropologist James Scott, Kelley notes that these "transcripts" are part of what Scott refers to as "infrapolitics." As Scott puts it, "the circumspect struggle waged daily by subordinate groups is, like infrared rays, beyond the visible end of the spectrum. That it should be invisible . . . is in large part by design—a tactical choice born of a prudent awareness of the balance of power."[9] Kelley observes that "one measure of the power and importance of the informal infrapolitics of the oppressed is the response of those who dominate traditional politics."[10] In Memphis, infrapolitics combined with more traditionally described acts of resistance could and did lead to lynchings. One of the reasons for these extreme responses by the dominant group was that in Memphis for the years under study the transcripts were often not hidden. The ever-increasing migration of African Americans into the city of Memphis served as a constant and obvious reminder to whites that regardless of their efforts to keep African Americans in their place, migrating African Americans continued to take up residence in Memphis. Moreover, as this essay will demonstrate, the response to legal, verbal, and physical attempts to keep African Americans in their "place" often was physical retaliation. Scott writes,

> In the day-to-day conduct of class relations, these . . . transcripts are never in direct contact. Only at rare moments of historical crisis are these transcripts and the actions they imply brought into a direct confrontation. When they are, it is often assumed that there has come into being a new consciousness, a new anger, or a new ideology that has transformed class relations. It is far more likely, however, that this new "consciousness" was already there in the unedited transcript and it is the situation that has changed in a way that allows or requires one or both parties to act on that basis.[11]

Kelley is correct to see the analogy between Scott's descriptions of peasant cultures and the African American working classes and to implore historians to look for the hidden transcripts of resistance. However, Kelley's article does not go far enough. In Memphis, "resistance" not only meant physical resistance, labeled "criminal" by the white and African American elites and white working classes; it was also represented in the African American desire to choose where they wanted to live. And as Scott acknowledges in the quotation cited above, "consciousness" had not really changed—these working-class people in Memphis were the same people from the countryside whose resistance there had generated the thousands of lynchings over this time period. What was different now was the "social and cultural spaces" involved—they were in urban areas but they were the same people. They had left the countryside in the hope of a better situation in an urban setting and, as the following material indicates, they were not going to be denied without a struggle.

The first major attempt to keep African Americans and their community in their "place" had come with the Memphis Riot in 1866. Immediately after the fall of Memphis to Union troops in 1862, thousands of newly freed African Americans poured into the city from western Tennessee, eastern Arkansas, and northern Mississippi. The African American migrants had created a personal, social, and gender leveling. African Americans in Memphis and Shelby County demanded the right to vote, to serve on juries even when whites were being tried, and to be treated as equals. One of the best descriptions of the state of mind of Memphians (white) about their fate following the war was expressed by Elizabeth Meriwether in memoirs she penned after the Civil War. The Meriwether family was one of Memphis's most distinguished. Elizabeth Meriwether's husband was a colonel in the Confederate Army. Like many members of her class, she and her family abandoned Memphis when the Union troops moved into Tennessee. Mrs. Meriwether describes her return to Memphis and the conditions she found:

> As we drew near Memphis the country around seemed more and more desolate and dreary; the farm houses on the road side had been either deserted or burned to the ground. No inhabitants of the country did we see, nor human beings of any kind excepting occasionally we saw troops of Union soldiers marching northward. Some of these troops were black. . . . I was worried . . . at the thought of what might happen were some of those soldiers to leave Fort Pickering at night and wander in my direction. Men but lately released from slavery, men

> but a degree removed from savagery sometimes do terrible things when suddenly entrusted with power. . . . Next morning on the way into town I saw things which both distressed and alarmed me—blue uniformed negro soldiers swaggering, and sometimes staggering, on the streets; some of these soldiers were playing the gallant to ugly, gaudily dressed negro women. During my walks from one shop to another I sometimes had to get off the sidewalk into the street in order to make way for these negro soldiers—they walked four or five abreast and made not the slightest effort to let white women pass. Had I not voluntarily gotten out into the street they would have elbowed me off the sidewalk. As it was, when I stepped off the curb to let them pass they gave me insolent looks and laughed and sometimes one of them would say: "we's all ekal now. Git o' our way, white woman!" . . . Any stranger, seeing those negroes would have supposed the Blacks, not the Whites, were masters in the South. As I saw those things that first morning in Memphis, and as I myself experienced them the thought came to me that, if negroes were to dominate life in the South, it would be better for us to emigrate.[12]

The social leveling brought about by the Union occupation of Memphis would cause a great many Memphians (white) to rethink just what had happened to them and what would become of them.[13]

For Memphians (white) and some of their African American collaborators, these conditions became intolerable. A minor police skirmish would lead to a race riot headed by the mayor, aldermen, police, and white citizens against selected African Americans. Long-term African American residents of Memphis were not attacked; the attacks were directed instead at the African American migrant community, civilian and military. During the Civil War, Memphis underwent significant demographic change as the number of blacks increased from 3,872 in 1860 to 15,471 by 1870. Memphis also experienced a significant increase in the Irish population during the 1850s, assuring mounting tensions between the two groups as they competed for jobs as draymen, boatmen, and construction and railway workers. Tensions between black and white Memphians finally escalated into a race riot on May 1, 1866. During three days of rioting, black Memphians were beaten and robbed, and African American women were raped. A Joint Congressional Committee to Investigate the Memphis Riots came to the city on May 22. The Committee found that forty-six blacks and two whites had been killed in the riot.[14] Yet despite this brutal assault by whites, the African American community in Memphis continued to grow.

By 1890, Memphis was an "improved" southern city. It led the South in the lumber market and was the world's largest inland cotton market. Because of its strategic location on the Mississippi River, Memphis became a major distribution center, boasting ten railroads and convenient water routes for steamships heading north or south. Additionally, the city's debt and tax rate had declined significantly, sanitation conditions had improved, and the population had risen to more than 64,000, with 44.5 percent being African American.[15]

Memphians (white) who had ties to the plantation economy expected obsequious behavior by black tenant farmers and sharecroppers who were generally financially dependent on planters. When rural black migrants began moving into Memphis, therefore, whites expected them to display social and economic insecurities and remain in their place in relation to whites. But black migrants were inspired by the opportunities urban life offered. Even the yellow fever epidemics did not hinder the growth of the black community. Between 1870 and 1890, the proportion of African Americans living in Memphis increased from 39 percent to 44 percent.[16]

The black population comprised an achievement-oriented community. There were several black schools, churches, businesses, and community organizations. By the end of the 1880s, three black newspapers—the Memphis *Watchman*, the *Living Way*, and the *Free Speech and Headlight*—were published regularly in the city.[17] Black migrants to Memphis took advantage of various employment opportunities not related to plantation labor. They worked on the levee or the steamship lines and in the lumber industry. Denied opportunities in skilled labor, some were hired as cooks and maids and in other service occupations reserved specifically for them. However, the percentage of African Americans in the workforce in Memphis declined from 48 percent in 1900 to 39 percent by 1920, reflecting the continued migration into the city not only of African Americans but also of native-born whites from the fields and small towns of the region.[18] These white migrants were moving into traditional "negro" jobs and displacing African American workers; this displacement was one of the leading causes of the heightened racial tension in the city.

But the economic situation was not all bleak. The African American community continued to provide a clientele for African American professionals and business people. Besides the growth of African American businesses, there were increases in the number of doctors, lawyers, and teachers. African Americans also owned and operated grocery stores, barber and beauty shops, saloons, funeral homes, and other businesses that served the community.[19] It was the success of the black-owned People's Grocery Store that led to the 1892 lynchings of Tom Moss, Calvin McDowell, and William Stewart.

This physical attack, known as the "curve riot," was part of a larger trend to impose a new "accommodation" of segregation and disenfranchisement on African American communities all over the South. Howard Rabinowitz, in *Race Relations in the Urban South,* has posited that by 1890, when white southerners could be sure that the North would not interfere with their race relations practices, whites "driven by fears about the consequences if blacks were not 'kept in their place' established a new pattern of race relations in the urban South by 1890."[20] And, indeed, conditions in Memphis appeared to confirm this thesis.

The attempt in Memphis to put into place the "final settlement" centered around the lynching of the owner-operators of the People's Grocery. The People's Grocery was a symbol of strength and vitality for the Memphis African American community. The "aristocrats of color" in Memphis had organized the grocery store to take advantage of the growing African American community in and around Memphis. The store, located in south Memphis (where the heaviest concentration of African Americans, particularly migrants, lived), was an immediate success and drew customers away from a white-owned grocery store owned and operated by W. H. Barret just across the street. Barret got the Shelby County grand jury to indict the People's Grocery on a nuisance charge. The African American community was outraged. This was *their* store, and at community meetings, members rallied around and vowed to defend the store against "white trash." Barret used these meetings to gain indictments against the speakers for "conspiring against white people." On March 6, 1892, Barret had a message delivered to the People's Grocery that a mob was going to attack them. Several armed members of the community then went to defend the store. Waiting until after nightfall, Barret, with nine deputies in civilian clothes, approached the store to serve arrest warrants. Those in the grocery took the deputies and Barret for the mob, and firing commenced. Spreading the cry of "race riot," armed white supporters of Barret assisted the deputies in rounding up more than thirty African American males, including Moss, McDowell, and Stewart. Four days later in the early morning hours, sheriff's deputies seized Moss, McDowell, and Stewart, took them about a mile from the jail and shot them. This lynching had been committed not in the rural South but in a city of over 64,000!

Although the entire African American community was outraged, their response to this lynching was varied. Between 2,000 and 6,000 African Americans chose to leave the city. Some went to the Oklahoma territory, which was just opening up; others went to Kansas to follow the "exodusters"; and one minister with his congregation emigrated to California. Many of those who migrated were members of the Du Bois's "talented tenth." The lynching of some of Memphis's "best and brightest" had convinced them that no matter what their education, no matter what their attainment, being black in the American South meant that your life would always be endangered.[21] This was to be the lesson of the lynching. Accept your place; accept segregation in hotels, theaters, and railroad cars; and do not challenge the economic arrangements or the white power structure of the South will be brought against you. Yet although some did migrate, the vast majority of African American Memphians stayed, including Ida B. Wells, already a staunch advocate of activism and self-defense for the African American community, who was prompted by the 1892 lynchings and other incidents to begin her great work against lynching.[22]

But if a "new accommodation" had been imposed by 1892, why did a lynching occur in 1917? We believe the key to answering that question lies in part in the not-so-hidden transcripts of one segment of the Memphis African American community—the working class.

Memphians (white) were concerned about black community growth and economic competition. Each black business that was developed and each black vote that was cast in a political election exemplified black self-determination and challenged the notion of white supremacy. Tennessee extended African Americans the right to vote in 1867. In the 1870s, African Americans, with the support of some ethnic groups, secured seats on the city council, twelve in all during the decade. They demanded and received some city and county patronage. This patronage included the selection of African Americans as assistant attorneys general (assistant DAs), wharfmasters, coal inspectors, and other minor political offices. It should be noted that the ethnic support for African American candidates did not come out of a sense of racial "fraternity" but rather out of a political coalition to keep native-born elite whites (Redeemers) out of office. This fractious and fragile coalition, white ethnics and African Americans, was also aided by the periodic white evacuations of the city caused by the yellow fever epidemics.

Redemption came in Memphis in about 1879, when a group of wealthy old elites (ex-Confederates) convinced the state legislature to revoke the city's charter. They claimed that the ethnic/African American coalition through partisan/ward politics had mismanaged the city and had driven it into enormous debt. Actually, the coalition had not mismanaged the city; epidemics and subsequent population declines that made tax and revenue collection almost impossible prompted borrowing and debt. But the state bought the old elite's argument, and in January 1879, Memphis surrendered its city charter and became a taxing district. Black representation in the city government was "conveniently destroyed." In 1889, the Tennessee legislature passed a registration law, secret ballot law, and separate ballot-box law that significantly reduced black political influence. A subsequent poll tax also helped reduce black voting participation. Yet black citizens continued to vote in spite of the efforts by whites to curtail their political activities.[23] In essence, the increase of black in-migration and the further development of black social and economic institutions aroused fears and prejudices among Memphians (whites). But more than just economic competition and assertive community building were at work. Although there are various explanations for lynching, we agree with Joel Williamson when he writes in *The Crucible of Race* that "the crime [lynching] was seen to be part of a larger interracial conflict. White society was endangered. Negroes were becoming 'uppity,' presumptuous, and bumptious" and "symbolically, the lynching was often seen as an act against the whole black community."[24]

Although some sources claim race relations in Tennessee and Memphis were "moderate," actual events indicate a different story.[25] On December 10, 1908, for example, Bill Latura walked into a Beale Street saloon and killed four black customers for no apparent reason. Even though Latura admitted to the shooting, he was acquitted by an all-white jury.[26] By 1916, with urban migration growing, the animosity and violence by whites toward African Americans, particularly migrants, increased. Whites complained of farm or labor shortages and streetcar congestion.[27] Because of the large number of rural migrants coming into Memphis, police officers had "orders to stop all suspicious-looking negroes and search them." The police department even hired "negro spotters," who were local African Americans who ostensibly could identify the city's new black residents and point them out to officers.[28]

But the extent of how far race relations had really deteriorated was evident in the newspapers. These news articles demonstrated how some elements within the African American community rejected the "new accommodations" and how the white community would in 1917, as in 1866 and 1892, again try violently to impose its "new order." Rabinowitz, in his conclusion, noted that before the attempt to impose the "new" racial accommodations in the 1890s, "most threatening was the tendency among lower-class blacks to use physical means to resist local white policemen."[29] In Memphis, the resort to physical violence by the police, by white civilians, and by African Americans well into the twentieth century was a clear indication that racial politics in Memphis were not settled, and that there was still a contesting over the rules of the game.

Examples of the increased violence were the killings and attacks on African Americans by the police during the second decade of the twentieth century. For example, in September 1915, three African American women out walking at night were ordered by a police officer to stop. When they did not, he drew his gun and shot one, Fannie Shepherd. Fannie Shepherd bled to death when the first ambulance called would not transport her to the hospital. When later questioned by the police, the driver said he did not want to "bloody the car."[30] In August 1916, a headlined story read, "Officer Shoots Negro." As the *CA* reported, when Noble Conley, "negro," refused to halt after he had been commanded three times to do so, Officer L. J. (Red) Daine shot and wounded him.[31] In November 1916, a headlined story read, "Officer Kills Negro." Albert Moody, an African American, refused to stop for the police and was simply shot. Two weeks later, a police officer attempted to stop another "suspicious-looking negro." Officer Mayers ordered him to stop and drew his gun. The African American drew faster and shot the police officer.[32] What was going on here? Clearly, these African Americans resisted the authority of the police. African American men and women in these cases were not being intimidated, even by the imminent use of physical force. It should be noted, given the reputation of the Memphis police, that this was also a matter of survival.

African Americans living in Memphis had great reason to have little or no respect for the police. Police shootings provided the most dramatic examples, but harassment occurred daily. In December 1916, for example, police arrested a group of forty-six people, most likely migrants, who were unable or unwilling to tell the police why they were in Memphis. In addition to those who were "suspicious looking," the police were also "to take into custody all negroes of questionable character and all strange blacks whose presence is not explained in a logical manner."[33] This policy was really harassment of migrants in an attempt to drive them out of the city. Sometimes when large groups were questioned by the police, they did not simply go quietly to jail. In October 1916, "When Capt. Couch and his squad raided Jim Kiname's place [to see why so many people had congregated together], Officer Rutt was attacked by a negro with a butcher knife."[34]

The time period included in this study, 1890-1920, may have been, as Rayford Logan noted, the "nadir" of American race relations but it was not the "nadir" of black agency.[35] African Americans were not just passive victims; this was part of an interracial struggle that was taking place in Memphis. When they perceived injury or insult, African Americans resisted or sought redress even to the extent of using physical violence. In April 1915, Thomas Brooks, a twenty-year-old African American, shot and killed two white men who attacked him for no reason.[36] Tragically, two weeks later, the *Chicago Defender* finished the story under the headline "Another Member of the Race Is Lynched." The sheriff had locked Brooks up in Sommerville, a small town outside of Memphis. When the sheriff attempted to return Brooks to Memphis for trial, "a mob of white gentlemen (blue hoods) strung [Brooks] upon a railroad trestle where his body was riddled with bullets."[37] In August 1916, officers attempted to arrest seven African Americans (men and women) on trumped-up charges. One person drew a gun, fired, and killed the officer. With the help of his friends, he escaped. While being captured, he was shot ten times and died in the hospital. His friends immediately went into hiding. Eventually, all but one person were captured and the police had shot each one.[38]

But the streetcars of Memphis were the real racial battlegrounds. Even though "officially" segregated, they were one of the few places where African Americans and whites came into direct and regular contact with each other. On August 14, 1916, Charley Parks, an African American, and Tom Lee, white, got into a verbal argument over seating arrangements. Parks stabbed Lee, then jumped off the train. Lee, who was not seriously injured, grabbed the switchman's bar and along with the other white passengers ran after Parks, caught him, and beat him "into insensibility."[39] This incident should not be seen as an example of the old "victimization" model, but rather as an example of the contestation over racial rules and etiquette that was still taking place in Memphis at this time. Just because a white person told an African American to give up his or her seat was no guarantee that the seat would be given up. This incident is noteworthy because there was not simply one white person involved. The other whites on the streetcar immediately took up his cause. Everyone, it seems, was aware of the nature and stakes of the contest. A loss for one might be a loss for all.

Another example of African American resistance took place in October 1916, when John Knox, an African American, and an unidentified white passenger got into an argument over a streetcar seat. The verbal confrontation led to Knox drawing his gun, while a white passenger drew his, which caused several other white male passengers to draw their guns as well. When the shooting was over, four African Americans, including Knox, had been shot. The other three African Americans had not been part of the argument, nor had they participated in the shooting. They had been shot simply for being "in the wrong place at the wrong time." No whites were injured in this gun battle.[40] A third interracial altercation on the streetcars took place in November 1916, just after the reelection of President Woodrow Wilson. After an African American supposedly insulted Wilson, a newsboy, "Red Eye" Sanders, took umbrage at the remark and grabbed the switchman's bar to hit Wilson's insulter. Another African American intervened, grabbing the bar, and the "near race riot" was on. People started jumping off the cars and a group of whites surrounded a lone African American and beat him unconscious.[41] In December 1916, when a white conductor demanded more fare from an African American passenger who had already paid, the two got into an argument. The conductor grabbed the passenger, at which point the passenger drew a knife and stabbed the conductor in the arm. He then leaped off the trolley and made good his escape.[42]

Memphis was a racial battleground. Memphis-born and migrant whites who were attempting to enforce white supremacy were fighting an African American population made up of Memphis-born residents who knew but still challenged the "accommodation" as well as African American migrants to Memphis who, if they knew the new rules, often chose to defy them. Not all the resistance was physical; some was psychological as well. An example was the case of *Lula McLaurin (C) [Colored] v. Memphis Street Railway Company.* This case was heard in Shelby County Circuit Court on May 13, 1915. Lula had boarded a streetcar with three other friends—Pearl Allen, India Allen, and Mattie Cooper. Pearl, who boarded the streetcar first, paid for the others. Lula boarded last, and not knowing that Pearl had already paid, paid her nickel. Pearl saw Lula paying her fare and told Lula that she should get her money back from the conductor. When Lula tried to get her money back, the conductor said, "Go on and sit down nigger bitch" (Lula and her friends' testimony), or "Go up front of the car and sit down" (the conductor's testimony). When Lula continued to press her case, the conductor then threatened to throw her off the car, to which she replied, "No, I guess you won't do anything like that." The conductor also claimed he thought she was white and that he had not known she was with the other women, hence (according to him) his telling her to "go up front," and if that was true he certainly would not called a white woman a "nigger bitch." Lula's friends were called to testify. When Mattie was asked by Lula's attorney to describe Lula's demeanor throughout this encounter, she replied that "Mrs. McLaurin" had always been polite and soft-spoken. The attorney said, "Are you referring to Lula?" Mattie replied, "No, I'm talking about *Mrs. McLaurin.*" Mattie, Pearl, and India never called Mrs. McLaurin "Lula" and never responded to the prosecution or the defense when they referred to Mrs. McLaurin by her first name. By the time the trial was over, both attorneys were calling Mrs. McLaurin "Mrs." Mrs. McLaurin was awarded $300 for "insult" by the conductor. Not only were these women not intimidated by the streetcar conductor and the court system, they were not going to allow the customary practice by whites in and out of court to refer to African American women (particularly married individuals) by their first name. In defending Mrs. McLaurin, they of course were defending their own dignity as well.[43]

Another source of psychological resistance was, of course, the black church. As Evelyn Brooks Higginbotham has noted, "In the closed society of Jim Crow, the church afforded African-Americans an interstitial space in which to critique and contest white America's racial domination."[44] Nowhere was this more apparent in Memphis than with the Church of God in Christ (COGIC), which emerged as the largest religious denomination in the city during the early twentieth century. Labeled as a representation of the "Sanctified Church," COGIC, through emotional and spirit-driven services, attracted a significant number of Memphis's black working-class community. This population desired a source of religious inspiration and community empowerment, which COGIC tried to offer. As Cheryl Townsend Gilkes reveals, the "Sanctified Church" offered some unique "resistance strategies." COGIC members who had been "sanctified" (redeemed through the removal of all sin) defined themselves as "Saints" to challenge white stereotypes that depicted African Americans as criminals. Additionally, "Saints" were encouraged to use first initials instead of their proper names. This protocol served as a subtle means of pressing whites to refer to COGIC members with proper and respectful titles and not the traditional practice of calling all African Americans, male and female, old and young, by their first names. Because of this type of cultural resistance, religion served an even greater role in the process of self-determination.[45]

For some migrants, Memphis was but a way station on the track north. For others, Memphis was or turned out to be their final destination. Almost all the literature on migrants describes those who went north, but one suspects there was little difference between those who left and those who stayed. Thus, the migrants to Memphis were probably young people, generally twenty-five to thirty-five, more male than female, likely to have made an intermittent move to another town or locality and to have some resources, financial and intellectual. And this generation had not been "schooled" in the old ways.[46] Despite the bravado in the "humorous" descriptions of African Americans in the newspapers, these people were not "Hambones." They were not acting as "negroes" should have been acting. They talked back. They fought back. They shot back. These were not the submissive and deferential darkies perpetuated by Old South mythology.

A clear detractor of the notion that migrants who remained in the South were deferential and submissive was black sociologist Charles Johnson. Writing in 1923 about the effects of the Great Migration (1910-1920), Charles Johnson noted, "All Negroes are not uniformly sensitive to their social environment." He looked at migration for the six decades following emancipation and concluded that while migration to northern cities was important, intraregional migration to western southern states and states deeper south was more significant. Over 130 years, from the 1790s to the 1920s, the center of African American population had shifted from Dinwiddie County, Virginia, to northern Alabama. The series of migrations that brought about the shift, Johnson believes, resulted from the fact that tenancy and sharecropping were not yielding returns in proportion to population growth. Buttressing the argument, he noted that counties in general that suffered loss of African Americans also suffered loss of whites. Johnson noted as well that many of the southern counties African Americans migrated to were places that had experienced a large number of lynchings. Applying this information to the migration north and being careful to note that these were just tentative proposals, he wrote,

> It means that the Negroes who left the South were motivated more by the desire to improve their economic status than by fear of being manhandled by unfriendly whites. . . . Persecution plays its part. . . . But when the whole of the migration of Southern Negroes is considered, the part seems to be limited.[47]

If Johnson is correct for the Great Migration, and if that argument can apply to the early period 1890-1910, then the continued migration into Memphis becomes understandable. The migrants were not intimidated by the lynchings in Memphis. They wanted to improve their economic lot and, for black people, the mid-South area of Memphis was probably the best place to make that effort.

As was attempted in 1866 and in 1892, whites, many of them migrants themselves, attempted in 1917 to drive the lesson home again: Either play by the new rules of accommodation or face physical retribution. The lesson was delivered in the form of a lynching.

And as was the case in previous attempts, lynching was almost a last resort in response to the infrapolitics of this African American community. Laws, persuasion, co-optation of African American leaders, and physical violence administered by civilians and the police had not been enough to keep African Americans in "their place." Well into the 1920s, African Americans would be working to keep the Ku Klux Klan from taking over the city. They would still be exercising the franchise. They would still be building their community and defending themselves. And they would still be physically challenging whites. Immediately after the 1917 lynching of Ell Persons, Johnson Burlingame, an African American chauffeur, protested by tearing up the American flag and suggesting African Americans join the Germans. He was arrested by federal authorities. Roscoe Conkling Simmons, former member of the Tuskegee machine and nephew by marriage to Booker T. Washington, on July 1917, addressed a large and enthusiastic crowd at Church Park in Memphis; to raucous cheers he declared, "You may burn me, but you cannot burn away my record of undying loyalty." And, in a possible precursor to the later Double V campaign conducted by African Americans during World War II, Simmons stated, "After that job is well done we will make the United States safe for the Negroes."[48]

African Americans were still resisting the police. In September 1917, George Parker, an African American, refused to halt for the police and was shot.[49] In October 1919, an African American farmer shot a white farmer who had insulted him.[50] In March 1918, the police followed two African American males into a cafe. The African Americans sat at the end of the counter. The officer said to the first man, "What's in the grips, nigger?" The second man replying for the first said, "I ain't got nothing in it boss." The patrolman then pulled out his billy club and was about to strike them when one of the African Americans pulled his gun and shot the officers. Both black men ran out of the cafe and were chased by the white customers but escaped.[51] Apparently, some people were willing to show only so much deference. The very next month, when the police had come up with a new scheme to extort money out of African American individuals, they came upon someone who would not go along. On April 5, Mrs.

George Williams was told to confess to the theft of $300 from a farmer. She refused. The police beat her. She still refused to confess. It turned out the police had split the money with the real robber, a white man, and wanted to get Mrs. Williams to cover for them. She not only resisted but was able to get enough pressure brought to bear on this incident that a grand jury handed down indictments against the police.[52]

The streetcars were no quieter after May 1917. Approximately a year later, a white man desiring a seat tried to pull an African American out of his seat on the streetcar. Several other white men joined the fight, one attempting to throw the African American off. The African American drew his gun, the whites drew theirs, and the fighting commenced. When it was over, a white woman had been killed. Also, in May 1918, Patrolman Bryant (white) instantly killed Lee Turner, a locomotive fireman. Bryant ordered Turner, an African American, to throw up his hands and Turner refused. Bryant then shot him.[53]

Even accommodators seemed more assertive. Dr. Robert Moton appeared to be catching whatever spirit was around in May 1920 when he addressed an interracial group of Memphians who were meeting in Clarksville, Tennessee, to discuss how to improve race relations. It should be noted that Clarksville was the county that abuts Greene County, Arkansas, which proudly proclaimed in the *CA* that "no negroes live here," because all African Americans had been lynched and the rest driven out. The first speaker was C. P. J. Mooney, editor of the *CA*. Mooney proclaimed that if African Americans (although he used the word "darkies") would just behave there would be no problem. The next speaker was Dr. Moton, who "scolded" Mooney for referring to African Americans as darkies or niggers. The *Chicago Defender* noted, "The Tuskegee educator made the audience gasp when he flayed the Memphis editor."[54]

Clearly, the physical attacks against the Memphis African American community were not as threatening as whites had anticipated. Migration then became the most effective "unhidden transcript" used by African Americans. To be sure, the African American community in Memphis was divided. There certainly were some who were silenced verbally, physically, and mentally by the lynching of Ell Persons. And there were certainly some who were "eager" to show that they bore no malice for the horrible past, but the majority of African Americans continued to live their lives in dignity and self-respect. When someone attempted to interfere with their ability to do so, African Americans created and vigorously employed both hidden and unhidden transcripts. Although Rabinowitz asserts that "blacks had to be disciplined" in most of the urban South, and though "this job had largely been accomplished in Southern cities by 1890," Memphis, we believe, was *an* exception.[55]

Notes

1. Memphis *Commercial Appeal,* May 22, 1917, hereafter cited *CA.*
2. Ibid.
3. Throughout this essay, where appropriate, we have referred to white Memphians as Memphians (white). This was a practice adopted by the African American press as a protest to the white press's continually noting after the name of African Americans that they were (colored) or (negro).
4. Memphis *News Scimitar,* May 3, May 4, May 7, May 8, May 22, 1917; *CA,* May 8, 1917.
5. "Lynching at Memphis," *The Crisis* (August 1917), 185-88. Meetings of the Board of Directors, Records of Annual Conferences, Major Speeches and Special Reports, 1909-1950, NAACP Papers, part 1, frame 556-59.
6. Memphis *News Scimitar,* May 8, 1917, May 22, 1927; *CA,* May 8, May 22, 1917. See also the supplement to *The Crisis* 14 (July 1917), 1-4.
7. Memphis *News Scimitar,* May 22, 1917; *The Crisis* (July 1917), 2-3.
8. Robin D. G. Kelley, " 'We Are Not What We Seem': Rethinking Black Working-Class Opposition in the Jim Crow South," *Journal of American History* 80 (June 1993), 75-112.
9. James C. Scott, *Domination and the Arts of Resistance: Hidden Transcripts* (New Haven, 1990), 183.
10. Ibid., 76-77.
11. James C. Scott, *Weapons of the Weak: Everyday Forms of Peasant Resistance* (New Haven, 1985), 288.
12. Elizabeth Avery Meriwether, *Recollections of 92 Years, 1824-1916* (Nashville, 1959), 164-67.
13. Kevin R. Hardwick, " 'Your Old Father Abe Lincoln Is Dead and Damned': Black Soldiers and the Memphis Race Riot of 1866," *Journal of Social History* 27 (Fall 1993), 109-28.
14. See Bobby L. Lovett, "Memphis Riots: White Reaction in Memphis: May 1865-July 1866," *Tennessee Historical Quarterly* 38 (Spring 1979), 9-10, 20-23, 30; Memphis *Appeal,* January 22, 1890; Gerald M. Capers, *The Biography of a River Town: Memphis, Its Heroic Age* (Chapel Hill, 1939), 205.

15. Roger Biles, *Memphis in the Great Depression* (Knoxville, 1986), 15, 34; William D. Miller, *Memphis During the Progressive Era* (Memphis, 1957), 92-93.

16. Gloria Melton, "Blacks in Memphis, Tennessee, 1920-1955: A Historical Study" (Ph.D. diss., Washington State University, 1982), 22.

17. W. E. B. Du Bois was so impressed with the African American middle class in Memphis that he made it a base for his first newspaper, *The Moon Illustrated Weekly*, which was directed toward the middle class in Memphis and Atlanta. See Paul G. Partington, "The Moon Illustrated Weekly—Precursor of the *Crisis*," *Journal of Negro History* 48 (July 1963), 212-13.

18. There are two studies that provide invaluable information on Memphis's black community. See Gloria Brown Melton, "Blacks in Memphis, Tennessee, 1920-1955," 25; Kathleen Berkeley, " 'Like a Plague of Locusts': Immigration and Social Change in Memphis, Tennessee, 1850-1880" (Ph.D. diss., University of California at Los Angeles, 1980). Black employment and black newspapers are discussed respectively in Joel Roitman, "Race Relations in Memphis, Tennessee, 1880-1905" (master's thesis, Memphis State University, 1964), 71, 74, 75; David M. Tucker, "Miss Ida B. Wells and Memphis Lynching," *Phylon* 32 (Summer 1971), 113, 116.

19. Denoral Davis, "Against the Odds: Postbellum Growth and Development in a Southern Urban Black Community, 1865-1900" (Ph.D. diss., State University of New York at Binghampton, 1987), 50-55.

20. Howard Rabinowitz, *Race Relations in the Urban South: 1865-1890* (New York, 1978), 30. See also William Cohen, *At Freedom's Edge: Black Mobility and the Southern White Quest for Racial Control, 1861-1915* (Baton Rouge, 1991), 217.

21. Indianapolis *Freeman*, May 7, 1892. Details of the 1892 lynching have been covered in various publications. See David M. Tucker, "Miss Ida B. Wells and Memphis Lynching"; Fred Hutchins, *What Happened in Memphis* (Memphis, 1965), 37-39; Alfreda Duster, ed., *Crusader for Justice: The Autobiography of Ida B. Wells* (Chicago, 1970), 35-54.

22. Duster, ed., *Crusader for Justice*, 18-20.

23. Joseph H. Cartwright, *The Triumph of Jim Crow: Tennessee Race Relation in the 1880s* (Knoxville, 1976), 137-39.

24. Joel Williamson, *The Crucible of Race* (New York, 1984), 187. See also Arthur Raper, *The Tragedy of Lynching* (Montclair, 1969), 28-29. The following works also provide invaluable information on lynching in the United States: James Harmon Chadburn, *Lynching and the Law* (Chapel Hill, 1933); James E. Cutler, *Lynch Law: An Investigation Into the History of Lynching in the United States* (New York, 1905); Ralph Ginzburg, *100 Years of Lynching* (New York, 1962); Jacquelyn Dowd Hall, *Revolt Against Chivalry: Jessie Daniel Ames and the Woman's Campaign Against Lynching* (New York, 1979); National Association for the Advancement of Colored People, *Thirty Years of Lynching in the United States, 1889-1918* (New York, 1919); Howard Smead, *Blood Justice* (New York, 1980); George C. Wright, *Racial Violence in Kentucky, 1865-1940, Lynchings, Mob Rule, and "Legal Lynchings"* (Baton Rouge, 1990); Robert L. Zangrando, *The NAACP Crusade Against Lynching, 1909-1950* (Philadelphia, 1980); Edward L. Ayers, *Vengeance and Justice: Crime and Punishment in the 19th-Century American South* (New York, 1984), 241; W. Fitzhugh Brundage, *Lynching in the New South: Georgia and Virginia, 1880-1930* (Urbana, 1993).

25. Gerald Capers, Jr. writes: "The Negroes, as long as they observed a certain unwritten code regarding their relations with the whites, were allowed to live their own lives without interference." See the following studies, which suggest race relations were moderate in Tennessee and Memphis. *The Biography of a River Town, Memphis*, 330; Joel Roitman, "Race Relations in Memphis," 144.

26. *CA*, December 11, 1908, February 12, 1909.

27. Lester Lamon, *Black Tennesseans, 1900-1930* (Knoxville, 1977), 2, 18, 231.

28. *CA*, August 10, November 11, 1916.

29. Rabinowitz, *Race Relations*, 266. This conflict between police and African Americans is the continuation of a trend identified in Rabinowitz, "The Conflict Between Blacks and Police in the Urban South, 1965-1900," *The Historian* 39 (November 1976), 62-76.

30. *Chicago Defender*, September 11, 1915. We have by no means included all of the police/African American and/or white/African American incidents. We have included only those incidents in which no criminal accusation of wrongdoing was made. The incidents we have included refer to incidents when African Americans were responding to racial insults or were being stopped or harassed by the police for being "suspicious."

31. *CA*, August 14, 1916, 5.

32. Ibid., November 11, 1916, 6.

33. Ibid., December 4, 1916, 20.

34. Ibid., October 29, 1916, 20.

35. Rayford Logan, *The Negro in American Life and Thought: The Nadir, 1877-1901* (New York, 1954).

36. *Chicago Defender*, April 24, 1915, 1.

37. Ibid., May 8, 1915, 1.

38. Ibid., August 6, 1916, 5.

39. Ibid., August 14, 1916, 5.

40. Ibid., October 22, 1916, 1.

41. Ibid., November 10, 1916, 10.

42. Ibid., December 4, 1916, 7; Melton, "Blacks in Memphis," 21; U.S. Department of Interior, Bureau of Education, *The Public School System of Memphis, Tennessee*, Bulletin No. 50, 1919 (Washington, D.C., 1920), 13; U.S. Department of Commerce, Bureau of the Census, *Negroes in the United States, 1920-1932* (Washington, D.C., 1935), 3. We believe that migrants were well represented in the incidents cited above. In 1918, for example, the U.S. Bureau of Education surveyed the parents of white and African American students in the public school system. They found that less than 2 percent of the 11,871 white parents and less than 5 percent of the 3,801 African American parents had been born in Memphis. In addition, the 1920 census shows that 53.6 percent of the African American population in Memphis was born outside of Tennessee. In the state's next largest city, Nashville, only 8.4 percent were born outside of Tennessee. It is very likely that migrants were heavily involved in these incidents.

43. *Lula McLaurin (c) v. Memphis Street Railway Company*, Circuit Court of Shelby County, Tennessee, Division 3, No. 25 589 T.D., filed May 29, 1915.

44. Evelyn Brooks Higginbotham, *Righteous Discontent: The Women's Movement in the Black Baptist Church, 1880-1920* (Cambridge, 1993), 10.

45. Cheryl Townsend Gilkes, " 'Together and in Harness': Women's Traditions in the Sanctified Church," *Signs: Journal of Women in Culture and Society* 10 (1985), 678-99; David Tucker, *Black Pastors and Leaders: Memphis, 1819-1972* (Memphis, 1975), chap. 7.

46. Carole Marks, *"Farewell—We're Good and Gone": The Great Black Migration* (Bloomington, 1989), 35-44.

47. Charles Johnson, "How Much Is the Migration a Flight From Persecution," *Opportunity: A Journal of Negro Life* 1 (September 1923), 272-74. Currently, sociologists are revisiting Johnson's original thesis. See, for example, E. M. Beck and Stewart E. Tolnay, "The Killing Fields of the Deep South: The Market for Cotton and the Lynching of Blacks, 1882-1930," *American Sociological Review* 55 (August 1990), 526-39;

Tolnay and Beck, "Black Flight: Lethal Violence and the Great Migration, 1900 to 1930," *Social Science History* 14 (Fall 1990), 347-70; Tolnay et al., "Black Lynchings: The Power Threat Hypothesis Revisited," *Social Forces* 67 (March 1989), 605-23; Tolnay and Beck, "Racial Violence and Black Migration in the South, 1910-1930," *American Sociological Review* 57 (February 1992), 105-16. See also Cohen, *At Freedom's Edge*, chap. 10.

48. *Chicago Defender*, May 26, 1917, 1, September 9, 1917, 1.
49. Ibid., September 9, 1917, 1.
50. Ibid., October 13, 1917, 1.
51. Ibid., March 2, 1918, 1.
52. Ibid., April 6, 1918, 1.
53. Ibid., May 22, 1920, 1.
54. Ibid., October 7, 1920, 1.
55. Rabinowitz, *Race Relations in the Urban South*, 30.

6

Domination and Resistance

The Politics of Wage Household Labor in New South Atlanta

Tera W. Hunter

> Relations of domination are, at the same time, relations of resistance. Once established, domination does not persist of its own momentum. Inasmuch as it involves the use of power to extract work, production, services, taxes against the will of the dominated, it generates considerable friction and can be sustained only by continuous efforts at reinforcement, maintenance, and adjustment.
>
> James C. Scott, *Domination and the Arts of Resistance: Hidden Transcripts* (1990)

Washerwomen in Atlanta organized a massive strike in the summer of 1881. Over the course of a two-week period in July, they summoned 3,000 supporters through the neighborhood networks they

Author's Note: This chapter originally appeared in *Labor History*, *34*, Spring-Summer, 1993, pp. 205-220, and is reprinted with permission. Thanks to Stephanie Shaw, Robin D. G. Kelley, Elsa Barkley Brown, Leon Fink, Jacquelyn Hall, and Elizabeth Faue for

had been building since emancipation. The strike articulated economic as well as political grievances: The women demanded higher fees for their services and fought to maintain the distinctive autonomy of their trade. When city officials threatened the "washing amazons" with the possibility of levying an exorbitant tax on each individual member of the Washing Society (the group responsible for the strike), the women issued a warning of their own: "We, the members of our society, are determined to stand our pledge . . . we mean business this week or no washing."[1]

Southern household workers, who are often stereotyped as passive victims of racial, sexual, and class oppression, displayed a profound sense of political consciousness through the organization of this strike. Moreover, they initiated it at the dawn of the New South movement, an effort by ambitious businessmen to change the course and fortunes of regional economic development. To promote the goals of industrial capitalism and to attract northern capital below the Mason-Dixon Line, proponents of the New South heralded an image of all southern workers as artless by nature and indifferent to class struggle. But these working-class women stridently scorned this agenda.

The protest in Atlanta was not unique in the postslavery era. Washerwomen in Jackson, Mississippi, struck in 1866. And on the heels of the Great Strike of 1877, laundresses and other household workers in Galveston stopped work as well.[2] Both of these boycotts articulated goals for a living wage and autonomy, yet neither matched the proportions and the affront the Atlanta women posed to the emergent New South ideology. The Atlanta strike was unusual; domestic workers rarely organized strikes. But they did find a multitude of other ways to oppose oppression, usually in the form of surreptitious and quotidian resistance.[3]

---

incisive critiques. I am also grateful to Julius Scott, III, David Montgomery, and Nancy Cott for their insightful readings of the many incarnations of the larger project from which this essay is drawn. The writing was supported by a grant from the Institute for Research in Social Science of the University of North Carolina, Chapel Hill.

Household workers often resorted to covert tactics of resistance because they were frequently the only options available within a system of severe constraints. The magnitude of seemingly unassuming gestures looms large if we realize that workers sometimes transformed them into collective dissent or used them as building blocks for the occasional large-scale outburst. Nonetheless, it is a testament to the potency of the forces dominating women workers in the South that defiance would assume this form and that these forces were powerful enough to cover up the expression of opposition. The importance of strikes such as that by the Atlanta washerwomen in part is that they have generated a precious few documents straight from the mouths of working-class women in the form of letters and petitions to municipal officials and reports from journalists who witnessed mass meetings and rallies. In the main, little direct testimony exists from household workers about their activities and the motivations that prompted them. But there is another way to scout out working-class women's discontent and dissent. Evidence from employers and their proxies in public authority positions unwittingly expose the resilience and creativity of African American household workers' efforts to counter domination.

This essay is an effort to understand resistance by looking at the character of domination and the attempts to counter it from Reconstruction to World War I. Domination is defined here as the process of exercising power over the dispossessed by whatever means necessary, but without overt conflict where possible. Conversely, resistance is defined as any act, individual or collective, symbolic or literal, intended by subordinates to deny claims, to refuse compliance with impositions made by superordinates, or to advance claims of their own.[4] This essay outlines examples of African American women domestics combating injustice, and it analyzes the responses of employers and public officials. As household workers struggled to negate conditions of abject servitude, their employers worked even harder to repress and contain these workers. The subsequent contests reveal how structures of inequality were reproduced and challenged in daily interactions; their public airing suggests that wage household labor had broader social and political implications beyond its significance to private homes. Atlanta is a fitting place to begin exploring the larger ramifications of wage household labor. Young, white, upwardly mobile businessmen in the years after the Civil War began cultivating an image of the city as the vanguard of a "New South." As the ideas of these urban boosters were instituted, it became all too clear that "modernization" of the social, political, and economic order included racial segregation and political disfranchisement. From this perspective, Atlanta did not simply embody the contradictions of life under Jim Crow; the conscious leadership role it assumed in the region also made it instrumental in creating and perpetuating them.[5]

This self-proclaimed model of the New South held the distinction of employing one of the highest per capita numbers of domestic workers in the nation during the period of this study.[6] Such a repute was not coincidental to the seeming contradiction between the goals of modernization and the advocacy of a retrogressive system such as segregation. One might expect that a modernizing economy would shirk old-fashioned manual household labor in favor of up-to-date mechanized and commercial production. Yet manual household work furthered the goals of the advocates of the New South in restricting black workers' social and economic opportunities. African American women who migrated to Atlanta following emancipation were segregated into household labor. Virtually no other options were available to them, yet wage work was essential to the sustenance of their livelihoods from childhood to death. And in Atlanta, as in other southern cities, the disproportionate sex ratio among blacks made wage work all the more imperative for women, especially for single, divorced, or widowed mothers saddled with the sole responsibility for taking care of their families. And the low wages paid to black men meant that even married women could rarely escape outside employment and worked in far greater numbers than their white counterparts.[7]

Yet despite this occupational confinement, black women managed to assert some preferences for the particular kind of domestic labor they performed. Single and younger women accepted positions as general maids or child-nurses more often, for example, whereas married women usually chose positions as laundresses. Washerwomen represented the largest single category of waged household workers in Atlanta, and by 1900 their total numbers exceeded all other domestics combined.[8] Laundresses picked up loads of dirty clothes from their patrons on Monday; washed, dried, and ironed throughout the week; and returned the finished garments on Saturday. This labor process encumbered their already cramped living quarters with the accoutrements of the trade, but it exempted workers from employer supervision, yielded a day "off," allowed workers to care for their children and to perform other duties intermittently, incorporated family members into the work routine, and facilitated communal work among adult women.[9]

Regardless of the specific domestic job black women chose, the majority insisted on living in their own homes rather than with their employers. Elsewhere in the country, where immigrant European and native-born white women were more numerous, live-in domestic work predominated, but for recently freed slaves, living with their own families was foremost to approximating independence.[10] Above all, living on their own meant for the former slaves breaking the physical chains of bondage and reestablishing the kinship ties scattered and torn asunder by the caprice of fluctuating fortunes or the ill will of owners. It also meant preventing employers from exercising unmitigated control over their entire lives. Some employers accepted a live-out arrangement; perhaps, because it coincided with their own ambivalence about continuing the intimacy that prevailed between master and slaves. But many employers resented the loss of control that resulted.[11]

Black women's priorities in the post-Civil War years demonstrated that economic motivations alone did not influence their decisions about wage labor. They sought instead to balance wage-earning activities with other needs and obligations. Consequently, they moved in and out of the labor market as circumstances in their personal lives demanded and switched jobs frequently. Domestic workers quit in order to buy time off for a variety of reasons; among them participation in special functions, such as religious revivals, or taking care of family members who became ill. The workers also resorted to quitting to make clear their discontent over unfair practices when other efforts to obtain satisfactory redress failed. Quitting did not necessarily guarantee a better situation elsewhere—indeed, it often did not improve the workers' situation—but it reinforced workers' desire for self-determination and deprived employers of the ascendancy to which they were accustomed as slaveholders.

Consequently, quitting made it difficult for employers to find "good" servants and, especially, to keep them—the single most oft-repeated complaint from Reconstruction onward.[12] Quitting violated employers' expectations of the ideal worker: one who conformed to relentless hours of labor, made herself available at beck and call, and showed devoted loyalty throughout her entire life. In 1866, as the clamor among employers demanding relief quickly rose to a high pitch, the Atlanta City Council interceded on their behalf by passing a law to nullify free labor's most fundamental principle. To obstruct the liberties essential to authentic independence, to hinder the ease and frequency of workers changing jobs, the law required employers of domestics to obtain recommendations from the previous employer before hiring them.[13]

The 1866 law is instructive of the general crisis of free labor in the South following the Civil War. As African Americans showed a marked determination to make their new status live up to their needs and expectations, planters and urban employers rejected the ideals of the free-labor system that conflicted with the safekeeping of white supremacy. In 1865, during the brief reign of Presidential Reconstruction under Andrew Johnson, Democrats in state legislatures in the South instituted the Black Codes, laws designed, among other things, to diminish blacks' rights in labor contracts.[14] The 1866 law was strongly reminiscent of this mechanism, and its passage signaled the increasing role of the state in relationships formerly governed entirely by individual masters. Black women workers would still be vulnerable to arbitrary personal power although its exercise would be tempered by the Thirteenth Amendment. Nonetheless, employers would try to coerce workers with the aid of the state. The enactment of the law in 1866 provided concrete evidence that household workers' refusals to acquiesce to unrelenting physical exertion forced employers to procure outside intervention.

Employers' augmentation of their authority with municipal power, however, proved ineffective in part because of their ambivalent attitude toward the law. Frustrated employers were often willing to employ almost any black woman in their ever-illusive search for individuals whose personal characteristics and occupational behavior coincided with the traits of good servants. Despite the employers' dissatisfaction with the way the system worked, and in defiance of the law passed for their own protection, they preferred to hire workers without the requisite nod from former bosses rather than face the unthinkable possibility of no servants at all.

Black women's active opposition to the law also helped to defeat it as they continued to quit work at will. Quitting was an effective strategy of resistance precisely because it could not be quelled outside a system of bound labor. Although some women workers may have openly confronted their employers before departing, quitting as a tactic thrived because it did not require such direct antagonism. Workers who had the advantage of living in their own homes could easily make up excuses for leaving, or leave without notice at all—permitting small and fleeting victories for individuals to accumulate into bigger results as domestics throughout Atlanta and the urban South repeated these actions over and over again. The instability created in the labor market strengthened the bargaining position of domestic workers because employers persisted in thinking of the pool as scarce, although in absolute numbers, the supply of domestic workers available to the employing population in Atlanta was virtually endless. The incongruence between the perception of a dearth and the reality of an abundance suggests that black women's self-assertion had indeed created a shortage of workers with the attributes employers preferred.[15]

Quitting and other forms of everyday struggle continued for many decades long after Reconstruction. In 1912, an Atlanta mayoral candidate offered an extreme, if novel, solution to the menacing problem of restraining domestic workers' self-assertions. George

Brown, a physician, supported a public health reform that encompassed the concerns of white employers. The candidate promised pure drinking water, free bathing facilities, improved sanitary provisions at railroad stations, and a (white) citizenry protected from exposure to contagious germs.[16] The latter proposal had direct implications for black domestics whom employers and health officials accused of spreading tuberculosis through the food they cooked, the houses they cleaned, and the clothes they washed. Laundry workers were the most vociferously attacked objects of scorn. The freedom they enjoyed from direct white supervision permitted them to operate more as contractors than as typical wage workers, which made them vulnerable to scrutiny of their labor and personal lives.[17]

Brown and like-minded individuals heightened the fear that domestic workers were the primary emissaries of physical contagions and impressed on white minds that black women were the harbingers of social disease as well. The attribution of pestilence to domestics unveils deeper frictions that lay bare a central paradox about Jim Crow, which by then was firmly in place. The social and political geography of Atlanta bolstered the exploitation and containment of black bodies and their spatial separation from upper-class whites. African Americans were segregated in the worst areas of the city and had the least access to the municipal resources essential to good health; services such as street pavings, proper waste disposal, and potable water were provided to Atlantans on the basis of both racial and class privileges. By the late nineteenth century, upper-class whites in large numbers had moved out to ostentatious suburbs and had begun to escape regular interaction with the unattractive sites that the inequitable distribution of city resources typically bred.[18] Yet these white suburbanites continued to hire black household workers from such malodorous neighborhoods. White anxieties about the contaminating touch of black women reflected the ambivalence of a tension between revulsion and attraction to the worker who performed the most intimate labor, taking care, for example, of children.

Brown proposed to wipe out the public health problem and to diminish the ubiquitous "servant problem" in one sweeping measure. He proposed the creation of a city-run servant bureau invested with broad discretionary judiciary powers that would require domestics to submit to rigorous physical examination and to offer detailed personal and employment histories before obtaining prerequisite licenses for work. Brown sought to reinstitute "absolute control" of servants and to relieve white fears by criminalizing presumed carriers of disease; he promised to punish domestics who impeded efforts to keep the scourge away from the doorsteps of their white bosses.

And the mayoral hopeful went further: he called for disciplinary measures to be used against workers who exercised the conventional liberties of wage work. Quitting for reasons employers did not consider "just" or displaying other forms of recalcitrance would constitute sufficient grounds for arrest, fines, incarceration, or labor on the chain gang.[19] As a candidate outside the inner circle of New South politicos, Brown hardly had a chance to win the election, but his campaign is noteworthy for its dissemination of pejorative images of domestics that further legitimized their subordination as a source of cheap labor.

The Brown campaign is also suggestive about the changing constitution of domination in response to household workers' agency. The prominence of the disease issue, even beyond the mayoral campaign, showed signs of a shift in the "servant problem" discourse from an emphasis on so-called inherent deficiencies of black women, such as laziness and the lack of a proper work ethic, to a more powerful critique of domestic workers as the bearers of deadly organisms.[20] Worker mobility and other acts of defiance undoubtedly took their toll on employers' patience, but the prospects of contracting tuberculosis or other communicable diseases provided new and greater rationalizations for establishing comprehensive mechanisms of control over black females. The ostensible concern with public health, however, falters as an adequate explanation for these exacerbated prejudices, if we consider that proposals like Brown's were based on the faulty assumption that disease traveled solely on one-way tickets from blacks to whites. The servant bureau of Brown's imagination would not have alleviated the propagation of germs, but it would have stripped household workers of important rights. Carried to their logical conclusion, the punitive measures could have conveniently led to a convict labor system for domestic workers, forcing them to work at the behest of employers without compensation and under the threat of physical brutality.

Several of the issues raised in George Brown's run for mayor reverberated in another infamous campaign. Joseph M. Brown, son of the former Confederate governor and unrelated to George, ran for the U.S. Senate against the incumbent Hoke Smith in 1914. The two Brown men shared the view that domestic workers' defiance posed an ample threat to social stability in the New South that justified state intervention. Both men berated the large numbers of household workers who participated in benevolent and mutual aid associations, also known as secret societies, and both believed that it was imperative to dismantle the workers' capacity to bolster clandestine resistance through such institutions.[21]

From Reconstruction onward, black women led and joined secret societies to pool their meager resources to aid the sick, orphaned, widowed, or unemployed and to create opportunities for personal enrichment as well as broader race advancement. The number of such organizations with explicit labor-related goals were few, but groups that brought working-class women together for other expressed purposes were known to transform themselves on the spur of the moment and operate as quasi-trade unions when necessary.[22]

George Brown had entreated white men to put him in the mayor's seat so that he could direct the cleansing mission of his servant bureau toward eradicating these organizations that debilitated "helpless" white housewives.[23] "Little Joe" Brown followed suit in his bid for the Senate two years later by rebuking African American domestics for devising "blacklists" in secret societies that deprived errant employers. This tactic was especially unnerving to him (and others) because it shrouded a collective act by relying on individuals to quietly refuse to work, leaving behind perplexed housewives with the sudden misfortune of not being able to find willing workers. Joe Brown preyed on white southern fears to dramatize the urgent need to eliminate these quasi-union activities, and he tried to race-bait his opponent Hoke Smith, no stranger to this ploy himself. Brown accused the black mutual aid groups of conspiring with white labor unions in an interracial syndicate, a charge that white labor leaders quickly rebutted.[24] Brown forewarned the voters against choosing Smith and of the consequences of failing to elect him and neglecting to outlaw the institutional basis of African American women's dissent: "Every white lady in whose home negro servants are hired then becomes subservient to these negroes," he stated.[25] Brown lost the Senate race, yet his devotion to assailing household workers' resistance had unintended consequences; it acknowledged its effect.

Schemes designed to thwart household workers' agency reached a peak as the Great Migration intensified during World War I. In May 1918, Enoch Crowder, the selective service director, issued a "work or fight" order aimed at drafting unemployed men into the armed forces. The order stressed the nation's need for labor's cooperation in contributing to the war effort through steady, gainful work or military service. Trade unions immediately protested the potential abuses that could result from such a directive, having heard of abuses perpetrated against striking British workers under a similar law. Newton D. Baker, the secretary of war, made assurances to the contrary, but striking machinists in Bridgeport, Connecticut, were threatened with Crowder's order.[26] Southern legislatures and city councils deliberately designed their own work-or-fight laws to break the will of black workers in order to maintain white supremacy in a time of rapid change and uncertainty. Similar to the logic used by white Progressives in antivagrancy campaigns during the same period, work-or-fight laws were rationalized as a solution to alleged crime and moral depravity that resulted when blacks filled all or part of their day with pursuits other than gainful work. Atlanta had one of the highest per capita arrest records in the country in the early twentieth century, largely because of vagrancy and other misdemeanor convictions; the individuals apprehended were often gainfully employed and always disproportionately black.[27] The relative scarcity of labor produced by the war prompted southern lawmakers to manipulate Crowder's order and use it to clamp down on African

Americans at the very moment when the war opened new opportunities for employment and increased their bargaining positions in existing jobs.

White southerners abandoned the original intention of the federal measure to fill the army with able-bodied men by making the conscription of women central to its provisions.[28] As opportunities for black women expanded in the sewing trades, commercial laundries, and, less rapidly, in small manufacturing plants, the number available for household work declined, giving an edge to those who remained in negotiating for better terms.[29] Employers of domestics resented this new mobility and sought to contain it by using work-or-fight laws to punish black women who vacated traditional jobs.

Individuals arrested under the laws' provisions included black housewives, defined as "idle" and unproductive, and other self-employed black women such as hairdressers. A group of self-described "friends" of the Negro race in Macon, Georgia, iterated some of the assumptions behind such enforcement. Black women should not withdraw from wage work in general and household labor in particular, no matter what the circumstances; the Macon group argued that patriotic duty required that black women not "sit at home and hold their hands, refusing to do the labor for which they are specially trained and otherwise adapted." Black women's domestic work was essential to the war effort, insisted the Macon group, because it exempted white women "from the routine of housework in order that they may do the work which negro women cannot do."[30] In Atlanta, two seventeen-year-old girls experienced the encroachment of this notion of patriotism firsthand. "You cannot make us work," Nellie Atkins and Ruth Warf protested upon arrest and proceeded to break windows to vent their anger at the injustice, which doubled the sentence to sixty days each in the prison laundry.[31] Warf and Atkins were relatively fortunate, however; other women were tarred and feathered and violently attacked by vigilantes.[32]

African Americans in Atlanta took the lead in organizing what eventually became a regional assault against racist and sexist implementation of work-or-fight laws. They enlisted the national office of the NAACP, which in turn launched an investigation and supported local chapters in the South in order to stop the passage of the abusive laws. The NAACP discovered that employers not only used the laws to conscript nondomestics but also used them against employed household workers who demanded higher wages to meet the rising costs of living, organized protests, quit work because of unfair treatment, or took time out for other activities.[33] Over a half century after the Jackson washerwomen's strike, for example, all the household workers in that city organized and established a six-day work week, with Sundays off. But employers launched a counteroffensive, forcing the workers to return to an unforgiving seven-day schedule or face prosecution.[34]

Blacks in Atlanta successfully lobbied Governor Hugh Dorsey to veto discriminatory work-or-fight legislation passed by the Georgia House and Senate. Fearing the intensification of the Great Migration and the loss of black laborers, Dorsey responded to their demands. The Atlanta branch of the NAACP similarly appealed to the city council and managed to preempt legislation at the local level and eliminated de jure discrimination through a wartime measure. Police and vigilantes, however, found other methods of abusing black women with impunity.[35]

The blatantly unjust harassment of household workers during World War I revealed another variation on a familiar theme—the New South's unabashed disdain for the privileges of free labor. Yet the physical brutality and legal coercion rationalized by state work-or-fight laws also signaled the breakdown of the authority of the elite in controlling a workforce whose hallmark was supposedly servility. Like similar proposals to regulate domestic workers in previous years, work-or-fight laws uncovered an effort by employers to eliminate black women's ongoing resistance. The abusive legislation also uncloaked the impact of the Great Migration. As African Americans left the South en masse to pursue freedom in northern industrial towns, white southern employers struggled to maintain power over those who stayed.

Work-or-fight laws and the other efforts to control domestic workers are interesting in part because they evidence struggle and contestation that till now had been obscured. Although in many of the instances noted above the household workers' collective consciousness may have been out of sight, it was not out of mind. The washerwomen's strike in the summer of 1881 reveals how working-class women's resistance could and did take a different form, as they openly proclaimed the usually "hidden transcript" of opposition in a profound way.[36] The strike displayed an astute political consciousness among black working-class women who made so-called private labor a public issue and insisted on autonomy and a living wage.[37]

The communal character and self-organization of laundry work proved critical to this mobilization as it facilitated the creation of a relatively autonomous space that had already nurtured the foundation of working-class women's solidarity. The Atlanta laundresses built on this tightly knit system, extended it through an intensive door-to-door recruitment of adherents to their cause, and sustained it through mass or decentralized ward meetings held nightly. Their capacity to rise to this occasion demonstrates why washerwomen were the most outspoken leaders in domestic workers' strikes documented in the South. It is no accident that—as incidents in later years would indicate—employers often combined forces to repress this particular group.

White city leaders put their full weight behind employers' attempts to annihilate a strike. At least one landlord threatened to raise the rent of his washerwoman if she raised the fees for her work. A businessman scoffed "at the colored people's stupidity in not seeing that they were working their own ruin" and warned that if they persisted they would be faced with a harsh winter without white charity.[38] The police arrested several street organizers for "disorderly conduct," charging them with disruptive and violent behavior as they canvassed their neighborhoods. Leading capitalists raised funds for a state-of-the-art steam laundry and offered to employ "smart Yankee girls" to buttress the counteroffensive and requested a tax exemption from the city council to subsidize the costs. Meanwhile, municipal authorities proposed a scheme to regulate the trade and destroy the workers' independence: Councilmen suggested that each member of any washerwomen's organization pay an exorbitant business tax of $25.00.[39] In the end, however, the city council rejected the license fee; the councilmen may have been daunted by the continued determination of women who refused to buckle under to threats and who vowed to reappropriate the license fee and city regulation to gain the benefits of private enterprise. As the women themselves stated in an open letter to the mayor, "We have agreed, and are willing to pay $25 or $50 for licenses as a protection so we can control the washing for the city."[40]

Not only did the washerwomen's spirit of rebellion frustrate the actions of their opponents, it set an example for other black workers. Waiters at the National Hotel followed on the women's coattails and won demands for better wages and working conditions previously rejected by management. Cooks, maids, and child nurses also were inspired to begin organizing for better wages. Even the *Atlanta Constitution,* ardent ally of the employers, begrudgingly admitted that the "amazons" had shown remarkable organization.[41]

The most telling piece of evidence about the strike's impact appeared several weeks after the event had apparently subsided, when an unidentified source divulged to the newspaper that the washerwomen were threatening to call a second potentially more perilous general strike of all domestics during the upcoming International Cotton Exposition. Although there were no further reports to suggest that this rumor ever came to fruition, the mere threat of a second strike at such a critical moment is quite telling. The laundry workers were clearly conscious of the significance of this event, which had been touted as the debut of the New South movement and as a showcase for Atlanta, an upstart metropolis eager to be emulated. A strike held at that particular time not only would have spoiled the image of docile labor that New Southerners were carefully projecting to attract northern capital, it would have wreaked havoc on a city already anxious about its capacity to host the thousands of visitors who would require the services of cooks, maids, child nurses, and laundresses. The newspaper forewarned white housewives to "prepare for the attack before it is made," and they did.[42]

The actual outcome of the washerwomen's strike is inconclusive, although it is curious that reports on the protest petered out in the medium that had openly flaunted its partisanship against it. Whether or not some or all the washerwomen were able to gain higher wages we may never know; however, they continued to maintain a modicum of independence in their labor not enjoyed by other domestics. The strike speaks volumes symbolically about African American working-class women's consciousness of their racial, class, and gender position. Domestic work was synonymous with black women in freedom as it was in slavery, and the active efforts by whites to exploit labor clearly circumscribed black lives. Yet black women fought for dignity, to be treated with respect, and for a fair chance to earn the necessary resources for making a decent living. The women identified autonomy as vital to freedom and to making decisions about wage work most commensurate with their nonwage responsibilities as mothers, sisters, daughters, and wives.

The employers could not fathom the motivations that inspired domestic workers to act in these ways. But employers knew they could not afford to take a pacified workforce for granted. They used coercion, repression, and violence and sought support from the state to extract compliance to their wishes, which helped to determine the form that resistance would take. Domination and resistance were always defined in dynamic relationship to one another, thus it is not surprising that strikes were atypical events. Domestic workers developed other ways to articulate their grievances and assert their own demands, however, and in return their actions influenced the character of domination itself. The illusive quality of the black women's surreptitious actions made them difficult to control by individual employers and kept them vigilant. Domination was not a project that could be erected in full form and left to operate on its own momentum; it required ongoing efforts of surveillance and reconstitution to guarantee its effect.[43] At times, this meant that domestic workers won small gains and moments of relief, as when they quit work. At other times, their resistance led to greater repression, as during the period of World War I with the implementation of work-or-fight laws.

The contested character of wage household labor between Reconstruction and World War I also highlights another important point. Far from functioning as "separate spheres," the so-called public sphere of politics and business and the private sphere of family and home infiltrated one another in complex ways. It should be noted, however, that employers sometimes displayed an ambivalence about the relationship between their prerogatives as managers of labor and the intervention of public authorities, literally, on their home turfs. Municipalities and legislatures often stopped short of imposing legislation; recall, for example, that the Atlanta City Council failed to impose the business tax on individual laundry workers during the 1881 strike. African American women's opposition may have thwarted employers' efforts to subdue them, but other factors may have also hindered employers from realizing the optimal balance between compulsion and free labor. In an economy moving toward modernization, even in the constrained version of southern capitalism, the issue of state power versus individual employer authority was never consistently resolved. Waged household labor played an important role in the economic, social, and political life of the New South. The women who performed the labor, and the women and men who employed them, were consummate political actors all. Further theoretical speculation and empirical research of the issues raised in this essay will advance our understanding on the development of New South capitalism beyond what we already know about social relations in agriculture and industry.

Notes

1. *Atlanta Constitution,* August 3, 1881.

2. See *Jackson Daily Clarion,* June 24, 1866, reprinted in Philip S. Foner and Ronald Lewis, eds., *The Black Worker: A Documentary History From Colonial Times to the Present* (Philadelphia, 1978-1984), vol. 2, 345; *Galveston Daily News,* 1, 2, 5, 7, and 16, August 1877. For a full account of all the strikes, see Tera W. Hunter, "Household Workers in the Making: Afro-American Women in Atlanta and the New South, 1861 to 1920" (Ph.D. diss., Yale University, 1990).

3. This essay relies on the following works on resistance: James C. Scott, *Weapons of the Weak: Everyday Forms of Peasant Resistance* (New Haven, 1985); *Domination and the Arts of Resistance: The Hidden Transcripts* (New Haven, 1990); Rosalind O'Hanlon, "Recovering the Subject: Subaltern Studies and Histories of Resistance in Colonial South Asia," *Modern Asian Studies* 22 (1988), 189-224: John Fiske, *Understanding Popular Culture* (Boston, 1989).

4. See O'Hanlon, "Recovering the Subject," 199-200; Scott, *Weapons of the Weak,* 289-303.

5. On Atlanta as a leading city in the New South, see James Michael Russell, *Atlanta, 1847-1890: City Building in the Old South and the New* (Baton Rouge, 1988), passim; Don H. Doyle, *New Men, New Cities, New South: Atlanta, Nashville, Charleston, Mobile, 1860-1910* (Chapel Hill, 1990), passim; Howard N. Rabinowitz, *Race Relations in the Urban South, 1865-1890* (New York, 1978), passim; C. Vann Woodward, *Origins of the New South, 1877-1913* (Baton Rouge, 1951), 124.

6. For a comparison of rates of employment of domestic workers in various cities, see David Katzman, *Seven Days a Week: Women and Domestic Service and Industrializing America* (New York, 1978), 61, 286.

7. On rates of married women in the workforce, see Joseph A. Hill, *Women in Gainful Occupations, 1870 to 1920* (Washington, D.C., 1929), 334-36.

8. U.S. Department of Commerce and Labor, Bureau of the Census, *Special Reports: Occupations of the Twelfth Census* (Washington, D.C., 1904), 486-89.

9. On laundry work, see Sarah Hill, "Bea, the Washerwoman," Federal Writer's Project Papers, Southern Historical Collection, University of North Carolina, Chapel Hill; Jasper Battle, "Wash Day in Slavery," in George P. Rawick, ed., *The American Slave: A Composite Autobiography* (Westport, 1972-1978), vol. 2, pt. 1, 70; Katzman, *Seven Days a Week*, 72, 82, 124; Daniel Sutherland, *Americans and Their Servants: Domestic Service in the United States From 1800 to 1920* (Baton Rouge, 1981), 92; Faye E. Dudden, *Serving Women: Household Service in Nineteenth Century America* (Middletown, 1983), 224-25; Patricia E. Malcolmson, *English Laundresses: A Social History, 1850-1930* (Urbana, 1986), 11-43.

10. On live-out arrangements, see Katzman, *Seven Days a Week*, 87-91.

11. See, for example, testimony of Albert C. Danner, U.S. Senate, Committee on Education and Labor, *Report Upon the Relations Between Labor and Capital* (Washington, D.C., 1885), 105 (hereafter cited as *Labor and Capital*).

12. See Myrta Lockett Avary, *Dixie After the War: An Exposition of Social Conditions Existing in the South During the Twelve Years Succeeding the Fall of Richmond* (Boston, 1906; reprinted 1937), 192 entries for June 17 through December 2, 1866, Samuel P. Richard Diary, Atlanta Historical Society; entries for May 1865, Ella Gertrude Clanton Thomas Journal, Duke University Archives; Emma J. S. Prescott, "Reminiscences of the War," 49-55, Atlanta Historical Society.

13. Alexa Wynell Benson, "Race Relations in Atlanta, as Seen in a Critical Analysis of the City Council Proceedings and Other Related Works, 1865-1877" (master's essay, Atlanta University, 1966), 43-44.

14. On the crisis of free labor, see Eric Foner, *Politics and Ideology in the Age of the Civil War* (New York, 1978), 97-125. On Black Codes, see Eric Foner, *Reconstruction: America's Unfinished Business, 1863-1877* (New York, 1988), 109-202.

15. David Katzman speculates on the basis of the ratio of workers to employers that there were enough laundresses in Atlanta for every white household and even some black. See Katzman, *Seven Days a Week*, 91-92 and table 2-6.

16. On George Brown's campaign, see *Atlanta Constitution*, September 8, September 15, September 28, and September 29, 1912.

17. See, for example, H. McHatton, "Our House and Our Servant," *Atlanta Journal-Record of Medicine* 5 (July 1903), 212-19; *Atlanta Constitution*, December 19, 1909; William Northen, "Tuberculosis Among Negroes," *Journal of the Southern Medical Association* 6 (October 1909), 415; H. L. Sutherland, "Health Conditions of the Negro in the South: With Special Reference to Tuberculosis," *Journal of the Southern Medical Association* 6 (October 1909), 399-407; *Daily Times*, n.p., September 7, 1912, in Tuskegee Institute News Clip file (hereinafter TINF).

18. On the social and political implications of Atlanta's geography, see James M. Russell, "Politics, Municipal Services, and the Working Class in Atlanta, 1865 to 1890," *Georgia Historical Quarterly* 66 (1982), 467-91; Jerry Thornberry, "The Development of Black Atlanta 1865-1885" (Ph.D. diss., University of Maryland, 1977); Dana F. White, "The Black Sides of Atlanta: A Geography of Expansion and Containment, 1870-1970," *Atlanta Historical Journal* 26 (Summer/Fall 1982-1983), 199-225.

19. *Atlanta Constitution*, September 15, 1912.

20. For other discussions associating domestic workers with disease and proposals to regulate them, see *Atlanta Constitution*, February 11, March 11, March 12, March 25, 1910, October 2, 1912; and *Atlanta Independent*, February 19, 1910.

21. *Atlanta Constitution*, September 15, 1912; 1914 campaign literature, Joseph M. Brown Papers, Atlanta Historical Society.

22. See, for example, *Atlanta Constitution*, March 31, 1910; Ruth Reed, *Negro Women of Gainesville, Georgia* (Athens, Ga., 1921), 46. Canadian working-class women's mutual aid organizations operated similarly; see Varpu Lindström-Best, *Defiant Sisters: A Social History of Finnish Immigrant Women in Canada* (Toronto, 1988), 56-60.

23. *Atlanta Constitution*, September 15, 1912.

24. See 1914 campaign literature, Joseph M. Brown Papers, Atlanta Historical Society. Also see *Atlanta Constitution*, March 31, 1910. The white trade unionists denied the charges by reminding their supporters that "the 'nigger' question is generally the last and most desperate resort of demagogues to win votes." Although they admitted the importance of black workers organizing in separate unions to prevent undercutting white workers, they opposed integration and social equality. 'Little Joe' knows that there is not a single white labor unionist in Georgia, or the South, who would stand for that sort of thing," they insisted. *Journal of Labor*, July 24, 1914.

25. See 1914 campaign literature, Joseph M. Brown Papers, Atlanta Historical Society: Dewey Grantham, *Hoke Smith and the Politics of the New South* (Baton Rouge, 1959), 270-73.

26. David M. Kennedy, *Over Here: The First World War and American Society* (New York, 1980), 269; David Montgomery, *Workers' Control in America* (Cambridge, 1979), 127-34.

27. Charles Crowe, "Racial Violence and Social Reform: Origins of the Atlanta Riot of 1906," *Journal of Negro History* 52 (1968), 247; John Dittmer, *Black Georgia in the Progressive Era 1900-1920* (Urbana, 1977), 87-88.

28. Walter F. White, " 'Work or Fight' in the South," *The New Republic*, 18 (March 1, 1919), 144-46.

29. See U.S. Department of the Interior, Bureau of the Census, *Report of the Population of the United States at the Eleventh Census: 1890* (Washington, D.C., 1897), pt. 2, 634-35; U.S. Dept. of Commerce and Labor, Bureau of the Census, *Special Reports: Occupations at the Twelfth Census* (Washington, D.C., 1904), 486-89; idem, *Thirteenth Census of the United States Taken in the Year 1910*, vol. 4, "Population Occupational Statistics" (Washington, D.C., 1914), 536-37; idem, *Fourteenth Census of the United States Taken in the Year 1920*, vol. 4, "Population, Occupations" (Washington, D.C., 1923), 1053-55.

30. Macon *News*, October 18, 1918, in TINF. The federal government also made similar appeals to black women through war propaganda. See, for example, the Portsmouth, Virginia, *Star*, October 21, 1918, in TINF.

31. Quoted in *Baltimore Daily Herald*, September 10, 1918, Group 1, Series C, Administrative Files, Box 417, National Association for the Advancement of Colored People Papers, Library of Congress (hereinafter NAACP, LC).

32. For instances of violence against women, see Walter F. White, "Report of Conditions Found in Investigation of 'Work or Fight' Laws in Southern States," Group 1, Series C, Administrative Files, Box 417, NAACP, LC.

33. Walter F. White, "Report of Conditions Found in Investigation of 'Work or Fight' Laws in Southern States," NAACP, LC; Chicago *Defender*, July 13, 1918, and *New York Age*, November 19, 1918, in TINF.

34. *New York Age*, November 19, 1918, in TINF.

35. One outcome of the NAACP's involvement in this campaigning was that it increased the interests of black southerners in joining the organization. Thus, work-or-fight laws became a critical galvanizing issue for the growth of local NAACP chapters in the South. See, for example, *Atlanta Constitution*, July 10-August 25, 1918; Reverend P. J. Bryant, Remarks to the 10th Annual Conference of the NAAP, June 24, 1919, Group 1, Series B, Annual Conference Files, Box 2, NAACP, LC.

36. On "hidden transcripts," see Scott, *Domination and the Arts of Resistance*, passim.

37. See Rabinowitz, *Race Relations in the Urban South*, 74-76; Katzman, *Seven Days a Week*, 196-97; William H. Harris, *The Harder We Run: Black Workers Since the Civil War* (1982), 37; Dudden, *Serving Women*, 232; Dorothy Sterling, *We Are Your Sisters: Black Women in the Nineteenth Century* (New York, 1984), 357-58; Jacqueline Jones, *Labor of Love, Labor of Sorrow: Black Women, Work, and the Family From Slavery to Freedom* (New York, 1985), 148-49; Donna Van Raaphorst, *Union Maids Not Wanted: Organizing Domestic Workers, 1870-1940* (New York, 1988), 200. My own interpretation is closest to the only other study that considers most of the available evidence: Thornberry, "The Development of Black Atlanta," 215-20. Rabinowitz's account has prevailed as the definitive one, often cited uncritically by other historians. But in the haste to force the event to conform to a thesis that emphasizes white attitudes and black inefficacy in the face of white power, he ignores significant evidence and overstates the known reprisals made against the women.

38. *Atlanta Constitution*, August 3, 1881.

39. Ibid., July 24, 1881.

40. Ibid., August 3, 1881. The women may have been counting on resources from individual savings and mutual aid organizations to help defray the costs of the fees. Nonetheless, the cost still would have been exorbitant.

41. Ibid., July 21, 1881.

42. Ibid., September 6, 1881.

43. See epigram above. Scott, *Domination and the Arts of Resistance*, 45.

7

"We Are Not What We Seem"

Rethinking Black Working-Class Opposition in the Jim Crow South

Robin D. G. Kelley

> Each day when you see us black folk upon the dusty land of the farms or upon the hard pavement of the city streets, you usually take us for granted and think you know us, but our history is far stranger than you suspect, and we are not what we seem.
>
> Richard Wright,
> *Twelve Million Black Voices* (1940)

> The Negro, in spite of his open-faced laughter, his seeming acquiescence, is particularly evasive. You see we are a polite people and we do not say to our questioner, "Get out of here!" We smile and tell him or her something that satisfies the white person because, knowing so little about us, he doesn't know what he is missing.
>
> Zora Neale Hurston,
> *Mules and Men* (1935)

On the factory floor in North Carolina tobacco factories, where women stemmers were generally not allowed to sit or to talk with one another, it was not uncommon for them to break out in song. Not only did singing in unison reinforce a sense of collective identity in these black workers; also, the songs themselves—most often religious hymns—ranged from veiled protests against the daily indignities of the factory to utopian visions of a life free of difficult wage work.[1]

Throughout the urban South in the early twentieth century, black women household workers were accustomed to staging so-called incipient strikes, quitting or threatening to quit just before important social affairs to be hosted by their employers. The strategy's success often depended on a collective refusal on the part of other household workers to fill in.[2]

In August 1943, on the College Hills bus line in Birmingham, Alabama, black riders grew impatient with a particularly racist bus driver who within minutes twice drew his gun on black passengers, intentionally passed one black woman's stop, and ejected a black man who complained on the woman's behalf. According to a bus company report, "The negroes then started ringing the bell for the entire block and no one would alight when he stopped."[3]

These daily, unorganized, evasive, seemingly spontaneous actions form an important yet neglected part of African American political history. By ignoring or belittling such everyday acts of

---

Author's Note: This chapter originally appeared in the *Journal of American History, 80*(1), June, 1993, pp. 75-112, and is reprinted here with the permission of the Organization of American Historians. I want to send a shout out to Diedra Harris-Kelley, Edward L. Ayers, Eileen Boris, Elsa Barkley Brown, Geoff Eley, Elizabeth Faue, Jacquelyn Dowd Hall, Robert L. Harris, Michael Honey, Tera Hunter, Robert Korstad, Cliff Kuhn, Earl Lewis, Nelson Lichtenstein, Peter Linebaugh, George Lipsitz, August Meier, Bruce Nelson, David Roediger, Armstead Robinson, Joe W. Trotter, Jr., Victoria Wollcott, and the anonymous readers for the *Journal of American History* for their comments and suggestions. I am especially grateful to David Thelen for his direct but encouraging critique and Susan Armeny for her editing and effort. I want to dedicate this essay to Herbert Aptheker in recognition of the fiftieth anniversary of *American Negro Slave Revolts*, a pioneering work that inspired this essay, and to the late Brenda McCallum, a brilliant and imaginative historian of black working-class culture who passed unexpectedly and prematurely in September of 1992.

resistance and privileging the public utterances of black elites, several historians of southern race relations concluded, as Lester C. Lamon did in his study of Tennessee, that black working people "remained silent, either taking the line of least resistance or implicitly adopting the American faith in hard work and individual effort."[4] But as Richard Wright, Zora Neale Hurston, and countless cases like those recounted above suggest, the appearance of silence and accommodation was not only deceiving but frequently intended to deceive. Beneath the veil of consent lies a hidden history of unorganized, everyday conflict waged by African American working people. Once we explore in greater detail those daily conflicts and the social and cultural spaces where ordinary people felt free to articulate their opposition, we can begin to ask the questions that will enable us to rewrite the political history of the Jim Crow South to incorporate such actions and actors.

Drawing examples from recent studies of African Americans in the urban South, mostly in the 1930s and 1940s, I would like to sketch out a research agenda that might allow us to render visible hidden forms of resistance; to examine how class, gender, and race shape working-class consciousness; and to bridge the gulf between the social and cultural world of the "everyday" and political struggles.[5] First and foremost, my thoughts grow out of rereading Herbert Aptheker's *American Negro Slave Revolts*, a pioneering study that is celebrating its fiftieth anniversary this year. In it, Aptheker gave us a framework to study the hidden and disguised, not only locating acts of resistance and plans for rebellion among slaves but also showing how their opposition shaped all of antebellum southern society, politics, and daily life.[6] Second, I am indebted to scholars who work on South Asia, especially the political anthropologist James C. Scott. Scott and other proponents of subaltern studies maintain that despite appearances of consent, oppressed groups challenge those in power by constructing a "hidden transcript," a dissident political culture that manifests itself in daily conversations, folklore, jokes, songs, and other cultural practices. One also finds the hidden transcript emerging "on stage" in spaces controlled by the powerful, although almost always in disguised forms. The submerged social and cultural worlds of oppressed people frequently surface in everyday forms of resistance—theft, footdragging, the destruction of property—or, more rarely, in open attacks on individuals, institutions, or symbols of domination. Together, the hidden transcripts that are created in aggrieved communities and expressed through culture and the daily acts of resistance and survival constitute what Scott calls "infrapolitics." As he puts it, "The circumspect struggle waged daily by subordinate groups is, like infrared rays, beyond the visible end of the spectrum. That it should be invisible . . . is in large part by design—a tactical choice born of a prudent awareness of the balance of power."[7]

Like Scott, I use the concept of infrapolitics to describe the daily confrontations, evasive actions, and stifled thoughts that often inform organized political movements. I am not suggesting that the realm of infrapolitics is any more or less important or effective than what we traditionally consider politics. Instead, I want to suggest that the political history of oppressed people cannot be understood without reference to infrapolitics, for these daily acts have a cumulative effect on power relations. Although the meaning and effectiveness of acts differ according to circumstances, they make a difference, whether they were intended to or not. Thus, one measure of the power and historical importance of the informal infrapolitics of the oppressed is the response of those who dominate traditional politics. Daily acts of resistance and survival have had consequences for existing power relations, and the powerful have deployed immense resources in response. Knowing how the powerful interpret, redefine, and respond to the thoughts and actions of the oppressed is just as important as identifying and analyzing opposition. The policies, strategies, or symbolic representations of those in power—what Scott calls the "official" or "public" transcript—cannot be understood without examining the infrapolitics of oppressed groups. The approach I am proposing will help illuminate how power operates, how effective the southern power structure was in maintaining social order, and how seemingly innocuous, individualistic acts of survival and opposition shaped southern urban politics, workplace struggles, and the social order generally. I take the lead of the ethnographer Lila Abu-Lughod, who argues that everyday forms of resistance ought to be "diagnostic" of power. Instead of seeing these practices primarily as examples of the "dignity and heroism of resistors," she argues that they can "teach us about the complex interworkings of historically changing structures of power."[8]

An infrapolitical approach requires that we substantially redefine our understanding of politics. Too often politics is defined by *how* people participate rather than *why;* by traditional definition the question of what is political hinges on whether or not groups are involved in elections, political parties, grassroots social movements. Yet the how seems far less important than the why because many of the so-called real political institutions have not proved effective for, or even accessible to, oppressed people. By shifting our focus to what motivated disenfranchised black working people to struggle and what strategies they developed, we may discover that their participation in "mainstream" politics—including their battle for the franchise—grew out of the very circumstances, experiences, and memories that impelled many to steal from an employer, to join a mutual benefit association, or to spit in a bus driver's face. In other words, those actions all reflect, to varying degrees, larger political struggles. For southern blacks in the age of Jim Crow, politics was not separate from lived experience or the imagined world of what is possible. It was the many battles to roll back constraints, to exercise power over, or create space within, the institutions and social relationships that dominated their lives.[9]

Using this revised framework for understanding power, resistance, and politics, the following explores three sites of urban black working-class opposition in the American South in the early twentieth century: the semiprivate/semipublic spaces of community and household, the workplace, and public space. My remarks are intended to be interrogations that may lead to new ways of understanding working-class politics.

At Home, at Play, at Prayer

Several southern labor and urban historians have begun to unveil the hidden social and cultural world of black working people and to assess its political significance. They have established that during the era of Jim Crow, black working people carved out social space and constructed what George Lipsitz calls a "culture of opposition" through which to articulate the hidden transcript free from the watchful eye of white authority or the moralizing of the black middle class. Those social spaces constituted a partial refuge from the humiliations and indignities of racism, class pretensions, and waged work. African American communities often created an alternative culture emphasizing collectivist values, mutuality, and fellowship. There were vicious, exploitative relationships within southern black communities, particularly across class and gender lines, and the tentacles of Jim Crow touched even black institutions. But the social and cultural institutions and ideologies that ultimately informed black opposition placed more emphasis on communal values and collective uplift than the prevailing class-conscious, individualist ideology of the white ruling classes. As Earl Lewis so aptly put it, African Americans turned segregation into "congregation."[10]

Ironically, segregation facilitated the creation and maintenance of the unmonitored, unauthorized social sites in which black workers could freely articulate the hidden transcript. Jim Crow ordinances ensured that churches, bars, social clubs, barbershops, beauty salons, even alleys, remained "black" space. When southern white ruling groups suspected dissident activities among African Americans, they tried to monitor and sometimes to shut down black social spaces —usually swiftly and violently. During World War II, as Howard Odum observed, mere rumors of black uprisings made any black gathering place fair game for extralegal, often brutal invasions. More significant, employers and police officials actively cultivated black stool pigeons to maintain tabs on the black community. Clearly, even if historians have underestimated the potential threat that rests within black-controlled spaces, the southern rulers did not.[11]

Grassroots black community organizations such as mutual benefit societies, church groups, and gospel quartets were crucial to black people's survival. Through them, African Americans created and sustained bonds of community, mutual support networks, and a collectivist ethos that shaped black working-class political struggle.

As Elsa Barkley Brown points out in her work on Richmond, Virginia, mutual benefit societies, like many other black organizations, "institutionalized a vision of community based on notions of collectivity and mutuality even as [they] struggled with the practical problems of implementing and sustaining such a vision." Although the balance of power in these organizations was not always equal, with males and the middle class sometimes dominant, Brown demonstrates that within benevolent societies all members played some role in constructing a vision of the community.[12]

Yet we need to acknowledge intraracial class tensions. Mutual disdain, disappointment, and even fear occasionally found their way into the public transcript. Some middle-class blacks, for example, regarded the black poor as lazy, self-destructive, and prone to criminal behavior. Geraldine Moore, a black middle-class resident of Birmingham, Alabama, wrote that many poor blacks in her city knew "nothing but waiting for a handout of some kind, drinking, cursing, fighting and prostitution." On the other hand, in his study of a small Mississippi town, the sociologist Allison Davis found that "lower-class" blacks often "accused upper-class persons (the 'big shots,' the 'Big Negroes') of snobbishness, color preference, extreme selfishness, disloyalty in caste leadership ('sellin' out to white folks'), and economic exploitation of their patients and customers."[13]

To understand the significance of class conflict among African Americans, we need to examine how specific communities are constructed and sustained rather than to presume the existence (until recently) of a tight-knit, harmonious black community. This romantic view of a "golden age" of black community—an age when any elder could beat a misbehaving child, when the black middle class mingled with the poor and offered themselves as "role models," when black professionals cared more about their downtrodden race than about their bank accounts—is not just disingenuous; it has deterred serious historical research on class relations within African American communities. As a dominant trope in the popular social science literature on the so-called underclass, it has hindered explanations of the contemporary crisis in the urban United States by presuming a direct causal relationship between the disappearance of middle-class role models as a result of desegregation and the so-called moral degeneration of the black jobless and underemployed working class left behind in the cities.[14]

Such a reassessment of African American communities would also require us to rethink the role of black working-class families in shaping ideology and strategies of resistance. Social historians and feminist theorists have made critical contributions to our understanding of the role of women's (and, to a lesser degree, children's) unpaid work in reproducing the labor power of male industrial workers and maintaining capitalism.[15] Nevertheless, we still know very little about power relations and conflicts within black working-class families, the role of family life in the development of class consciousness (especially among children), and how these things shape oppositional strategies at the workplace and in neighborhoods. For instance, if patriarchal families enabled exploited male wage earners to control and exploit the labor of women and children, then one might find a material basis to much intrafamily conflict, as well as hidden transcripts and resistance strategies framed within an ideology that justifies the subordinate status of women and children.[16] We might, therefore, ask how conflicts and the exploitation of labor power in the family and household shape larger working-class politics.

Indeed, in part because most scholarship privileges the workplace and production over the household and reproduction, the role of families in the formation of class consciousness and in developing strategies of resistance has not been sufficiently explored. The British women's historian Carolyn Steedman, for example, points out that radical histories of working people have been slow "to discuss the *development* of class consciousness (as opposed to its expression)" and to explore "it as a learned position, *learned* in childhood, and often through the exigencies of difficult and lonely lives." Likewise, Elizabeth Faue asks us to look more carefully at the formation of class, race, and gender identities long before young people enter the wage labor force. She adds that "focusing on reproduction would give meaning to the relationship between working class family organization and behavior and working class collective action and labor organization."[17]

Such a reexamination of black working-class families should provide insights into how the hidden transcript informs public, collective action. We might return, for example, to the common claim that black mothers and grandmothers in the age of Jim Crow raised their boys to show deference to white people. Were black working-class parents "emasculating" potential militants, as several black male writers argued in the 1960s, or were they arming their boys with a sophisticated understanding of the political and cultural terrain of struggle?[18] And what about black women's testimony that their mothers taught them values and strategies that helped them survive and resist race, class, and gender oppression? Once we begin to look at the family as a central (if not *the* central) institution where political ideologies are formed and reproduced, we may discover that households hold the key to understanding particular episodes of black working-class resistance. Elsa Barkley Brown has begun to search for the sources of opposition in black working-class households. In an essay on African American families and political activism during Reconstruction, she not only demonstrates the central role of black women (and even children) in Republican Party politics, despite the restriction of suffrage to adult males, but also persuasively argues that newly emancipated African Americans viewed the franchise as the collective property of the whole family. Men who did not vote according to the family's wishes were severely disciplined or ostracized from community institutions.[19]

Black workers, therefore, participated in or witnessed oppositional politics—whether in community institutions or households—before they entered the workplace or the labor movement. Average black workers probably experienced greater participatory democracy in community- and neighborhood-based institutions than in the interracial trade unions that claimed to speak for them. Anchored in a prophetic religious ideology, these collectivist institutions and practices took root and flourished in a profoundly undemocratic society. For instance, Tera Hunter demonstrates that benevolent and secret societies constituted the organizational structures through which black washerwomen organized strikes. In separate studies, Michael Honey and Robert Korstad suggest that black religious ideology and even some churches were key factors in the success of the Food, Tobacco, Agricultural, and Allied Workers union in Memphis, Tennessee, and Winston-Salem, North Carolina. Brenda McCallum illustrates that black gospel quartet circuits were crucial to the expansion and legitimation of the Congress of Industrial Organizations (CIO) in Birmingham, Alabama. My book argues that black working people enveloped the Alabama Communist Party with a prophetic religious ideology and collectivist values that had grown out of black communities. The subcultures of working people do not always or automatically suffuse formal working-class organizations. The relationship is dialectical; the political culture that permeated labor organizations, including radical left-wing movements, often conflicted with aspects of working-class culture. The question historians might explore is whether certain interracial labor organizations were unable to mobilize sufficient black support because they failed to work through black community institutions or to acknowledge, if not to embrace, the cultural values of the African American working class.[20]

Much of southern black working-class culture falls outside "conventional" labor history, in part because historians have limited their scope to public action and formal organization: Part of the problem is that those who frequented the places of rest, relaxation, recreation, and restoration rarely maintained archives or recorded the everyday conversations and noises that filled the bars, dance halls, blues clubs, barbershops, beauty salons, and street corners of the black community. Nevertheless, folklorists, anthropologists, oral historians, musicians, and writers fascinated by "Negro life" preserved cultural texts that allow scholars access to the hidden transcript. Using those texts, pioneering scholars and critics, including Amiri Baraka, Lawrence Levine, and Sterling Stuckey, have demonstrated that African American working people created an oppositional culture that represents at least a partial rejection of the dominant ideology and that was forged in the struggle against class and racial domination. The challenge for southern labor historians is to determine how this rich expressive culture—which was frequently at odds with formal working-class institutions—shaped and reflected black working-class opposition.[21]

Even modes of leisure could undergird opposition. Of course, black working-class popular culture was created more to give pleasure than to challenge or explain domination. But people thought before they acted, and what they thought shaped, and was shaped by, cultural production and consumption. Moreover, for members of a class whose long workdays were spent in backbreaking, low-paid wage work in settings pervaded by racism, the places where they played were more than relatively free spaces in which to articulate grievances and dreams. They were places that enabled African Americans to take back their bodies, to recuperate, to be together. Two of the most popular sites were dance halls and blues clubs. Despite opposition from black religious leaders and segments of the black middle class, as well as many white employers, black working people of both sexes shook, twisted, and flaunted their overworked bodies, drank, talked, flirted and—in spite of occasional fights—reinforced their sense of community. Whether it was the call and response of a blues man's lyrics or the sight of hundreds moving in unison on a hardwood dance floor, the form and content of such leisure activities were unmistakably collective.[22]

Much African American popular culture can be characterized as alternative rather than oppositional.[23] Most people went to parties, dances, and clubs to escape from the world of assembly lines, relief lines, and color lines and to leave momentarily the individual and collective battles against racism, sexism, and material deprivation. But their search for the sonic, visceral pleasures of music and fellowship, for the sensual pleasures of food, drink, and dancing, was not just about escaping the vicissitudes of southern life. They went with people who had a shared knowledge of cultural forms, people with whom they felt kinship, people with whom they shared stories about the day or the latest joke, people who shared a vernacular whose grammar and vocabulary struggled to articulate the beauty and burden of their lives. Places of leisure allowed freer sexual expression, particularly for women, whose sexuality was often circumscribed by employers, family members, the law, and the fear of sexual assault in a society with few protections for black women. Knowing what happens in these spaces of pleasure can help us understand the solidarity black people have shown at political mass meetings, illuminate the bonds of fellowship one finds in churches and voluntary associations, and unveil the *conflicts* across class and gender lines that shape and constrain these collective struggles.

When we consider the needs of employers and the dominance of the Protestant work ethic in American culture, these events were resistive, although not consciously. Speaking of the African diaspora in general, and that in Britain in particular, cultural critic Paul Gilroy argues that black working people who spent time and precious scarce money at the dance halls, blues clubs, and house parties "see waged work as itself a form of servitude. At best, it is viewed as a necessary evil and is sharply counterposed to the more authentic freedoms that can be enjoyed only in nonwork time. The black body is here celebrated as an instrument of pleasure rather than an instrument of labor. The nighttime becomes the right time, and the space allocated for recovery and recuperation is assertively and provocatively occupied by the pursuit of leisure and pleasure."[24]

In southern cities where working-class blacks set Friday and Saturday nights aside for the "pursuit of leisure and pleasure," some of the most intense skirmishes between such blacks and authority erupted during and after weekend gatherings. During World War II in Birmingham, for example, racial conflicts on public transportation on Friday and Saturday nights were commonplace; many of the incidents involved black youths returning from dances and parties. The young men and women who rode public transportation in groups were energized by a sense of social solidarity rooted in a shared culture, common friends, and generational identity, not to mention naivete as to the possible consequences of "acting up" in white-dominated public space. Leaving social sites that had reinforced a sense of collectivity, sometimes feeling the effects of alcohol and reefer, many young black passengers were emboldened. On the South Bessemer line, which passed some of the popular black dance halls, white passengers and operators dreaded the "unbearable" presence of large numbers of African Americans who "pushed and shoved" white riders at will. As one conductor noted, "Negroes are rough and boisterous when leaving down town dances at this time of night."[25]

The nighttime also afforded black working people the opportunity to become something other than workers. In a world where clothes signified identity and status, "dressing up" was a way of shedding the degradation of work and collapsing status distinctions between themselves and their oppressors. As one Atlanta domestic worker remembers, the black business district of Auburn Avenue was "where we dressed up, because we couldn't dress up during the day. We'd dress up and put on our good clothes and go to the show on Auburn Avenue, and you were going places. It was like white folks' Peachtree."[26] Seeing oneself and others dressed up was important to constructing a collective identity based on something other than wage work, presenting a public challenge to the dominant stereotypes of the black body, and shoring up a sense of dignity that was perpetually under assault. In these efforts to re-present the body through dress, African Americans wielded a double-edged sword, because the styles they adopted to combat racism all too frequently reinforced, rather than challenged, bourgeois notions of respectability. Yet by their dress as by their leisure, black people took back their bodies.

Clothing, as a badge of oppression or an act of transgression, is crucial to understanding opposition by subordinate groups. Thus, black veterans were beaten and lynched for insisting on wearing their military uniforms in public. A less-known but equally potent example is the zoot suit, which became popular during World War II. Although the suit itself was not created and worn as a direct political statement, the language and culture of zoot suiters emphasized ethnic identity and rejected subservience. Young black males created a fast-paced, improvisational language that sharply contrasted with the passive stereotype of the stuttering, tongue-tied Sambo, and whereas whites commonly addressed them as "boy," zoot suiters made a fetish of calling each other "man." The zoot suiters constructed an identity in which their gendered and racial meanings were inseparable; they opposed racist oppression through public displays of masculinity. Moreover, because fabric rationing regulations instituted by the War Productions Board forbade the sale and manufacturing of zoot suits, wearing the suit (which had to be purchased through informal networks) was seen by white servicemen as a pernicious act of anti-Americanism—a view compounded by the fact that most zoot suiters were able-bodied men who refused to enlist or found ways to dodge the draft. A Harlem zoot suiter interviewed by black social psychologist Kenneth Clark declared to the scholarly audience for whom Clark's research was intended: "By [the] time you read this I will be fighting for Uncle Sam, the bitches, and I do not like it worth a dam. I'm not a spy or a saboteur, but I don't like goin' over there fightin' for the white man—so be it." It is not a coincidence that whites who assaulted black and Chicano zoot suiters across the country during the fateful summer of 1943 took great pains to strip the men or mutilate the suits.[27]

Although no one, to my knowledge, has investigated zoot suiters in the South, they undoubtedly were a presence on the wartime urban landscape. As Howard Odum observed during the early 1940s, the mere image of these draped-shape-clad hipsters struck fear into the hearts of many white southerners. On Birmingham's already overcrowded buses and streetcars during World War II, some of these zooted "baaad niggers" put on outrageous public displays of resistance that left witnesses in awe, although their transgressive acts did not lead directly to improvements in conditions, nor were they intended to. Some boldly sat down next to white female passengers and challenged operators to move them, often with knife in hand. Others refused to pay their fares or simply picked fights with bus drivers or white passengers. Nevertheless, like the folk hero himself, the Stagolee-type rebel was not always admired by other working-class black passengers. Some were embarrassed by his actions; the more sympathetic feared for his life. Black passengers on the Pratt-Ensley streetcar in 1943, for example, told a rebellious young man who was about to challenge the conductor to a fight "to hush before he got killed." Besides, black hipsters were hardly social bandits.

Some were professional hustlers whose search for pleasure and avoidance of waged labor often meant exploiting the exploited. Black hustlers took pride in their ability to establish parasitical relationships with women wage earners or sex workers, and those former hipsters who recorded memories in print wrote quite often of living off women, in many cases by outright pimping. The black male hipsters of the zoot suit generation remind us that the creation of an alternative culture can simultaneously challenge *and* reinforce existing power relations.[28]

Last, I want to briefly leap from the "bad," lawless, secular world to the sacred—a realm of practice to which historians have paid great attention. Despite the almost axiomatic way the church becomes central to black working-class culture and politics, religion is almost always treated simply as culture, ideology, and organization. We need to recognize that the sacred and the spirit world were also often understood and invoked by African Americans as weapons to protect themselves or to attack others. How do historians make sense of, say, conjure as a strategy of resistance, retaliation, or defense in the daily lives of some working-class African Americans? How do we interpret divine intervention, especially when one's prayers are *answered?* How does the belief that God is by one's side affect one's willingness to fight with police, leave an abusive relationship, stand up to a foreman, participate in a strike, steal, or break tools? Can a sign from above, a conversation with a ghost, a spell cast by an enemy, or talking in tongues unveil the hidden transcript? If a worker turns to a root doctor or prayer rather than to a labor union to make an employer less evil, is that "false consciousness"? These are not idle questions. Most of the oral narratives and memoirs of southern black workers speak of such events or moments as having enormous material consequences.[29] Of course, reliance on the divine or on the netherworlds of conjure was rarely, if ever, the only resistance or defense strategy used by black working people, but in their minds, bodies, and social relationships this was real power—power of which neither the CIO, the Populists nor the National Association for the Advancement of Colored People (NAACP) could boast. With the exception of Vincent Harding, no historian that I know of since W. E. B.

Du Bois has been bold enough to assert a connection between the spirit and spiritual world of African Americans and political struggle. Anticipating his critics, Du Bois in *Black Reconstruction* boldly considered freed people's narratives of divine intervention in their emancipation and, in doing so, gave future historians insight into an aspect of African American life that cannot be reduced to "culture": "foolish talk, all of this, you say, of course; and that is because no American now believes in his religion. Its facts are mere symbolism; its revelation vague generalities; its ethics a matter of carefully balanced gain. But to most of the four million black folk emancipated by civil war, God was real. They knew Him. They had met Him personally in many a wild orgy of religious frenzy, or in the black stillness of the night."[30]

At the Point of Production

Nearly a quarter of a century ago, as Herbert Gutman was poised to lead a revolution in the study of labor in the United States, George Rawick published an obscure article that warned against treating the history of working-class opposition as merely the history of trade unions or other formal labor organizations. If we are to locate working-class resistance, Rawick insisted, we need to know "how many man hours were lost to production because of strikes, the amount of equipment and material destroyed by industrial sabotage and deliberate negligence, the amount of time lost by absenteeism, the hours gained by workers through the slowdown, the limiting of the speed-up of the productive apparatus through the working class's own initiative." Unfortunately, few southern labor historians have followed Rawick's advice. Missing from most accounts of southern labor struggles are the ways unorganized working people resisted the conditions of work, tried to control the pace and amount of work, and carved out a modicum of dignity at the workplace.[31]

Not surprisingly, studies that seriously consider the sloppy, undetermined, everyday nature of workplace resistance have focused on workers who face considerable barriers to traditional trade union organization. Black domestic workers devised a whole array of creative strategies, including slowdowns, theft (or "pan-toting"), leaving work early, or quitting, in order to control the pace of work, increase wages, compensate for underpayment, reduce hours, and seize more personal autonomy. These individual acts often had a collective basis that remained hidden from their employers. Black women household workers in the urban South generally abided by a code of ethics or established a blacklist so they could collectively avoid employers who had proved unscrupulous, abusive, or unfair. In the factories, such strategies as feigning illness to get a day off, slowdowns, sometimes even sabotage often required the collective support of co-workers. Studies of black North Carolina tobacco workers by Dolores Janiewski and Robert Korstad reveal a wide range of clandestine, yet collective, strategies to control the pace of work or to strike out against employers. When black women stemmers had trouble keeping up with the pace, black men responsible for supplying tobacco to them would pack the baskets more loosely than usual. Among black women who operated stemmer machines, when one worker was ill, other women would take up the slack rather than call attention to her inability to handle her job, which could result in lost wages or dismissal.[32]

Theft at the workplace was a common form of working-class resistance, and yet the relationship between pilfering—whether of commodities or of time—and working-class opposition has escaped the attention of most historians of the African American working class, except in slavery studies and the growing literature on domestic workers.[33] Any attempt to understand the relationship between theft and working-class opposition must begin by interrogating the dominant view of "theft" as deviant, criminal behavior. From the vantage point of workers, as several criminologists have pointed out, theft at the workplace is a strategy to recover unpaid wages or to compensate for low wages and mistreatment. Washerwomen in Atlanta and other southern cities, Hunter points out, occasionally kept their patrons' clothes "as a weapon against individual employers who perpetuated injustices or more randomly against an oppressive employing class." In the tobacco factories of North Carolina, black workers not only stole cigarettes and chewing tobacco (which they usually sold or bartered at the farmers' market) but, in Durham at least, also figured out a way to rig the clock in order to steal time. In the coal mines of Birmingham and Appalachia, miners pilfered large chunks of coke and coal for their home ovens. Black workers sometimes turned to theft as a means of contesting the power public utilities had over their lives. During the Great Depression, for example, jobless and underemployed working people whose essential utilities had been turned off for nonpayment stole fuel, water, and electricity: They appropriated coal, drew free electricity by tapping power lines with copper wires, illegally turned on water mains, and destroyed vacant homes for firewood.[34]

Unfortunately, we know very little about black workplace theft in the twentieth-century South and even less about its relationship to working-class opposition. Historians might begin to explore, for example, what Michel de Certeau calls "wigging," employees' use of company time and materials for their own purposes (e.g., repairing or making a toy for one's child or writing love letters). By using part of the workday in this manner, workers not only take back precious hours from their employers but resist being totally subordinated to the needs of capital. The worker takes some of that labor power and spends it on herself or her family. One might imagine a domestic who seizes time from work to read books from her employer's library. In a less creative, although more likely, scenario, washerwomen wash and iron their families' clothes along with their employers'.[35]

A less elusive form of resistance is sabotage. Although the literature is nearly silent on industrial sabotage in the South, especially acts committed by black workers, it existed. Korstad's study of tobacco workers in Winston-Salem introduces us to black labor organizer Robert Black, who admitted using sabotage to counter speedups:

> These machines were more delicate, and all I had to do was feed them a little faster and overload it and the belts would break. When it split you had to run the tobacco in reverse to get it out, clean the whole machine out and then the mechanics would have to come and take all the broken links out of the belt. The machine would be down for two or three hours and I would end up running less tobacco than the old machines. We had to use all kinds of techniques to protect ourselves and the other workers.

Historians provide ample evidence that domestic workers adopted sabotage techniques more frequently than industrial workers. There is evidence of household workers scorching or spitting in food, damaging kitchen utensils, and breaking household appliances, but employers and white contemporaries generally dismissed these acts as proof of black moral and intellectual inferiority. Testifying on the "servant problem" in the South, a frustrated employer remarked: "The washerwomen . . . badly damaged clothes they work on, iron-rusting them, tearing them, breaking off buttons, and burning them brown; and as for starch!—Colored cooks too, generally abuse stoves, suffering them to get clogged with soot, and to 'burn out' in half the time they ought to last."[36]

These examples are rare exceptions, however, for workplace theft and sabotage in the urban South has been all but ignored by labor historians. Given what we know of the pervasiveness of these strategies in other parts of the world and among slaves as well as rural African Americans in the postbellum period, the absence of accounts of similar clandestine activity by black industrial workers is surprising.[37] Part of the reason, I think, lies in southern labor historians' noble quest to redeem the black working class from racist stereotypes. In addition, company personnel records, police reports, mainstream white newspaper accounts, and correspondence have left us with a somewhat serene portrait of black folks who only occasionally deviate from what I like to call the "cult of true Sambohood." The safety and ideological security of the South required that pilfering, slowdowns, absenteeism, tool breaking, and other acts of black working-class resistance be turned into ineptitude, laziness, shiftlessness, and immorality. But rather than reinterpret these descriptions of black working-class behavior, sympathetic labor historians are often too quick to invert the images, remaking the black proletariat into the hardest working, thriftiest, most efficient labor force around. Historians too readily naturalize the Protestant work ethic and project onto black working people as a whole the ideologies of middle-class and prominent working-class blacks. But if we regard most work as alienating, especially work done amid racist and sexist oppression, then a crucial aspect of black working-class struggle is to minimize labor with as little economic loss as possible. Let us recall one of Du Bois's many beautiful passages from *Black Reconstruction:* "All observers spoke of the fact that the slaves were slow and churlish; that they wasted material and malingered at their work. Of course they did. This was not racial but economic. It was the answer of any group of laborers forced down to the last ditch. They might be made to work continuously but no power could make them work well."[38]

Traditional documents, if used imaginatively, can be especially useful for reconstructing the ways in which workers exploited racial stereotypes to control the pace of work. Materials that describe "unreliable," "shiftless," or "ignorant" black workers should be read as more than vicious, racist commentary on African Americans; in many instances these descriptions are employers', foremen's, and managers' social reconstruction of the meaning of working-class self-activity, which they not only misunderstood but were never supposed to understand. Fortunately, many southern black workers understood the cult of true Sambohood all too well, and at times they used the contradictions of racist ideology to their advantage. In certain circumstances, their inefficiency and penchant for not following directions created havoc and chaos for industrial production or the smooth running of a household, and all the while the appropriate grins, shuffles, and "yassums" served to mitigate potential punishment.[39]

Among workers especially, the racial stereotypes associated with industrial disruption were also gendered. As David Roediger has demonstrated in a penetrating essay, Covington Hall and the Brotherhood of Timber Workers (BTW) in Louisiana understood sabotage as a direct, militant confrontation with the lumber companies rather than an evasive strategy. As a native southern white leader of the working class, born of privilege, Hall sought to use appeals to "manhood" to build biracial unity. His highly gendered rhetoric, which insisted that there were no "Niggers" or "white trash"—only *men*—had the effect of turning clandestine tactics into direct confrontation. Roediger writes, "It is hard to believe the zeal with which [sabotage] was propagandized was not intensified by the tremendous emphasis on manhood, in part as a way to disarm race, in BTW thinking. . . . Hall's publications came to identify sabotage with the improbable image of the rattlesnake, not the black cat symbolizing the tactic elsewhere."[40]

Yet despite Hall's efforts, employers and probably most workers continued to view what black male workers in the lumber industry were doing as less than manly—indeed, as proof of their inferiority at the workplace and evidence that they should be denied upward mobility and higher wages. Thus, for some black male industrial workers, efficiency and the work ethic were sometimes more effective as signifiers of manliness than sabotage and foot dragging. As Joe W. Trotter's powerful new book on African Americans in southern West Virginia reminds us, theft, sabotage, and slowdowns were two-edged weapons that more often than not reinforced the subordinate position of black coal miners in a racially determined occupational hierarchy. As he explains, "Job performance emerged as one of the black miners' most telling survival mechanisms. To secure their jobs, they resolved to provide cooperative, efficient, and productive labor." Their efficient labor was a logical response to a rather limited struggle for job security and advancement because their subordination to specific tasks and pay scales were based, at least ostensibly, on race alone. More than a few black workers seemed to believe that a solid work record would eventually topple the racial ceiling on occupational mobility. Obviously, efficiency did not always lead to improved work conditions, nor did sabotage and foot dragging always go unnoticed or unpunished. What we need to know is why certain occupations seemed more conducive to particular strategies. Was efficiency more prevalent in industries where active, interracial trade unions at least occasionally challenged racially determined occupational ceilings (e.g., coal mining)? Did extensive workplace surveillance deter sabotage and theft? Were black workers less inclined toward sabotage when disruptions made working conditions more difficult or dangerous for fellow employees? Were evasive strategies more common in service occupations? These questions need to be explored in greater detail. They suggest, as British labor historian Richard Price has maintained, that to understand strategies of resistance thoroughly we need to explore with greater specificity the character of subordination at the workplace.[41]

Nevertheless, the relative absence of resistance at the point of production does not mean that workers acquiesced or accommodated to the conditions of work. On the contrary, the most pervasive form of black protest was simply to leave. Central to black working-class infrapolitics was mobility, for it afforded workers relative freedom to escape oppressive living and working conditions and power to negotiate better working conditions. Of course, one could argue that in the competitive context of industrial capitalism—North and South—some companies clearly benefited from such migration because wages for blacks remained comparatively low no matter where black workers ended up. But the very magnitude of working-class mobility weakens any thesis that southern black working-class politics was characterized by accommodationist thinking. Besides, there is plenty of evidence to suggest that a significant portion of black migrants, especially black emigrants to Africa and the Caribbean, were motivated by a desire to vote, to provide a better education for their children, or to live in a setting in which Africans or African Americans exercised power. The ability to move represented a crucial step toward empowerment and self-determination; employers and landlords understood this, which explains why so much energy was expended limiting labor mobility and redefining migration as "shiftlessness," "indolence," or a childlike penchant to wander.[42]

Gender, Race, Work, and the Politics of Location

Location plays a critical role in shaping workplace resistance, identity, and—broadly speaking—infrapolitics. By location, I mean the racialized and gendered social spaces of work and community, as well as black workers' position in the hierarchy of power, the ensemble of social relations. Southern labor historians and race relations scholars have established the degree to which occupations and, in some cases, work spaces were segregated by race. But only recently has scholarship begun to move beyond staid discussions of such labor market segmentation and inequality to an analysis of how spatial and occupational distinctions helped create an oppositional consciousness and collective action. Feminist scholarship on the South and some community histories have begun to examine how the social spaces in which people work (in addition to the world beyond work, which was also divided by race and, at times, sex) shaped the character of everyday resistance, collective action, and domination.[43]

Earl Lewis offers a poignant example of how the racialized social locations of work and community formed black working-class consciousness and oppositional strategies. During World War I, the all-black Transport Workers Association (TWA) of Norfolk, Virginia, began organizing African American waterfront workers irrespective of skill. Soon thereafter, its leaders turned their attention to the ambitious task of organizing all black workers, most notably cigar stemmers, oyster shuckers, and domestics. The TWA resembled what might have happened if Garveyites had taken control of an Industrial Workers of the World (IWW) local: The ultimate goal seemed to be One Big Negro Union. What is important about the Norfolk story is the startling success of the TWA's efforts, particularly among workers who have been deemed unorganizable. Lewis is not satisfied with such simplistic explanations as the power of charismatic leadership or the primacy of race over class to account for the mass support for the TWA; rather, he makes it quite clear that the labor process, work spaces, intraclass power relations, communities and neighborhoods—indeed, class struggle itself—were all racialized. The result, therefore, was a "racialized" class consciousness. "In the world in which these workers lived," Lewis writes, "nearly everyone was black, except for a supervisor or employer. Even white workers who may have shared a similar class position enjoyed a superior social position because of their race. Thus, although it appears that some black workers manifested a semblance of worker consciousness, that consciousness was so embedded in the perspective of race that neither blacks nor whites saw themselves as equal partners in the same labor movement."[44]

A racialized class consciousness shaped black workers' relations with interracial trade unions as well. Black workers did not always resist segregated union locals (although black union leaders often did). Indeed, in some instances African American workers preferred segregated locals—if they maintained control over their own finances and played a leading role in the larger decision-making process. To cite one example, black members of the Brotherhood of Timber Workers in Louisiana, an IWW affiliate, found the idea of separate locals quite acceptable. However, at the 1912 BTW convention black delegates complained that they could not "suppress a feeling of taxation without representation" because their dues were in the control of whites, and demanded a "coloured executive board, elected by black union members and designed to work 'in harmony with its white counterpart.' "[45]

Although gender undoubtedly shaped the experiences, work spaces, and collective consciousness of all southern black workers, historians of women have been the most forthright and consistent in employing gender as an analytical category. Recent work on black female tobacco workers, in particular, has opened up important lines of inquiry. Not only were the dirty and difficult tasks of sorting and stemming tobacco relegated to black women, but those women had to do the tasks in spaces that were unbearably hot, dry, dark, and poorly ventilated. The coughing and wheezing, the tragically common cases of workers succumbing to tuberculosis, the endless speculation as to the cause of miscarriages among coworkers, were constant reminders that these black women spent more than a third of the day toiling in a health hazard. If some compared their work space to a prison or a dungeon, then they could not help but notice that all of the inmates were black women like themselves. Moreover, foremen referred to them only by their first names or changed their names to "girl" or something more profane and regarded their bodies as perpetual motion machines as well as sexual objects. Thus bonds of gender as well as race were reinforced by the common experience of sexual harassment. Recalled one Reynolds worker, "I've seen [foremen] just walk up and pat women on their fannies and they'd better not say anything." Women, unlike their black male coworkers, had to devise a whole range of strategies to resist or mitigate the daily physical and verbal abuse of their bodies, ranging from putting forth an "asexual" persona to posturing as a "crazy" person to simply quitting. Although these acts seem individualized and isolated, the experience of, and opposition to, sexual exploitation probably reinforced bonds of solidarity. In the tobacco factories, these confrontations usually took place in a collective setting, the advances of lecherous foremen were discussed among the women, and strategies to deal with sexual assault were observed, learned from other workplaces, or passed down (e.g., former domestics had experience staving off the sexual advances of male employers). Yet to most male union leaders, such battles were private affairs that had no place among "important" collective bargaining issues. Unfortunately, most labor historians have accepted this view, unable to see resistance to sexual harassment as a primary struggle to transform everyday conditions at the workplace. Nevertheless, out of this common social space and experience of racism and sexual exploitation, black female tobacco workers constructed "networks of solidarity." They referred to each other as "sisters," shared the same neighborhoods and community institutions, attended the same churches, and displayed a deep sense of mutuality by collecting money for coworkers during sickness and death and celebrating each other's birthdays. In fact, those networks of solidarity were indispensable for organizing tobacco plants in Winston-Salem and elsewhere.[46]

In rethinking workplace struggles, black women's work culture, and the politics of location, we must be careful not to assume that home and work were distinct. Although much of this scholarship and the ideas I am proposing directly challenge the "separate spheres" formulation, there is an implicit assumption that working-class households are separate from spaces in which wage labor takes place. Recent studies of paid homework remind us that working women's homes were often extensions of the factory. For African American women, in particular, Eileen Boris and Tera Hunter demonstrate that the decision to do piecework or to take in laundry grows out of a struggle for greater control over the labor process, out of a conscious effort to avoid workplace environments in which black women have historically confronted sexual harassment, and out of "the patriarchal desires of men to care for their women even when they barely could meet economic needs of their families or from women's own desires to care for their children under circumstances that demanded that they contribute to the family economy."[47] The study of homework opens up numerous possibilities for rethinking black working-class opposition in the twentieth century. How do homeworkers resist unsatisfactory working conditions? How do they organize? Do community- and neighborhood-based organizations protect their interest as laborers? How does the extension of capital-labor relations into the home affect the use and meaning of household space, labor patterns, and the physical and psychological well-being of the worker and her family? How does the presumably isolated character of their work shape their consciousness? How critical is female homework as a survival strategy for households in which male wage earners are involved in strikes or other industrial conflicts? Thanks to the work of Boris and Hunter, many of these questions have been explored with regard to northern urban working women and southern laundry workers. But aside from washerwomen and occasional seamstresses, what do we really know about black homeworkers in the Jim Crow South?

For many African American women, homework was a way to avoid the indignities of household service, for as the experience of black tobacco workers suggests, much workplace resistance centered around issues of dignity, respect, and autonomy. White employers often required black domestics to don uniforms, which reduced them to their identities as employees and ultimately signified ownership—black workers literally became the property of whoever owned the uniform. As Elizabeth Clark-Lewis points out, household workers in Washington, D.C., resisted wearing uniforms because they were symbols of live-in service. Their insistence on wearing their own clothes was linked to a broader struggle to change the terms of employment from those of a "servant" (i.e., a live-in maid) to those of a day worker. "As servants in uniform," Clark-Lewis writes, "the women felt, they took on the identity of the job—and the uniform seemed to assume a life of its own, separate from the person wearing it, beyond her control. As day workers, wearing their own clothes symbolized their new view of life as a series of personal choices rather than predetermined imperatives."[48]

But struggles for dignity and autonomy often pitted workers against other workers. Black workers endured some of the most obnoxious verbal and physical insults from white workers, their supposed "natural allies." We are well aware of dramatic moments of white working-class violence—the armed attacks on Georgia's black railroad firemen in 1909, the lynching of a black strikebreaker in Fort Worth, Texas, in 1921, the racial pogroms in the shipyards of Mobile, Alabama, during World War II, to mention only three—but these were merely explosive, large-scale manifestations of the verbal and physical violence black workers experienced on a daily basis. Without compunction, racist whites in many of the South's mines, mills, factories, and docks referred to their darker coworkers as "boy," "girl," "uncle," "aunt," and more commonly, just plain "nigger." Memphis United Auto Workers (UAW) organizer Clarence Coe recalls, "I have seen the time when a young white boy came in and maybe I had been working at the plant longer than he had been living, but if he was white I had to tell him 'yes sir' or 'no sir.' That was degrading as hell [but] I had to live with it." Occasionally, white workers kicked and slapped black workers just for fun or out of frustration. Without institutional structures to censure white workers for racist and sexist attacks, black workers took whatever opportunity they could to contest white insults and reaffirm their dignity, their indignation often exploding into fisticuffs at the workplace or after work. Black tobacco worker Charlie Decoda recalled working "with a cracker and they loved to put their foot in your tail and laugh. I told him once, 'You put your foot in my tail again ever and I'll break your leg.' " Even sabotage, a strategy usually employed against capital, was occasionally used in the most gruesome and reactionary intraclass conflicts. Michael Honey tells of George Holloway, a black UAW leader in Memphis, Tennessee, whose attempts to desegregate his local and make it more responsive to black workers' needs prompted white union members to tamper with his punch press. According to Honey, the sabotage "could have killed him if he had not examined his machine before turning it on." But as Honey also points out, personal indignities and individual acts of racist violence prompted black workers to take collective action, sometimes with the support of antiracist white workers. Black auto workers in Memphis, for example, staged a wildcat strike after a plant guard punched a black woman in the mouth.[49]

Intraclass conflict was not merely a manifestation of false consciousness or a case of companies' fostering an unwritten policy of divide and rule. Rather, white working-class consciousness was also racialized. The construction of a white working-class racial identity, as has been illustrated in the works of Alexander Saxton, David Roediger, and Eric Lott, registered the peculiar nature of class conflict where wage labor under capitalism and chattel slavery existed side by side. That work is especially important, for it maps the history of how European American workers came to see themselves as white and to manifest that identity politically and culturally. What whiteness and blackness signified for antebellum white workers need not concern us here. We need to acknowledge, however, that although racism was not always in the interests of southern white workers, it was nonetheless a very "real" aspect of white working-class consciousness. Racist attacks by white workers did not need instigation from wily employers. Because they ultimately defined their own class interests in racial terms, white workers employed racist terror and intimidation to help secure a comparatively privileged position within the prevailing system of wage dependency, as well as what Du Bois and Roediger call a "psychological wage." A sense of superiority and security was gained by being white and not being black. White workers sometimes obtained very real material benefits by institutionalizing their strength through white-controlled unions that used their power to enforce ceilings on black mobility and wages. The limited privileges afforded white workers as whites meant a subordinate status for African American workers. Hence, even the division of labor was racialized—black workers had to perform "nigger work." And without the existence of "nigger work" and "nigger labor," whiteness to white workers would be meaningless.[50]

Determining the social and political character of "nigger work" is therefore essential to understanding black working-class infrapolitics. First, by racializing the division of labor, it has the effect of turning dirty, physically difficult, and potentially dangerous work into humiliating work. To illustrate this point, we might examine how the meaning of tasks once relegated to black workers changed when they were done predominantly, if not exclusively, by whites. Among contemporary coal miners in Appalachia, where there are few black workers and racial ceilings have been largely (although not entirely) removed, difficult and dangerous tasks are charged with masculinity. Michael Yarrow found the miners believed that "being able to do hard work, to endure discomfort, and to brave danger" is an achievement of "manliness." Although undeniably an important component of the miner's work culture, "the masculine meaning given to hard, dangerous work [obscures] its reality as class exploitation." On the other hand, the black miners in Trotter's study were far more judicious, choosing to leave a job rather than place themselves in undue danger. Those black miners took pride in their work; they often challenged dominant categories of skill and performed what had been designated as menial labor with the pride of skilled craftsmen. But once derogatory social meaning is inscribed on the work (let alone the black bodies that perform the work), it undermines its potential dignity and worth—frequently rendering "nigger work" less manly.[51]

Finally, because black men and women toiled in work spaces in which both bosses and white workers demanded deference, freely hurled insults and epithets at them, and occasionally brutalized their bodies, issues of dignity informed much of black infrapolitics in the urban South. Interracial conflicts between workers were not simply diversions from some idealized definition of class struggle; white working-class racism was sometimes as much a barrier to black workers' struggle for dignity and autonomy at the workplace as the racial division of labor imposed by employers. Thus, episodes of interracial solidarity among working people and the fairly consistent opposition by most black labor leaders to Jim Crow locals are all the more remarkable. More important, for our purposes at least, the normative character of interracial conflict opens up another way to think about the function of public and hidden transcripts for *white* workers. For southern white workers openly to express solidarity with African Americans was a direct challenge to the public transcript of racial difference and domination. Indeed, throughout this period, leaders of southern biracial unions, with the exception of some left-wing organizers, tended to apologize for their actions, insisting that the union was driven by economic necessity or assuring the public of their opposition to "social equality" or "intermixing." Thus, even the hint of intimate, close relations between workers across the color line had consequences that cut both ways. Except for radicals and other bold individuals willing to accept ostracism, ridicule, and even violence, expressions of friendship and respect for African Americans had to remain part of the "hidden transcript" of white workers. White workers had to disguise and choke back acts and gestures of antiracism; when white workers were exposed as "nigger lovers" or when they took public stands on behalf of African Americans, the consequences could be fatal.[52]

On Buses, Streetcars, and City Streets

African American workers' struggle for dignity did not end at the workplace. For most white workers, public space—after intense class struggle—eventually became a "democratic space," where people of different class backgrounds shared city theaters, public conveyances, streets, and parks. For black people, white-dominated public space was vigilantly undemocratic and potentially dangerous. Jim Crow signs, filthy and inoperable public toilets, white police officers, dark bodies standing in the aisles of half-empty buses, black pedestrians stepping off the sidewalk or walking with their eyes turned down or away, and other acts of interracial social "etiquette" all reminded black people every day of their second-class citizenship. The sights, sounds, and experiences of African Americans in white-dominated public spaces challenge the notion that southern black working-class politics can be understood by merely examining labor organization, workplace resistance, culture, and the family.

Although historians of the civil rights movement have exhaustively documented the organized movement to desegregate the South, the study of unorganized, day-to-day resistance to segregated public space remains undeveloped. We know very little about the everyday posing, discursive conflicts, and small-scale skirmishes that not only created the conditions for the success of organized, collective movements but also shaped segregation policies, policing, and punishment.[53] By broadening our focus to include the daily confrontations and blatant acts of resistance—in other words, the realm of infrapolitics—we will find that black passengers, particularly working people, were concerned with much more than legalized segregation. A cursory examination of black working-class resistance on Birmingham buses and streetcars during World War II reveals that in most incidents the racial compartmentalization of existing space was not the primary issue. Rather, the most intense battles were fought over the deliberate humiliation of African Americans by operators and other passengers, shortchanging, the power of drivers to allocate or limit space for black passengers, and the practice of forcing blacks to pay at the front door and enter through the center doors. For example, half-empty buses or streetcars often passed up African Americans on the pretext of preserving space for potential white riders. It was not unusual for a black passenger who had paid at the front of the bus to be left standing while she or he attempted to board at the center door.

The design and function of buses and streetcars rendered them unique sites of contest. An especially useful metaphor for understanding the character of domination and resistance on public transportation might be to view the interior spaces as "moving theaters." Here I am using the word *theater* in two ways: as a site of performance and a site of military conflict. First, plays of conflict, repression, and resistance are performed in which passengers witness, or participate in, "skirmishes" that shape the collective memory of the passengers, illustrate the limits as well as the possibilities of resistance to domination, and draw more passengers into the "performance." The design of streetcars and buses—enclosed spaces with seats facing forward or toward the center aisle—gave everyday discursive and physical confrontations a dramaturgical quality. Second, theater as a military metaphor is particularly appropriate because all bus drivers and streetcar conductors in Birmingham carried guns and blackjacks and used them pretty regularly to maintain (the social) order. In August 1943, for example, when a black woman riding the South East Lake-Ensley line complained to the conductor that he had passed her stop, he followed her out of the streetcar and, in the words of the official report, "knocked her down with handle of gun. No further trouble." Violence was not a completely effective deterrent, however. In the twelve months beginning September 1941, there were at least eighty-eight cases of blacks occupying "white" space on public transportation, fifty-five of which were open acts of defiance in which African American passengers either refused to give up their seats or sat in the white section. But this is only part of the story; reported incidents and complaints of racial conflict totaled 176. These cases included at least eighteen interracial fights among passengers, twenty-two fights between black passengers and operators, and thirteen incidents in which black passengers engaged in verbal or physical confrontations over being shortchanged.[54]

Public transportation, unlike any other form of public space (e.g., a waiting room or a water fountain), was an extension of the marketplace. Because transportation companies depend on profit, any action that might limit potential fares was economically detrimental. This explains why divisions between black and white space had to be relatively fluid and flexible. With no fixed dividing line, black and white riders continually contested readjustments that affected them. The fluidity of the color line meant that their protestations often fell within the proscribed boundaries of segregationist law, thus rendering public transportation especially vulnerable to everyday acts of resistance. Furthermore, for African Americans, public transportation —as an extension of the marketplace—was also a source of economic conflict. One source of frustration was the all too common cheating or shortchanging of black passengers. Unlike the workplace, where workers entered as disempowered producers dependent on wages for survival and beholden, ostensibly at least, to their superiors, public transportation gave passengers a sense of consumer entitlement. The notion that blacks and whites should pay the same for "separate but equal" facilities fell within the legal constraints of Jim Crow, although for black passengers to argue publicly with whites, especially those in positions of authority, fell outside the limits of acceptable behavior. When a College Hills line passenger thought she had been shortchanged, she initially approached the driver in a very civil manner but was quickly brushed off and told to take her seat. In the words of the official report, "She came up later and began cursing and could not be stopped and a white passenger came and knocked her down. Officer was called and made her show him the money which was $.25 short, then asked her where the rest of the money was. She looked in her purse and produced the other quarter. She was taken to jail." The incident served as compelling theater, a performance that revealed the hidden transcript, the power of Jim Crow to crush public declarations swiftly and decisively, the role of white passengers as defenders of segregation, the degree to which white men—not even law enforcement officers—could assault black women without compunction. The play closes with the woman utterly humiliated, for all along, the report claims, she had miscounted her change.[55]

Although the available records are incomplete, it seems that black women outnumbered black men in incidents of resistance on buses and streetcars. In 1941-1942, nearly twice as many black women were arrested as black men, most of them charged with either sitting in the white section or cursing. Indeed, there is a long tradition of militant opposition to Jim Crow public transportation by black women, a tradition that includes such celebrated figures as Sojourner Truth, Ida B. Wells-Barnett, and of course, Rosa Parks.[56] More significant, however, black working women in Birmingham generally rode public transportation more often than men. Male industrial workers tended to live in industrial suburbs within walking distance of their places of employment, whereas most black working women were domestics who had to travel to relatively wealthy and middle-class white neighborhoods on the other side of town.

Unlike the popular image of Parks's quiet resistance, most black women's opposition tended to be profane and militant. There were literally dozens of episodes of black women sitting in the white section, arguing with drivers or conductors, and fighting with white passengers. The "drama" usually ended with the woman being ejected, receiving a refund for her fare and leaving on her own accord, moving to the back of the vehicle, or being hauled off to jail. Indeed, throughout the war, dozens of black women were arrested for merely cursing at the operator or a white passenger. In October 1943, for example, a teenager named Pauline Carth attempted to board the College Hills line around 8:00 p.m. When she was informed that there was no more room for colored passengers, she forced her way into the bus, threw her money at the driver, and cursed and spit on him. The driver responded by knocking her out of the bus, throwing her to the ground, and holding her down until police arrived. Fights between black women and white passengers were also fairly common. In March 1943, a black woman and a white man boarding the East Lake-West End line apparently got into a shoving match, which angered the black woman to the point where she "cursed him all the way to Woodlawn." When they reached Woodlawn she was arrested, sentenced to thirty days in jail, and fined fifty dollars.[57]

Although black women's actions were as violent or profane as men's, gender differences in power relations and occupation did shape black women's resistance. Household workers were in a unique position to contest racist practices on public transportation without significantly transgressing Jim Crow laws or social etiquette. First, transit company rules permitted domestics traveling with their white employers' children to sit in the section designated for whites. The idea, of course, was to spare white children from having to endure the Negro section. Although this was the official policy of the Birmingham Electric Company (owner of the city transit system), drivers and conductors did not always follow it. The rule enabled black women to challenge the indignity of being forced to move or stand while seats were available because their retaining or taking seats was sometimes permissible under Jim Crow. Second, employers intervened on behalf of their domestics, which had the effect of redirecting black protest into legitimate, "acceptable" avenues. Soon after a white employer complained that the Mountain Terrace bus regularly passed "colored maids and cooks" and therefore made them late to work, the company took action. According to the report, "Operators on this line [were] cautioned."[58]

Among the majority of black domestics who had to travel alone at night, the fear of being passed or forced to wait for the next vehicle created a sense of danger. Standing at a poorly lit, relatively isolated bus stop left them prey to sexual and physical assault by white and black men. As the sociologist Carol Brooks Gardner reminds us, in many neighborhoods the streets, particularly at night, are perceived as belonging to men, and women without escorts are perceived as available or vulnerable. In the South, that perception applied mostly (although not exclusively) to black women, because the ideology of chivalry obligated white men to come to the defense of white women —although not always working-class white women. To argue that black women's open resistance on the buses is incompatible with their fear when on the streets misses the crucial point that buses and streetcars, although sites of vicious repression, were occupied, lighted public spaces where potential allies and witnesses might be found.[59]

Such black resistance on Birmingham's public transit system conveyed a sense of dramatic opposition to Jim Crow before an audience. But discursive strategies, which may seem more evasive, also carry tremendous dramatic appeal. No matter how effective drivers, conductors, and signs were at keeping bodies separated, black voices flowed easily into the section designated for whites, constantly reminding riders that racially divided public space was contested terrain. Black passengers were routinely ejected and occasionally arrested for making too much noise, often by directing harsh words at a conductor or passenger or launching a monologue about racism in general. Such monologues or verbal attacks on racism make for excellent theater. Unlike passersby who can hurry by a lecturing street corner preacher, passengers were trapped until they reached their destination, the space silenced by the anonymity of the riders. The reports reveal a hypersensitivity to black voices rising from the back of the bus. Indeed, verbal protests or complaints registered by black passengers were frequently described as "loud," an adjective almost never used to describe the way white passengers articulated their grievances. One morning in August 1943, during the peak hours, a black man boarded an Acipco line bus and immediately began "complaining about discrimination against negroes in a very loud voice."[60] Black voices, especially the loud and profane, literally penetrated and occupied white spaces.

Cursing, a related discursive strategy, was among the crimes for which black passengers were most commonly arrested. Moreover, only black passengers were arrested for cursing. The act elicited police intervention, not because the state maintained strict moral standards and would not tolerate profanity, but because it represented a serious transgression of racial boundaries. Although scholars might belittle the power of resistive, profane noise as opposition, Birmingham's policing structure did not.

On the South Bessemer line in 1942, one black man was sentenced to six months in jail for cursing. In most instances, however, cursing was punishable by a ten dollar fine and court costs, and jail sentences averaged about thirty days.[61]

Some might argue that the hundreds of everyday acts of resistance in public spaces—from the most evasive to the blatantly confrontational—amount to very little because they were individualized, isolated events that almost always ended in defeat. Such an argument misses the unique, dramaturgical quality of these actions within the interior spaces of public conveyances; whenever passengers were present no act of defiance was isolated. Nor were acts of defiance isolating experiences. Because African American passengers shared a collective memory of how they were treated on a daily basis, both within and without the "moving theaters," an act of resistance or repression sometimes drew other passengers into the fray. An interesting report from an Avenue F line bus driver in October 1943 illustrates such a moment of collective resistance: "Operator went to adjust the color boards, and negro woman sat down quickly just in front of board that operator was putting in place. She objected to moving and was not exactly disorderly but all the negroes took it up and none of [the] whites would sit in seat because they were afraid to, and negroes would not sit in vacant seats in rear of bus."[62]

Most occupants sitting in the rear who witnessed or took part in the daily skirmishes learned that punishment was inevitable. The arrests, beatings, and ejections were intended as much for all the black passengers on board as for the individual transgressor. The authorities' fear of an incident escalating into collective opposition often meant that individuals who intervened in conflicts instigated by others received the harshest punishment. On the South Bessemer line one early evening in 1943, a young black man was arrested and fined twenty-five dollars for coming to the defense of a black woman who was told to move behind the color dividers. His crime was that he "complained and talked back to the officer." The fear of arrest or ejection could persuade individuals who had initially joined collective acts of resistance to retreat. Even when a single, dramatic act captured the imaginations of other black passengers and spurred them to take action, there was no guarantee that it would lead to sustained, collective opposition. To take one example, a black woman and man boarded the South East Lake-Ensley line one evening in 1943 and removed the color dividers, prompting all of the black passengers already on board and boarding to occupy the white section. When the conductor demanded that they move to their assigned area, all grudgingly complied except the couple who had initiated the rebellion. They were subsequently arrested.[63]

Spontaneous, collective protest did not always fizzle out at the site of contestation. Occasionally, the passengers approached formal civil rights organizations asking them to intercede or to lead a campaign against city transit. Following the arrest of Pauline Carth in 1943, a group of witnesses brought the case to the attention of the Birmingham branch of the NAACP, but aside from a perfunctory investigation and an article in the black-owned *Birmingham World*, no action was taken. The Southern Negro Youth Congress (SNYC), a left-wing organization based in Birmingham, attempted a direct-action campaign on the Fairfield bus line after receiving numerous complaints from black youth about conditions on public transportation. Mildred McAdory and three other SNYC activists attempted to move the color boards on a Fairfield bus in 1942, for which she was beaten and arrested by Fairfield police. As a result of the incident, the SNYC formed a short-lived organization, the Citizens Committee for Equal Accommodations on Common Carriers. However, the treatment of African Americans on public transportation was not a high-priority issue for Birmingham's black protest organizations during the war, and very few middle-class blacks rode public transportation. Thus, working people whose livelihood depended on city transit had to fend for themselves.[64]

The critical point here is that the actions of black passengers forced mainstream black political organizations to pay some attention to conditions on Jim Crow buses and streetcars. Unorganized, seemingly powerless black working people brought these issues to the forefront by their resistance, which was shaped by relations of domination as well as the many confrontations they witnessed on the stage of the moving theater. Their very acts of insubordination challenged the system of segregation, whether they were intended to or not, and their defiance in most cases elicited a swift and decisive response. Even before the war ended, everyday acts of resistance on buses and streetcars declined for two reasons. First, resistance compelled the transit company to "re-instruct" the most blatantly discourteous drivers and conductors, who cost the company precious profits by passing up black passengers or initiating unwarranted violence. Second, and more important, the acts of defiance led to an increase in punitive measures and more vigorous enforcement of segregation laws. An internal study by the Birmingham Transportation Department concluded, "Continued re-instruction of train men and bus operators, as well as additional vigilance on the part of our private police, has resulted in some improvement."[65]

The bitter struggles waged by black working people on public transportation, although obviously exacerbated by wartime social, political, and economic transformations, should force labor historians to rethink the meaning of public space as a terrain of class, race, and gender conflict. The workplace and struggles to improve working conditions are fundamental to the study of labor history. For southern black workers, however, the most embattled sites of opposition were frequently public spaces, partly because policing proved far more difficult in public spaces than in places of work. Not only were employees constantly under the watchful eye of foremen, managers, and employers, but workers could also be dismissed, suspended, or have their pay docked on a whim. In the public spaces of the city, however, the anonymity and sheer numbers of the crowd, whose movement was not directed by the discipline of work (and was therefore unpredictable), meant a more vigilant and violent system of maintaining social order. Arrests and beatings were always a possibility, but so was escape. Thus, for black workers public spaces both embodied the most repressive, violent aspects of race and gender oppression and, paradoxically, afforded more opportunities to engage in acts of resistance than the workplace itself.

Black Working-Class Infrapolitics and the Revision of Southern Political History

Shifting our focus from formal, organized politics to infrapolitics enables us to recover the oppositional practices of black working people who, until recently, have been presumed to be silent or inarticulate. Contrary to the image of an active black elite and a passive working class one generally finds in race relations scholarship, members of the most oppressed section of the black community always resisted, but often in a manner intended to cover their tracks. Given the incredibly violent and repressive forms of domination in the South, workers' dependence on wages, the benefits white workers derived from Jim Crow, the limited influence black working people exercised over white-dominated trade unions, and the complex and contradictory nature of human agency, evasive, clandestine forms of resistance should be expected. When thinking about the Jim Crow South, we need always to keep in mind that African Americans, the working class in particular, did not *experience* a liberal democracy. They lived and struggled in a world that resembled, at least from their vantage point, a fascist or, more appropriately, a colonial situation.

Whether or not battles were won or lost, everyday forms of opposition and the mere threat of open resistance elicited responses from the powerful that, in turn, shaped the nature of struggle. Opposition and containment, repression and resistance are inextricably linked. A pioneering study, Herbert Aptheker's *American Negro Slave Revolts,* illustrates the dynamic. The opening chapters, "The Fear of Rebellion" and "The Machinery of Control," show how slave actions and gestures and mere discussions of rebellion created social and political tensions for the master class and compelled southern rulers to erect a complex and expensive structure to maintain order. Furthermore, Aptheker shows us how resistance and the threat of resistance were inscribed in the law itself; thus, even when black opposition appeared invisible or was censored by the press, it still significantly shaped southern political and legal structures. The opening chapters of Aptheker's book (the chapters most of his harshest critics ignored) demonstrate what Stuart Hall means when he says "hegemonizing is hard work."[66]

Hegemonizing was indeed hard work, in part because African American resistance did make a difference. We know that southern rulers during this era devoted enormous financial and ideological resources to maintaining order. Police departments, vagrancy laws, extralegal terrorist organizations, the spectacle of mutilated black bodies—all were part of the landscape of domination surrounding African Americans. Widely publicized accounts of police homicides, beatings, and lynchings as well as of black protest against such acts of racist violence abound in the literature on the Jim Crow South.[67] Yet dramatic acts of racial violence and resistance represent only the tip of a gigantic iceberg. The attitudes of most working-class blacks toward the police were informed by an accumulation of daily indignities, whether experienced or witnessed. African Americans often endured illegal searches and seizures, detainment without charge, billy clubs, nightsticks, public humiliation, lewd remarks, loaded guns against their skulls. African American women endured sexual innuendo, molestation during body searches, and outright rape. Although such incidents were repeated in public spaces on a daily basis, they are rarely a matter of public record. Nevertheless, everyday confrontations between African Americans and police not only were important sites of contestation but also help explain why the more dramatic cases carry such resonance in black communities.[68]

We need to recognize that infrapolitics and organized resistance are not two distinct realms of opposition to be studied separately and then compared; they are two sides of the same coin that make up the history of working-class self-activity. As I have tried to illustrate, the historical relationships between the hidden transcript and organized political movements during the age of Jim Crow suggest that some trade unions and political organizations succeeded in mobilizing segments of the black working class because they at least partially articulated the grievances, aspirations, and dreams that remained hidden from public view. Yet we must not assume that all action that flowed from organized resistance was merely an articulation of a preexisting oppositional consciousness, thus underestimating collective struggle as a shaper of working-class consciousness.[69] The relationship between black working-class infrapolitics and collective, open engagement with power is dialectical, not a teleological, transformation from unconscious accommodation to conscious resistance. Hence, efforts by grassroots unions to mobilize southern black workers, from the Knights of Labor and the Brotherhood of Sleeping Car Porters to the Communist Party and the CIO, shaped or even transformed the hidden transcript. Successful struggles that depend on mutual support among working people and a clear knowledge of the "enemy" not only strengthen bonds of class (or race or gender) solidarity but also reveal to workers the vulnerability of the powerful and the potential strength of the weak. Furthermore, at the workplace as in public spaces, the daily humiliations of racism, sexism, and waged work embolden subordinate groups to take risks when opportunities arise, and their failures are as important as their victories, for they drive home the point that each act of transgression has its price. Black workers, like most aggrieved populations, do not decide to challenge dominant groups simply because of the lessons they have learned; rather, the very power relations that force them to resist covertly also make clear the terrible consequences of failed struggles.

In the end, whether or not African Americans chose to join working-class organizations, their daily experiences, articulated mainly in unmonitored social spaces, constituted the ideological and cultural foundations for constructing a collective identity. Their actions, thoughts, conversations, and reflections were not always, or even primarily, concerned with work, nor did they abide well with formal working-class institutions, no matter how well these institutions articulated aspects of the hidden transcript. In other words, we cannot presume that trade unions and similar labor institutions were the "real" standard bearers of black working-class politics; even for organized black workers they were probably only a small part of an array of formal and informal strategies by which people struggled to improve or transform daily life. Thus, for a worker to accept reformist trade union strategies while stealing from work, to fight streetcar conductors while voting down strike action in the local, to leave work early in order to participate in religious revival meetings or rendezvous with a lover, to attend a dance rather than a CIO mass meeting was not to manifest an "immature" class consciousness. Such actions reflect the multiple ways black working people live, experience, and interpret the world around them. To assume that politics is something separate from all these events and decisions is to balkanize people's lives and thus completely miss how struggles over power, autonomy, and pleasure take place in the daily lives of working people. People do not organize their lives around our disciplinary boundaries or analytical categories; they are, as Elsa Barkley Brown so aptly puts it, "polyrhythmic."[70]

Although the approach outlined above is still schematic and tentative (there is so much I have left out, including a crucial discussion of periodization), I am convinced that the realm of infrapolitics—from everyday resistance at work and in public spaces to the elusive hidden transcripts recorded in working-class discourses and cultures—holds rich insights into twentieth-century black political struggle. As recent scholarship in black working-class and community history has begun to demonstrate, to understand the political significance of these hidden transcripts and everyday oppositional strategies, we must think differently about politics and reject the artificial divisions between political history and social history. A "remapping" of the sites of opposition should bring us closer to "knowing" the people Richard Wright correctly insists are not what they seem.

Notes

1. Jacqueline Jones, *Labor of Love, Labor of Sorrow: Black Women, Work, and the Family, From Slavery to the Present* (New York, 1985), 140.

2. Tera W. Hunter, "Household Workers in the Making: Afro-American Women in Atlanta and the New South, 1861-1920" (Ph.D. diss., Yale University, 1990), 76-82.

3. "Report Involving Race Question," August 1943, p. 2, box 10, Cooper Green Papers (Birmingham Public Library, Birmingham, Alabama).

4. Lester C. Lamon, *Black Tennessee, 1900-1930* (Knoxville, 1977), 18. See also, for example, Neil McMillen, *Dark Journey: Black Mississippians in the Age of Jim Crow* (Urbana, 1989); John Dittmer, *Black Georgia in the Progressive Era, 1900-1920* (Urbana, 1977); Jack Temple Kirby, *Darkness at the Dawning: Race and Reform in the Progressive South* (Philadelphia, 1972); Paul D. Casdorph, *Republicans, Negroes, and Progressiveness in the South, 1912-1916* (University, Ala., 1981); Robert Haws, ed., *The Age of Segregation: Race Relations in the South, 1890-1954* (Jackson, Miss., 1978); Margaret Law Callcott, *The Negro in Maryland Politics, 1870-1912* (Baltimore, 1969); I. A. Newby, *Black Carolinians: A History of Blacks in South Carolina From 1895-1968* (Columbia, S.C., 1973); and George C. Wright, *Life Behind a Veil: Blacks in Louisville, Kentucky, 1865-1930* (Baton Rouge, 1985). For the nineteenth century, see Howard N. Rabinowitz, *Race Relations in the Urban South, 1865-1890* (New York, 1978).

5. My definition of *urban* includes mining towns and company suburbs that have a significant black proletariat. Because of space considerations, I chose not to include rural areas. On twentieth-century rural resistance, see, for example, William Bennett Bizzell, "Farm Tenantry in the United States" (Ph.D. diss., Columbia University, 1921), 267-68; Ray Stannard Baker, *Following the Color Line: American Negro Citizenship in the Progressive Era* (1908; New York, 1964), 76-77; Glenn N. Sisk, "Alabama Black Belt: A Social History, 1875-1917" (Ph.D. diss., Duke University, 1951), 277-87; Allison Davis, Burleigh B. Gardner, and Mary R. Gardner, *Deep South: A Social Anthropological Study of Caste and Class* (Chicago, 1941), 396-98; Arthur F. Raper, *Preface to Peasantry: A Tale of Two Black Belt Counties* (Chapel Hill, 1936), 173-74; and Albert C. Smith, " 'Southern Violence' Reconsidered: Arson as Protest in Black Belt Georgia, 1865-1910," *Journal of Southern History* 55 (November 1985), 627-64.

6. Herbert Aptheker, *American Negro Slave Revolts* (New York, 1943). See also his essays and pamphlets, intended for the kind of working-class audiences he had been organizing: Herbert Aptheker, *Negro Slave Revolts in the United States, 1526-1860* (New York, 1939); Herbert Aptheker, *The Negro in the Civil War* (New York, 1938); and Herbert Aptheker, "Maroons Within the Present Limits of the United States," *Journal of Negro History* 24 (April 1939), 167-84. For an appraisal of *American Negro Salve Revolts* and a history of responses from the historical profession, see Herbert Shapiro, "The Impact of the Aptheker Thesis: A Retrospective View of *American Negro Salve Revolts*," *Science and Society* 48 (Spring 1984), 52-73.

7. James C. Scott, *Dominations and the Arts of Resistance: Hidden Transcripts* (New Haven, 1990), especially 183. See also James C. Scott, *Weapons of the Weak: Everyday Forms of Peasant Resistance* (New Haven, 1985); and James C. Scott, *The Moral Economy*

of the Peasant: Rebellion and Subsistence in Southeast Asia (New Haven, 1976). For a useful critique, see Rosalind O'Hanlon, "Recovering the Subject: *Subaltern Studies* and Histories of Resistance in Colonial South Asia," *Modern Asian Studies* 22 (February 1988), 189-224. The precursors of Scott's work include E. P. Thompson, *The Making of the English Working Class* (New York, 1966); E. P. Thompson, "The Crime of Anonymity," in *Albion's Fatal Tree: Crime and Society in Eighteenth-Century England*, Douglas Hay, Peter Linebaugh, John G. Rule, E. P. Thompson, and Cal Winslow, eds. (New York, 1975), 255-344; Michel de Certeau, *The Practice of Everyday Life*, Steven Rendall, trans. (Berkeley, 1984); Aptheker, *American Negro Slave Revolts;* Raymond A. Bauer and Alice H. Bauer, "Day to Day Resistance to Slavery," *Journal of Negro History* 37 (October 1942), 388-419; Eugene Genovese, *Roll, Jordan, Roll: The World the Slaves Made* (New York, 1974), 599-621; Peter Kolchin, *Unfree Labor: American Slavery and Russian Serfdom* (Cambridge, Mass., 1987), 241-44; Alex Lichtenstein, " 'That Disposition to Theft, With Which They Have Been Branded': Moral Economy, Slave Management, and the Law," *Journal of Social History* 22 (Spring 1988), 413-40; Sterling Stuckey, "Through the Prism of Folklore: The Black Ethos in Slavery," *Massachusetts Review* 9 (Summer 1968), 417-37.

8. Lila Abu-Lughod, "The Romance of Resistance: Tracing Transformations of Power Through Bedouin Women," *American Ethnologist* 17, 1 (1990), 55. I am grateful to Victoria Wolcott for bringing this source to my attention.

9. My recasting of "the political" is partly derived from my reading of Geoff Eley, "Labor History, Social History, *Alltagsgeschichte:* Experience, Culture, and the Politics of the Everyday—A New Direction for German Social History?" *Journal of Modern History* 61 (June 1989), 297-343.

10. Earl Lewis, *In Their Own Interests: Race, Class, and Power in Twentieth Century Norfolk, Virginia* (Berkeley, 1991), 10; George Lipsitz, *A Life in the Struggle: Ivory Perry and the Culture of Opposition* (Philadelphia, 1988); Elsa Barkley Brown, "Womanist Consciousness: Maggie Lena Walker and the Independent Order of St. Luke," *Signs* 14 (Spring 1989), 610-33; Elsa Barkley Brown, " 'Not Alone to Build This Pile of Bricks': Institution Building and Community in Richmond, Virginia," paper presented at the conference "The Age of Booker T. Washington," University of Maryland, College Park, May 3, 1990 (in Robin D. G. Kelley's possession); Elsa Barkley Brown, "Uncle Ned's Children: Richmond, Virginia's Black Community, 1890-1930" (Ph.D. diss., Kent State University, 1995); Michael K. Honey, *Southern Labor and Black Civil Rights: Organizing Memphis Workers* (Urbana, 1993); Hunter, "Household Workers in the Making"; Robert D. G. Kelley, *Hammer and Hoe: Alabama Communists During the Great Depression* (Chapel Hill, 1990); Robert Korstad, " 'Daybreak of Freedom': Tobacco Workers and the CIO, Winston-Salem, North Carolina, 1943-1950" (Ph.D. diss., University of North Carolina, Chapel Hill, 1987); Joe W. Trotter, *Coal, Class, and Color: Blacks in Southern West Virginia, 1915-1932* (Urbana, 1990). Although Peter J. Rachleff's outstanding book and his even more prodigious dissertation are limited to the nineteenth century and therefore beyond the scope of this essay, he offers one of the most sophisticated discussions of the relationship between community, culture, work, and self-activity. Peter J. Rachleff, *Black Labor in the South: Richmond, Virginia, 1865-1890* (Philadelphia, 1984), 109-15; Peter J. Rachleff, "Black, White, and Gray: Race and Working-Class Activism in Richmond, Virginia, 1865-1890" (Ph.D. diss., University of Pittsburgh, 1981).

11. Scott, *Domination and the Arts of Resistance*, 120; Kelley, *Hammer and Hoe;* Herbert Shapiro, *White Violence and Black Response: From Reconstruction to Montgomery* (Amherst, 1988), 224-37; Howard Odum, *Race and Rumors of Race: Challenge to American Crisis* (Chapel Hill, 1943), 96-104; Dolores E. Janiewski, *Sisterhood Denied: Race, Gender,*

and Class in a New South Community (Philadelphia, 1985), 121. For the antebellum period, see Aptheker, *American Negro Slave Revolts*, 59. The success of stool pigeons often depended on the strategies black working people used to resist exploitation. Black informers had to maintain a low profile and don a mask in front of other black folk because they were less effective as spies without entry into the community of workers.

12. Elsa Barkley Brown, " 'Not Alone to Build This Pile of Bricks' "; Evelyn Brooks Higginbotham, *Righteous Discontent: The Women's Movement in the Black Baptist Church, 1880-1920* (Cambridge, Mass., 1993); Trotter, *Coal, Class, and Color;* Rachleff, *Black Labor in the South.*

13. Geraldine Moore, *Behind the Ebony Mask* (Birmingham, 1961), 15; Davis, Gardner, and Gardner, *Deep South*, 230.

14. The best-known advocate of this position is William Julius Wilson. See William Julius Wilson, *The Truly Disadvantaged: The Inner City, the Underclass, and Public Policy* (Chicago, 1987), 56-57. Recent historical literature and a much older sociological literature challenge Wilson's claim of a "golden age" of black community. See especially Lewis, *In Their Own Interests;* Trotter, *Coal, Class, and Color;* Kenneth Marvin Hamilton, *Black Towns and Profit: Promotion and Development in the Trans-Appalachian West, 1877-1915* (Urbana, 1991); Robin D. G. Kelley, "The Black Poor and the Politics of Opposition in a New South City," in *The Underclass Debate: Views From History*, Michael Katz, ed. (Princeton, 1993), 293-333; E. Franklin Frazier, *Black Bourgeoisie* (New York, 1957); St. Clair Drake and Horace Cayton, *Black Metropolis: A Study of Negro Life in a Northern City* (1945; 2 vols., New York, 1962), vol. 2, 526-63; Davis, Gardner, and Gardner, *Deep South*, 230-36.

15. For a sampling of historical studies of African American families, see Herbert Gutman, *The Black Family in Slavery and Freedom, 1750-1925* (New York, 1976); James Borchert, *Alley Life in Washington: Family, Community, Religion, and Folklife in the City, 1850-1970* (Urbana, 1980); Andrew Billingsley, *Black Families in White America* (Englewood Cliffs, 1968); Sharon Harley, "For the Good of Family and Race: Gender, Work, and Domestic Roles in the Black Community, 1880-1930," *Signs* 15 (Winter 1990), 336-49; and Jones, *Labor of Love, Labor of Sorrow.* On women's unpaid labor, the reproduction of male labor power, and the maintenance of capitalism, see Emily Blumenfeld and Susan Mann, "Domestic Labour and the Reproduction of Labour Power: Towards an Analysis of Women, the Family, and Class," in *Hidden in the Household: Women's Domestic Labour Under Capitalism*, Bonnie Fox, ed. (Toronto, 1980), 267-307; Jeanne Boydston, *Home and Work: Housework, Wages, and the Ideology of Labor in the Early Republic* (New York, 1990); Martha E. Gimenez, "The Dialectics of Waged and Unwaged Work: Waged Work, Domestic Labor and Household Survival in the United States," in *Work Without Wages: Domestic Labor and Self-Employment Within Capitalism*, Jane L.Collins and Martha E. Gimenez, eds. (Albany, 1990), 24-45; and Susan Strasser, *Never Done: A History of American Housework* (New York, 1982).

16. Heidi Hartmann, "The Family at the Locus of Gender, Class, and Political Struggle: The Example of Housework," *Signs* 6 (Spring 1981), 366-94; Lois Rita Helmbold, "Beyond the Family Economy: Black and White Working-Class Women During the Great Depression," *Feminist Studies* 13 (Fall 1987), 629-55; Susan Mann, "Slavery, Sharecropping, and Sexual Inequality," *Signs* 14 (Summer 1989), 774-98.

17. Carolyn Steedman, *Landscape for a Good Woman: A Story of Two Lives* (New Brunswick, 1986), 13; Elizabeth Faue, "Reproducing the Class Struggle: Perspectives on the Writing of Working Class History," paper presented at the meeting of the Social Science History Association, Minneapolis, October 19, 1990 (in Kelley's possession), 8. See also Elizabeth Faue, *Community of Suffering and Struggle: Women, Men and the*

Labor Movement in Minneapolis, 1915-1945 (Chapel Hill, 1991), 15; and Elizabeth Faue, "Gender, Class, and the Politics of Work in Women's History," paper presented at the annual meeting of the American Historical Association, Chicago, December 1991 (in Kelley's possession).

18. Most black male social scientists suggested that black mothers inflicted irreparable psychological damage on their sons, but feminist scholars understood that learning the dominant codes and social conventions of the South was necessary for survival. See, for example, Calvin Hernton, *Sex and Racism in America* (Garden City, 1965); William H. Grier and Price M. Cobbs, *Black Rage* (New York, 1968), 51; Paula Giddings, *When and Where I Enter: The Impact of Black Women on Race and Sex in America* (New York, 1984); Patricia Morton, *Disfigured Images: The Historical Assault on Afro-American Women* (Westport, 1991), 116; Janiewski, *Sisterhood Denied*, 45; and Jacquelyn Dowd Hall, *Revolt Against Chivalry: Jessie Daniel Ames and the Women's Campaign Against Lynching* (New York, 1979), 142.

19. On the political ramifications of how southern black mothers raised their daughters, see Elsa Barkley Brown, "Mothers of Mind," *Sage* 6 (Summer 1989), 3-10; and Elsa Barkley Brown, "African-American Women's Quilting: A Framework for Conceptualizing and Teaching African-American Women's History," *Signs* 14 (Summer 1989), 928-29. Elsa Barkley Brown, "To Catch the Vision of Freedom: Reconstructing Southern Black Women's Political History, 1865-1885," paper presented at the workshop Historical Perspectives on Race and Racial Ideologies, Center for Afro-American and African Studies, University of Michigan, November 22, 1991 (in Kelley's possession).

20. Brenda McCallum, "Songs of Work and Songs of Worship: Sanctifying Black Unionism in the Southern City of Steel," *New York Folklore* 14, 1 and 2 (1988), 19-20; Hunter, "Household Workers in the Making," 151-86; Robert Korstad and Nelson Lichtenstein, "Opportunities Found and Lost: Labor, Radicals, and the Early Civil Rights Movement," *Journal of American History* 75 (December 1988), 786-811; Korstad, " 'Daybreak of Freedom' "; Honey, *Southern Labor and Black Civil Rights;* Kelley, *Hammer and Hoe.*

21. Eley, "Labor History, Social History, *Alltagsgeschichte*," 311-12. Important studies that move discussions of working-class culture beyond the trade union and the sphere of production include George Lipsitz, *Rainbow at Midnight: Labor and Culture in the 1940's* (Urbana, 1994); Francis G. Couvares, *The Remaking of Pittsburgh: Class and Culture in an Industrializing City, 1877-1919* (Albany, 1984); Kathy Lee Peiss, *Cheap Amusements: Working Women and Leisure in Turn-of-the Century New York* (Philadelphia, 1986); Steven J. Ross, *Workers on the Edge: Work, Leisure, and Politics in Industrializing Cincinnati, 1788-1890* (New York, 1985); and Roy Rosenzweig, *Eight Hours for What We Will: Workers and Leisure in an Industrial City, 1870-1920* (New York, 1983); LeRoi Jones, *Blues People* (New York, 1963); Lawrence Levine, *Black Culture and Black Consciousness: Afro-American Folk Thought From Slavery to Freedom* (New York, 1977), xi; Sterling Stuckey, *Slave Culture: Nationalist Theory and the Foundations of Black America* (New York, 1987); Stuckey, "Through the Prism of Folklore"; Hazel Carby, " 'It Jus Be's Dat Way Sometime': The Sexual Politics of Women's Blues," *Radical America* 20, 4 (1987), 9-22; Charles P. Henry, *Culture and African-American Politics* (Bloomington, 1990); John W. Roberts, *From Trickster to Badman: The Black Folk Hero in Slavery and Freedom* (Philadelphia, 1989); McCallum, "Songs of Work and Songs of Worship," 9-33; Gladys-Marie Fry, *Night Riders in Black Folk History* (Knoxville, 1975).

22. On the social meaning of dance halls and blues clubs in southern black life, see Hunter, "Household Workers in the Making," 92-93; Lewis, *In Their Own Interests,*

99-100; Wright, *Life Behind a Veil*, 138; Katrina Hazzard-Gordon, *Jookin': The Rise of Social Dance Formations in African-American Culture* (Philadelphia, 1990).

23. My use of the term *alternative* cultures is borrowed from Raymond Williams, "Base and Superstructure in Marxist Cultural Theory," *Problems in Materialism and Culture* (London, 1980), 41-42.

24. Paul Gilroy, "One Nation Under a Groove: The Cultural Politics of 'Race' and Racism in Britain," in *Anatomy of Racism*, David Theo Goldberg, ed. (Minneapolis, 1990), 274.

25. "Report Involving Race Question," June 1943, p. 1, box 10, Green Papers.

26. Clifford M. Kuhn, Harlon E. Joye, and E. Bernard West, *Living Atlanta: An Oral History of the City, 1914-1948* (Atlanta, 1990), 39.

27. Kenneth B. Clark and James Barker, "The Zoot Effect in Personality: A Race Riot Participant," *Journal of Abnormal and Social Psychology* 40 (April 1945), 145. See also Robin D. G. Kelley, "The Riddle of the Zoot: Malcolm Little and Black Cultural Politics During World War II," in *Malcolm X: In Our Own Image*, Joe Wood, ed. (New York, 1992), 155-82; Stuart Cosgrove, "The Zoot-Suit and Style Warfare," *History Workshop Journal* 18 (Autumn 1984), 77-91; Mauricio Mazon, *The Zoot-Suit Riots: The Psychology of Symbolic Annihilation* (Austin, 1984); Eric Lott, "Double V, Double-Time: Bebop's Politics of Style," *Callalloo* 11, 3 (1988), 597-605; Kobena Mercer, "Black Hair/Style Politics," *New Formations* 3 (Winter 1987), 49; Bruce M. Tyler, "Black Jive and White Repression," *Journal of Ethnic Studies* 16, 4 (1989), 31-66.

28. Odum, *Race and Rumors of Race*, 77-79; "Incidents Reported," September 1, 1941-August 31, 1942, pp. 4, 8, box 10, Green Papers: "Report Involving Race Question," February 1943, pp. 2, 3, ibid.; "Report Involving Race Question," September 1943, p. 1, ibid. On Stagolee folklore and the political implications of "baaad niggers" for African Americans, see Roberts, *From Trickster to Badman*, 171-215; Levine, *Black Culture and Black Consciousness*, 407-19. "Report Involving Race Question," March 1943, p. 3, box 10, Green Papers. See Julius Hudson, "The Hustling Ethic," in *Rappin' and Stylin' Out: Communication in Urban Black America*, Thomas Kochman, ed. (Urbana, 1972), 414-16; Kelley, "Riddle of the Zoot," 167-72. For postwar examples, see Elliot Liebow, *Tally's Corner: A Study of Negro Streetcorner Men* (Boston, 1967); 137-44; and Christina Milner and Richard Milner, *Black Players: The Secret World of Black Pimps* (New York, 1973).

29. See Nell Irvin Painter, *The Narrative of Hosea Hudson: His Life as a Negro Communist in the South* (Cambridge, Mass., 1979), 347-51; Jane Maguire, *On Shares: Ed Brown's Story* (New York, 1975), 125-33; Theodore Rosengarten, *All God's Dangers: The Life of Nate Shaw* (New York, 1974), 189, 192, 238-40.

30. See Vincent Harding, *There Is a River: The Black Struggle for Freedom in America* (New York, 1981). W. E. B. Du Bois, *Black Reconstruction in America: An Essay Toward a History of the Part Which Black Folk Played in the Attempt to Reconstruct Democracy in America, 1860-1880* (New York, 1935), 124.

31. George Rawick, "Working-Class Self-Activity," *Radical America* 3 (March-April 1969), 145. For examples of the voluminous literature on black urban workers, organized labor, and working-class politics in the South, see Eric Arnesen, *Waterfront Workers of New Orleans: Race, Class, and Politics, 1863-1923* (New York, 1991); Horace R. Cayton and George Mitchell, *Black Workers and the New Unions* (Chapel Hill, 1939); Philip Foner, *Organized Labor and the Black Worker, 1619-1981* (New York, 1982); Honey, *Southern Labor and Black Civil Rights;* Korstad, " 'Daybreak of Freedom' "; F. Ray Marshall, *Labor in the South* (Cambridge, Mass., 1967); Daniel Rosenberg, *New Orleans Dockworkers: Race, Labor, and Unionism, 1892-1923* (Albany, 1988); Richard

Straw, " 'This Is Not a Strike, It Is Simply a Revolution': Birmingham Miners Struggle for Power, 1894-1908" (Ph.D. diss., University of Missouri, Columbia, 1980); Philip Taft, *Organizing Dixie: Alabama Workers in the Industrial Era,* Gary Fink, ed. (Westport, 1981).

32. Hunter, "Household Workers in the Making," 76-82; Jones, *Labor of Love, Labor of Sorrow,* 123-33; Dolores Janiewski, "Sisters Under Their Skins: Southern Working Women, 1880-1950," in *Sex, Race, and the Role of Women in the South,* Joanne V. Hawks and Sheila L. Skemp, eds. (Jackson, Miss., 1983), 788. David Katzman, *Seven Days a Week: Women and Domestic Service in Industrializing America* (New York, 1978), 195-97; Korstad, " 'Daybreak of Freedom,' " 101; Janiewski, *Sisterhood Denied,* 98.

33. On this form of resistance in the unfree working class, see especially Genovese, *Roll, Jordan, Roll,* 599-621; Kolchin, *Unfree Labor,* 241-44; Lichtenstein, " 'That Disposition to Theft, With Which They Have Been Branded,' " 413-40; Bauer and Bauer, "Day to Day Resistance to Slavery," 388-419; Aptheker, *American Negro Slave Revolts,* 141-49; and Edward L. Ayers, *Vengeance and Justice: Crime and Punishment in the Nineteenth Century American South* (New York, 1984). The most sophisticated work on crime and resistance at the workplace continues to come from sociologists, radical criminologists, and historians of Europe and Africa. See, for example, Peter Linebaugh, *The London Hanged: Crime and Civil Society in the Eighteenth Century* (London, 1991); Hay et al., eds., *Albion's Fatal Tree;* Jeff Crisp, *The Story of an African Working Class: Ghanaian Miners' Struggles, 1870-1980* (London, 1984), 18, 26, 44-45, 68, 78; Bill Freund, "Theft and Social Protest Among the Tin Miners of Northern Nigeria," *Radical History Review* 26 (Spring 1982), 68-86; Charles Van Onselen, *Chibaro: African Mine Labour in Southern Rhodesia, 1900-1933* (London, 1976), 239-42; Charles Van Onselen, *Studies in the Social and Economic History of the Witwatersrand, 1886-1914: New Nineveh* (2 vols., New York, 1982), vol. 2, 171-95.

34. On workplace theft, see Alvin Ward Gouldner, *Wildcat Strike* (Yellow Springs, 1954). On British workers, see Steven Box, *Recession, Crime, and Punishment* (Totowa, 1987), 34. Jason Ditton, *Part-Time Crime: An Ethnography of Fiddling and Pilferage* (London, 1977); Richard C. Hollinger and J. P. Clark, *Theft by Employees* (Lexington, Mass., 1983). On the informal economy and working-class opposition, see Cyril Robinson, "Exploring the Informal Economy," *Crime and Social Justice* 15, 3 and 4 (1988), 3-16. Hunter, "Household Workers in the Making," 82-83; Jones, *Labor of Love, Labor of Sorrow,* 132; Korstad, " 'Daybreak of Freedom,' " 102; Janiewski, *Sisterhood Denied,* 124; Kelley, *Hammer and Hoe,* 20-21; Julia Kirk Blackwelder, "Quiet Suffering: Atlanta Women in the 1930's," *Georgia Historical Quarterly* 61 (Summer 1977), 119-20.

35. Certeau, *Practice of Everyday Life,* Rendall, trans., 25-26.

36. Korstad, " 'Daybreak of Freedom,' " 101; Hunter, "Household Workers in the Making," 85; Jones, *Labor of Love, Labor of Sorrow,* 131; John Dollard, *Caste and Class in a Southern Town* (New York, 1957), 108.

37. European and African labor historians have been more inclined than Americanists to study industrial sabotage. See, for example, Pierre DuBois, *Sabotage in Industry* (New York, 1979); Tim Mason, "The Workers' Opposition in Nazi Germany," *History Workshop* 11 (Spring 1981), 127-30; and Donald Quartaert, "Machine Breaking and the Changing Carpet Industry of Western Anatolia, 1860-1908," *Journal of Social History* 19 (Summer 1986), 473-89. On black members of the Industrial Workers of the World (IWW) and whether they practiced sabotage, see David Roediger, "Labor, Gender and the 'Smothering' of Race: Covington Hall and the Complexities of Class," 1992 (in Kelley's possession); and James Green, *Grass-Roots Socialism: Radical Movements in the Southwest, 1895-1943* (Baton Rouge, 1978), 219. Scholars have made

little effort to explore the question, partly because black workers have been treated by most IWW historians more as objects to be debated over than as subjects engaged in "the class struggle." See especially Paul Brissenden, *The IWW: A Study of American Syndicalism* (New York, 1920), 208; John S. Gambs, *The Decline of the IWW* (New York, 1932), 135, 198; Selig Perlman and Philip Taft *History of Labor in the United States* (New York, 1935), 247; Bernard A. Cook, "Covington Hall and Radical Rural Unionization in Louisiana," *Louisiana History*, 18 (1977), 230, 235; Melvyn Dubofsky, *We Shall Be All: A History of the IWW* (New York, 1969), 8-9, 210, 213-16; Merl E. Reed, "Lumberjacks and Longshoremen: The IWW in Louisiana," *Labor History* 13 (Winter 1972), 44-58; Philip Foner, "The IWW and the Black Worker," *Journal of Negro History* 55 (January 1970), 45-64; and Sterling Spero and Abram L. Harris, *The Black Worker: The Negro and the Labor Movement* (New York, 1931), especially 329-35.

38. My thinking is partly inspired by Sylvia Wynter, "Sambos and Minstrels," *Social Text* 1 (Winter 1979), 149-56. On the dominant assumptions about black criminality and laziness in the postbellum South, see George Fredrickson, *The Black Image in the White Mind: The Debate on Afro-American Character and Destiny, 1817-1914* (1971; Middletown, 1987), 251-52, 273-75, 287-88; Claude H. Nolen, *The Negro's Image in the South: The Anatomy of White Supremacy* (Lexington, Ky., 1967), 13-15, 25-27; and Ayers, *Vengeance and Justice*, 176-77; Du Bois, *Black Reconstruction*, 40.

39. Kelley, *Hammer and Hoe*, 101-3; Scott, *Domination and the Arts of Resistance*, 23-36. The mask of ignorance did not always work as a strategy to mitigate punishment. Some rural African Americans accused of stealing livestock or burning barns were lynched. See clippings in Ralph Ginzburg, ed., *100 Years of Lynchings* (New York, 1962), 92-93.

40. Roediger, "Labor, Gender, and the 'Smothering' of Race," 34.

41. Trotter, *Coal, Class, and Color*, 65, 108, 264-65; Richard Price, "The Labour Process and Labour History," *Social History* 8 (January 1983), 62-63.

42. See Trotter, *Coal, Class, and Color*, 68-85, 109; William Cohen, *At Freedom's Edge: Black Mobility and the Southern Quest for Racial Control* (Baton Rouge, 1991); Lewis, *In Their Own Interests*, 30-32, 168; James R. Grossman, *Land of Hope: Chicago, Black Southerners, and the Great Migration* (Chicago, 1989); Nell Irvin Painter, *Exodusters: Black Migration to Kansas After Reconstruction* (New York, 1976); Edwin S. Redkey, *Black Exodus: Black Nationalist and Back-to-Africa Movements, 1890-1910* (New Haven, 1969); and Joe W. Trotter, ed., *The Great Migration in Historical Perspective: New Dimensions of Race, Class, and Gender* (Bloomington, 1991). For examples of the ways in which black migration is redefined, see Jacqueline Jones, *The Dispossessed: America's Underclasses From the Civil War to the Present* (New York, 1992), 104-26; Dollard, *Caste and Class in a Southern Town*, 115-19; and Nolen, *Negro's Image in the South*, 186-88.

43. For rich descriptions of racially segregated work, one could go back as far as Charles Wesley, *Negro Labor in the United States, 1850-1925* (New York, 1927); Carter G. Woodson and Lorenzo Greene, *The Negro Wage Earner* (New York, 1930); and Spero and Harris, *Black Worker*. For sophisticated recent studies of the racialized dimension of industrial work and collective consciousness, see Arnesen, *Waterfront Workers of New Orleans;* Trotter, *Coal, Class, and Color*, especially 65-88, 102-11, 106. Feminist historians of southern labor have shown sensitivity to the relationship between work and collective consciousness. See especially Hunter, "Household Workers in the Making"; Janiewski, *Sisterhood Denied;* Janiewski, "Sisters Under Their Skins," 13-35; Janiewski, "Seeking 'a New Day and a New Way': Black Women and Unions in the Southern Tobacco Industry," in *"To Toil the Livelong Day": American Women at Work, 1780-1980*, Carol Groneman and Mary Beth Norton, eds. (Ithaca, 1987), 161-78; Korstad,

" 'Daybreak of Freedom' "; Julia Kirk Blackwelder, "Women in the Workforce: Atlanta, New Orleans, and San Antonio, 1930-1940," *Journal of Urban History* 4 (May 1978), 331-58. My ideas about a "politics of location" are derived from Adrienne Rich, "Notes Toward a Politics of Location," in *Women, Feminist Identity, and Society in the 1980's*, Myriam Diaz-Dioaretz and Iris M. Zavala, eds. (Philadelphia, 1985), 7-22; and Nina Gregg, "Women Telling Stories About Reality: Subjectivity, the Generation of Meaning, and the Organizing of a Union at Yale" (Ph.D. diss., McGill University, 1991).

44. Lewis, *In Their Own Interests*, 47-58, especially 58. See also McCallum, "Songs of Work and Songs of Worship," 14.

45. James R. Green, "The Brotherhood of Timber Workers: 1910-1913: A Radical Response to Industrial Capitalism in the Southern U.S.A.," *Past and Present* 60 (August 1973), 185. On the preference of black workers for segregated locals, see Bruce Nelson, "Class and Race in the Crescent City: The ILWU, From San Francisco to New Orleans," in *The CIO's Left-led Unions*, Steven Rosswurm, ed. (New Brunswick, 1992), 19-45.

46. Korstad, " 'Daybreak of Freedom,' " 90-91, especially 94; Janiewski, *Sisterhood Denied*, 97-109, passim; Janiewski, "Sisters Under Their Skins," 27-28; Janiewski, "Seeking 'a New Day and a New Way,' " 166; Beverly W. Jones, "Race, Sex, and Class: Black Female Tobacco Workers in Durham, North Carolina, 1920-1940, and the Development of Female Consciousness," *Feminist Studies* 10 (Fall 1984), 443-50; Jones, *Labor of Love, Labor of Sorrow*, 137-38. On sexual exploitation of domestic workers, see Hunter, "Household Workers in the Making," 116-17; and Kuhn, Joye, and West, *Living Atlanta*, 115. Works of labor history that attend to harassment are Vicki Ruiz, *Cannery Women, Cannery Lives: Mexican Women, Unionization, and the California Food Processing Industry, 1930-1950* (Albuquerque, 1987); and Mary Bularzik, "Sexual Harassment at the Workplace: Historical Notes," in *Workers' Struggles, Past and Present: A "Radical America" Reader*, James Green, ed. (Philadelphia, 1983), 117-35. And see Elsa Barkley Brown, " 'What Has Happened Here': The Politics of Difference in Women's History and Feminist Politics," *Feminist Studies* 18 (Summer 1992), 302-7, Jones; "Race, Sex, and Class," 449; Korstad, " 'Daybreak of Freedom.' "

47. Eileen Boris, "Black Women and Paid Labor in the Home: Industrial Homework in Chicago in the 1920's," in *Homework: Historical and Contemporary Perspectives on Paid Labor at Home*, Eileen Boris and Cynthia R. Daniels, eds. (Urbana, 1989), 47; Hunter, "Household Workers in the Making," 151-86. See also Eileen Boris, *Home to Work: Motherhood and the Politics of Industrial Homework in the United States* (Cambridge, England, forthcoming).

48. Elizabeth Clark-Lewis, " 'This Work Had a End': African-American Domestic Workers in Washington, D.C., 1910-1940," in *"To Toil the Livelong Day,"* Groneman and Norton, eds., 207.

49. John Michael Matthews, "The Georgia Race Strike of 1909," *Journal of Southern History* 40 (November 1974), 613-30; Hugh Hammet, "Labor and Race: The Georgia Railroad Strike of 1909," *Labor History* 16 (Fall 1975), 470-84; Foner, *Organized Labor and the Black Worker*, 105-7; Ginzburg, ed., *100 Years of Lynching*, 157-58; William H. Harris, *The Harder We Run: Black Workers Since the Civil War* (New York, 1982), 45-47; Shapiro, *White Violence and Black Response*, 338-39; Janiewski, *Sisterhood Denied*, 121; Michael Honey, "Black Workers Remember: Industrial Unionism in the Era of Jim Crow," paper presented to the Southern Labor History Conference, Atlanta, Georgia, October 9-12, 1991, pp. 13, 16, 18 (in Kelley's possession).

50. David Roediger, *The Wages of Whiteness: Race and the Making of the American Working Class* (London, 1991), 13-14; Alexander Saxton, *The Rise and Fall of the White Republic: Class Politics and Mass Culture in Nineteenth-Century America* (New York,

1990); Eric Lott, " 'The Seeming Counterfeit': Racial Politics and Early Blackface Minstrelsy," *American Quarterly* 43 (June 1991), 223-54; Arnesen, *Waterfront Workers of New Orleans*, 121-31; Herbert Hill, "Myth-Making as Labor History: Herbert Gutman and the United Mine Workers of America," *International Journal of Politics, Culture, Society* 2 (Winter 1988), 132-200; Robert J. Norrell, "Caste in Steel: Jim Crow Careers in Birmingham, Alabama," *Journal of American History* 73 (December 1986), 669-94; Horace Huntley, "Iron Ore Miners and Mine Mill in Alabama, 1933-1952" (Ph.D. diss., University of Pittsburgh, 1976), 110-69; Honey, "Black Workers Remember," 15; Roediger, "Labor in the White Skin," 287-308; Roediger, *Wages of Whiteness*, 43-87, passim.

51. Michael Yarrow, "The Gender-Specific Class Consciousness of Appalachian Coal Miners: Structure and Change," in *Bringing Class Back in: Contemporary and Historical Perspectives*, Scott G. McNall, Rhonda F. Levine, and Rick Fantasia, eds. (Boulder, 1991), 302-3; Trotter, *Coal, Class, and Color*, 109; Paul Willis, *Learning to Labour: How Working Class Kids Get Working Class Jobs* (New York, 1981), 133. See also Paul Willis, "Shop-Floor Culture, Masculinity, and the Wage Form," in *Working-Class Culture*, John Clarke, Charles Critcher, and Richard Johnson, eds. (New York, 1979), 185-98.

52. Scott, *Domination and the Arts of Resistance*, 113. On biracial unions in the South, a subject that deserves greater examination, see Eric Arnesen, "Following the Color Line of Labor: Black Workers and the Labor Movement Before 1930," *Radical History Review* 55 (Winter 1993), 53-87. See also Arnesen, *Waterfront Workers of New Orleans*; and Honey, *Southern Labor and Black Civil Rights*. Numerous white radicals and sympathizers in Alabama were severely beaten (and one lynched) for taking unpopular stands on African American rights. That they crossed the color line was far more important than that they were Communists; Communists in north Alabama, where the party was completely white and included ex-Klan members, faced virtually no violence until they began organizing black sharecroppers in the black belt counties. See Kelley, *Hammer and Hoe*, 47, 67-74, 130-31, 159-75, passim.

53. On early organized struggles against Jim Crow public transportation, see August Meier and Elliott Rudwick, "The Boycott Movement Against Jim Crow Streetcars in the South, 1900-1906," in *Along the Color Line: Explorations in the Black Experience*, August Meier and Elliott Rudwick, eds. (Urbana, 1976), 267-89; and Roger A. Fischer, "A Pioneer Protest: The New Orleans Street-Car Controversy of 1867," *Journal of Negro History* 53 (July 1968), 219-33. A few scholars briefly mention isolated incidents of day-to-day conflict on public transportation in the South. See Pete Daniel, "Going Among Strangers: Southern Reactions to World War II," *Journal of American History* 77 (December 1990), 906; Dittmer, *Black Georgia in the Progressive Era*, 16-19; Janiewski, *Sisterhood Denied*, 141; James A. Burran, "Urban Racial Violence in the South During World War II: A Comparative Overview," in *From the Old South to the New: Essays on the Transitional South*, Walter J. Fraser, Jr., and Winfred B. Moore, Jr., eds. (Westport, 1981), 171; Kuhn, Joye, and West, *Living Atlanta*, 77-82.

54. "Report Involving Race Question," August 1943, p. 1, box 10, Green Papers. For other examples, see "Race Complaints for Last Twelve Months," ibid.; and "Incidents Reported," September 1, 1941-August 31, 1942, ibid.; and James Armstrong interview by Cliff Kuhn, July 16, 1984, p. 9, Working Lives Oral History Collection (University of Alabama, Tuscaloosa). "Race Complaints for Last Twelve Months," September 1, 1941-August 31, 1942, box 10, Green Papers; "Incidents Reported," September 1, 1941-August 31, 1942, ibid.; "Analysis of Complaints and Incidents Concerning Race Problems on Birmingham Electric Company's Transportation System, 12 Months Ending August 31, 1942," ibid.

55. "Reports Involving Race Question," March 1943, pp. 4-5, box 10, Green Papers; "Reports Involving Race Question," November 1943, p. 2, ibid.

56. "Race Complaints for Last Twelve Months," September 1, 1941-August 31, 1942, ibid.; Willi Coleman, "Black Women and Segregated Public Transportation: Ninety Years of Resistance," in *Black Women in United States History*, Darlene Clark Hine, ed. (16 vols., Brooklyn, 1989), vol. 5, 295-301; Mary Ryan, *Women in Public: Between Banners and Ballots, 1825-1880* (Baltimore, 1990), 93-94.

57. "Race Complaints for Last Twelve Months," September 1, 1941-August 31, 1942, p. 1, box, 10, Green Papers; "Reports Involving Race Question," October 1943, p. 4, ibid.; *Birmingham World*, October 29, 1943; "Reports Involving Race Question," March 1943, p. 4, box 10, Green Papers.

58. For employee's complaints about the South Bessemer line, August 2, 1942, and the West End line, June 3, 1942, see "Race Complaints for Last Twelve Months," September 1, 1941-August 31, 1942, box 10, Green Papers. "Reports Involving the Race Question," December 1943, p. 2, ibid.; "Reports Involving the Race Question," September 1943, p. 2, ibid. See also "Reports Involving the Race Question," May 1944, p. 2, ibid.

59. Carol Brooks Gardner, "Analyzing Gender in Public Places: Rethinking Goffman's Vision of Everyday Life," *American Sociologist* 20 (Spring 1989), 42-56.

60. "Reports Involving Race Question," May 1994, pp. 2-3, box 10, Green Papers; "Reports Involving Race Question," August 1943, p. 1, ibid. "Loud-talking," according to linguistic anthropologist Claudia Mitchell-Kernan, is an age-old discursive strategy among African Americans that "by virtue of its volume permits hearers other than the addressee, and is objectionable because of this. Loud-talking requires an audience and can occur only in a situation where there are potential hearers other than the interlocutors." Moreover, loud-talking assumes "an antagonistic posture toward the addressee." Claudia Mitchell-Kerman, "Signifying, Loud-talking, and Marking," in *Rappin' and Stylin' Out*, Kochman, ed., 329, 331.

61. "Incidents Reported," September 1, 1941-August 31, 1942, pp. 4-6, box 10, Green Papers. On public cursing as a powerful symbolic act of resistance, see Scott, *Domination and the Arts of Resistance*, 215.

62. "Reports Involving Race Question," October 1943, p. 3, box 10, Green Papers.

63. "Reports Involving Race Question," March 1943, p. 1, ibid.; "Reports Involving Race Question," October 1943, p. 2, ibid.

64. *Birmingham World*, October 29, 1943; "Southern Negro Youth Congress—Forum," February 6, 1984, untranscribed tape, Oral History of the American Left (Tamiment Library, New York University, New York, New York); FBI Report, "Southern Negro Youth Congress, Birmingham, Alabama," June 14, 1943, p. 5. Headquarters File 100-82, Southern Negro Youth Congress Files (J. Edgar Hoover FBI Building, Washington, D.C.); James Jackson, "For Common Courtesy on Common Carriers," *Worker*, June 4, 1963.

65. N. H. Hawkins, Jr., Birmingham Electric Company Transportation Department, n.d., box 10, Green Papers.

66. Aptheker, *American Negro Slave Revolts*, 18-78, 140-61, passim. For Stuart Hall's statement, see George Lipsitz, "The Struggle for Hegemony," *Journal of American History* 75 (June 1988), 148. On the dialectic between southern black working-class self-activity and the response of employers, bureaucrats, social reformers, and the state, see Hunter, "Household Workers in the Making," 187-291, passim.

67. The literature on the late nineteenth and twentieth centuries alone is extensive. The best overview is Shapiro, *White Violence and Black Response*.

68. Kelley, "Black Poor and the Points of Opposition," 309, 311-12, 318-23, 327-29, 331-33; Kuhn, Joye, and West, *Living Atlanta*, 337-41; Wright, *Life Behind a Veil*, 254-57.

69. Du Bois, *Black Reconstruction;* Michael Schwartz, *Radical Protest and Social Structure: The Southern Tenant Farmers' Alliance and Cotton Tenancy, 1880-1890* (Chicago, 1976); Rick Fantasia, *Culture of Solidarity: Consciousness, Action, and Contemporary American Workers* (Berkeley, 1988); Lipsitz, *Rainbow at Midnight*.

70. Brown, " 'What Has Happened Here,' " 295-312.

Black Migration to the Urban Midwest

The Gender Dimension, 1915-1945

Darlene Clark Hine

The significance of temporal and spatial movement to a people, defined by and oppressed because of the color of their skin, among other things, defies exaggeration. Commencing with forced journeys from the interior of Africa to the waiting ships on the coast, over 11 million Africans began the trek to New World slave plantations that would, centuries later, land their descendants at the gates of the so-called promised lands of New York, Philadelphia, Chicago, Cleveland, Detroit, Milwaukee, and Indianapolis. The opening page of a privately published memoir of a black woman resident of Anderson, Indiana, captures well this sense of ceaseless movement on the part of her ancestors. D. J. Steans observed that "the backward trail of relatives spread from

Author's Note: This chapter previously appeared in *The Great Migration in Historical Perspective: New Dimensions of Race, Class, and Gender*, edited by Joe W. Trotter, © 1991, Indiana University Press, Bloomington, Indiana. Reprinted by permission.

Indiana to Mississippi, crisscrossing diagonally through several adjoining states. Whether the descendants came ashore directly from Africa to South Carolina or were detoured by way of islands off the coast of Florida is unknown."[1]

For half a millennium, black people in the New World have been, or so it seems, in continuous motion, much of it forced, some of it voluntary and self-propelled. Determined to end their tenure in the "peculiar institution," or die trying, thousands of blacks fled slavery during the antebellum decades, as the legendary exploits of Harriet Tubman and Frederick Douglass testify. Large numbers of blacks challenged, with their feet, the boundaries of freedom in the aftermath of the Civil War. Many moved west to establish new black towns and settlements in Kansas and Oklahoma in the closing decades of the nineteenth century. Others attempted to return to Africa.[2] To understand both the processes of black migration and the motivations of the individuals, men and women, who comprised this human tide is to approach a more illuminating portrait of American history and society. Central to all of this black movement was the compelling quest for that ever so elusive, but distinctly American, property: freedom and equality of opportunity.

Long a riveting topic, studies of the Great Migration abound. Indeed, recent histories of black urbanization, especially those focused on key midwestern cities and towns—Chicago, Cleveland, Detroit, Milwaukee, and Evansville, Indiana—pay considerable attention to the demographic transformation of the black population, a transformation that began in earnest in 1915 and continued through the World War II crisis.[3] As enlightening and pathbreaking as most of these studies are, there remains an egregious void concerning the experiences of black women migrants. This brief essay is primarily concerned with the gender dimension of black migration to the urban Midwest. It raises, without providing a comprehensive answer, the question, how is our understanding of black migration and urbanization refined by focusing on the experiences (similar to men in many ways, yet often unique) of the thousands of southern black women who migrated to the Midwest between the two world wars? A corollary question concerns the nature of the relations between those black women who migrated out of the lower Mississippi Valley states and those who stayed put. It is also important to shed light on the phenomenon of intraregional migration for there was considerable movement of women between the midwestern cities and towns.

By 1920, almost 40 percent of Afro-Americans residing in the North were concentrated in eight cities, five of them in the Midwest: Chicago, Detroit, Cleveland, Cincinnati, and Columbus, Ohio. The three eastern cities with high percentages of black citizens were New York, Philadelphia, and Pittsburgh. These eight cities contained only 20 percent of the total northern population. Two peaks characterized the first phase of the Great Black Migration: 1916-1919 and 1924-1925. These dates correspond to the passage of more stringent anti-immigration laws, and the years in which the majority of the approximately 500,000 southern blacks relocated northward.[4]

Clearly, the diverse economic opportunities in the midwestern cities served as the major "pull" factor in the dramatic black percentage increases registered between 1910 and 1920. Detroit's black population rose an astounding 611.3 percent. More precisely put, Detroit's Afro-Americans, attracted by the jobs available at the Ford, Dodge, Chrysler, Chevrolet, and Packard automobile plants, increased from 5,741 in 1910 to 120,066 in 1930. Home of the northern terminus of the Illinois Central Railroad, the *Chicago Defender*, meat packing, and mail order enterprises, Chicago was not outdone. The Windy City's black population, which in 1910 numbered 44,103, jumped to 233,903 in 1930.[5]

Drawn to midwestern jobs, throughout the World War I era, the numbers of black males far exceeded female migrants. Thus, the black population of midwestern cities, unlike in most eastern and southern cities, did not reflect a majority of females. Until the differences between the processes and consequences of migration to black men and to black women are fully researched, no comprehensive synthesis or portrait of the migrants is possible. Many unexplored dimensions of black migration and urbanization still beg attention. In addition to the good studies of political and cultural developments in Chicago, for example, more work needs to be done on the relation between migration and the development of black social, political, economic, and religious institutions in other midwestern cities. Although historians Peter Gottlieb and James Borchert have stressed the continuity between black life in the South and in northern cities, we do not yet understand fully the mechanisms by which this continuity was achieved, or its meaning.[6] It is perhaps fair to suggest that black women played a critical role in the establishment of an array of black institutions, especially the churches and mutual aid organizations that gave life in northern cities a southern flavor. Historian Elizabeth Clark-Lewis, although addressing Washington, D.C., directly, argues, "The growth of African-American churches in Washington, then, was a direct consequence of the steady influx of these working-class (former live-in) women. They strongly supported church expansion because their participation in the church activities further separated them from the stigma of servitude."[7]

Still in need of refinement is our understanding of the connection between migration and black social-class formation and between migration and the rise of protest ideologies that shaped the consciousness of the "New Negro," not only in Harlem but also in midwestern cities. We need studies of the relation between migration and family reorganization, and between migration and sex-role differentiation in the black communities, especially in terms of religious activity; and of the development of new types of community-based social welfare programs. Moreover, we need microstudies into individual lives, of neighborhoods, families, churches, and fraternal lodges in various cities. Examination of these themes makes imperative an even deeper penetration into the internal world of Afro-Americans. Perhaps even more dauntingly, to answer fully these questions requires that the black woman's voice and experience be researched and interpreted with the same intensity and seriousness accorded that of the black man.

Information derived from statistical and demographic data on black midwestern migration and urbanization must be combined with the knowledge drawn from the small, but growing, numbers of oral histories, autobiographies, and biographies of twentieth-century migrating women. Court records of legal encounters, church histories, black women's club minutes, scrapbooks, photographs, diaries, and histories of institutions ranging from old folks' homes, orphanages, businesses, and Phillis Wheatley Homes to local YWCAs yield considerable information on the lives of black women migrants to and within the middle western region. Actually, these sources, properly "squeezed and teased," promise to light up that inner world so long shrouded behind a veil of neglect, silence, and stereotype and will quite likely force a rethinking and rewriting of all of black urban history.

A perusal of the major studies of black urbanization reveals considerable scholarly consensus on several gender-related themes. Scholars generally acknowledge that gender did make a difference in terms of the reasons expressed for quitting the South and affected the means by which men and women arrived at their northern destinations. Likewise, scholars concur that men and women encountered radically divergent socioeconomic and political opportunities in midwestern cities. Gender and race stereotyping in jobs proved quite beyond their control and was intransigent in the face of protest. Scholars agree that black women faced greater economic discrimination and had fewer employment opportunities than black men. Their work was the most undesirable and least remunerative of all northern migrants. Considering that their economic condition or status scarcely improved or changed, for many women migrants were doomed to work in the same kinds of domestic service jobs they had held in the South, one wonders why they bothered to move in the first place. Of course, there were significant differences. A maid earning $7 a week in Cleveland perceived herself to be much better off than a counterpart receiving $2.50 a week in Mobile, Alabama. A factory worker, though the work was dirty and low status, could and did imagine herself better off than domestic servants who endured the unrelenting scrutiny, interference, and complaints of household mistresses.[8]

It is clear that more attention needs to be directed toward the noneconomic motives propelling black female migration. Many black women quit the South out of a desire to achieve personal autonomy and to escape from sexual exploitation both within and outside of their families and from sexual abuse at the hands of southern white as well as black men. The combined influence of domestic violence and economic oppression is key to understanding the hidden motivation informing major social protest and migratory movements in Afro-American history.[9]

That black women were very much concerned with negative images of their sexuality is graphically and most forcefully echoed in numerous speeches of the early leaders of the national organization of black women's clubs. Rosetta Sprague, the daughter of Frederick Douglass, declared in an address to the Federation of Afro-American Women in 1896:

> We are weary of the false impressions sent broadcast over the land about the colored woman's inferiority, her lack of noble womanhood. We wish to make it clear in the minds of your fellow country men and women that there are no essential elements of character that they deem worthy of cultivating that we do not desire to emulate that the sterling qualities of purity, virtue, benevolence and charity are no more dormant in the breast of the black woman than in the white woman.[10]

Sociologist Lynda F. Dickson cautions that "recognition of the major problem—the need to elevate the image of black womanhood —may or may not have led to a large scale club movement both nationally and locally."[11] It cannot be denied, however, that "the most important function of the club affiliation was to provide a support system that could continually reinforce the belief that the task at hand—uplifting the race, and improving the image of black womanhood was possible."[12] A study of the history of the early-twentieth-century black women's club movement is essential to the understanding of black women's migration to the middle-western towns and cities and the critical roles they played in creating and sustaining new black social, religious, political, and economic institutions. These clubs were as important as the National Urban League and the NAACP in transforming black peasants into the urban proletariat.[13]

This focus on the sexual and the personal impetus for black women's migration neither dismisses nor diminishes the importance of economic motives, a discussion of which I will return to later. Rather, I am persuaded by historian Lawrence Levine's reservations.

He cautions, "As indisputably important as the economic motive was, it is possible to overstress it so that the black migration is converted into an inexorable force and Negroes are seen once again not as actors capable of affecting at least some part of their destinies, but primarily as beings who are acted upon—southern leaves blown North by the winds of destitution."[14] It is reasonable to assume that many were indeed "southern leaves blown North" and that others were more likely self-propelled actors seeking respect, space in which to live, and a means to earn an adequate living.

Black men and women migrated into the Midwest in distinctive patterns. Single men, for example, usually worked their way north, leaving farms for southern cities, doing odd jobs, and sometimes staying in one location for a few years before proceeding to the next stop. This pattern has been dubbed "secondary migration." Single black women, on the other hand, as a rule, traveled the entire distance in one trip. They usually had a specific relative—or fictive kin—waiting for them at their destination, someone who may have advanced them the fare and who assisted with temporary lodging and advice on securing a job.[15] Amanda Jones-Watson, a fifty-year-old resident of Grand Rapids, Michigan, and three-time president of the still-functioning Grand Rapids Study Club, founded in 1904, migrated from Tennessee in 1936 in her thirties. She recalls asking her uncle, who had just moved to Grand Rapids, to send her a ticket. She exclaimed, "I cried when it came. I was kidding. My sister said, 'Amanda, what are you worried about? You can always come back if you don't like it.' " Jones-Watson was fortunate. Her uncle was headwaiter at the Pantlind Hotel. She continued, "I got a job as a maid and was written up in a local furniture magazine for making the best bed at the Pantlind."[16]

For Sara Brooks, a domestic, the idea that she should leave Alabama and relocate in Cleveland in 1940 originated with her brother. He implored his sister, "Why don't you come up here? You could make more here." Brooks demurred, "Well, I hadn't heard anything about the North because I never known nobody to come no further than Birmingham, Alabama, and that was my sister-in-law June, my husband's sister." A single mother of three sons and a daughter,

Brooks eventually yielded to her brother's entreaties, leaving her sons with her aging parents. She recalled, "But my brother wanted me to come up here to Cleveland with him, so I started to try to save up what little money I had. . . . But I saved what I could, and when my sister-in-law came down for me, I had only eighteen dollars to my name, and that was maybe a few dollars over enough to come up here. If I'm not mistaken it was about a dollar and fifteen cent over."[17]

The influence and pressure of family members played a substantial role in convincing many ambivalent young women to migrate. A not-so-young sixty-eight-year-old Melinda left her home in Depression-ridden rural Alabama to assist her granddaughter in child rearing in Anderson, Indiana. Even when expressing her plans to return home once her granddaughter was up and about, somehow Melinda knew that the visit would be permanent. Grounded largely in family folklore, D. J. Steans declared that Melinda had labored hard at sharecropping, besides taking in washing and ironing. Even after her sixty-second birthday, she was still going strong. Many weeks she earned less than fifty cents, but she was saving pennies a day for her one desire to travel north to visit her great-grandchildren.[18]

Some women simply seized the opportunity to accompany friends traveling north. Fired from her nursing job at Hampton Institute, in Hampton, Virginia, Jane Edna Hunter packed her bags determined to head for Florida. She never made it. According to Hunter, "En route, I stopped at Richmond, Virginia, to visit with Mr. and Mrs. William Coleman, friends of Uncle Parris. They were at church when I arrived; so I sat on the doorstep to await their return. After these good friends had greeted me, Mrs. Coleman said, 'Our bags are packed to go to Cleveland, Jane. We are going to take you with us.' " Jane needed little persuasion. She exclaimed, "I was swept off my feet by the cheerful determination of the Colemans. My trunk, not yet removed from the station, was rechecked to Cleveland."[19] Hunter arrived in the city on May 10, 1905, with $1.75 in her pockets, slightly more than Sara Brooks brought with her thirty-five years later.

The different migratory patterns of black males and females reflect gender conventions in the larger society. A woman traveling alone was surely at greater risk than a man. After all, a man could and did, with less approbation and threat of bodily harm, spend nights outdoors. More important, men were better suited to defend themselves against attackers. However, given the low esteem in which the general society held black women, even the courts and law officials would have ridiculed and dismissed assault complaints from a black female traveling alone, regardless of her social status. Yes, it was wise to make the trip all at once, and better still to have company.

Although greater emphasis has been placed on men who left families behind, black women, many of whom were divorced, separated, or widowed, too left loved ones, usually children, in the South when they migrated. Like married men, unattached or single black mothers sent for their families after periods of time ranging from a month to even several years. Actually, I suspect that a great number of women who migrated into the Midwest probably left children, the products of early marriages or romantic teenage liaisons, with parents, friends, and other relatives in the South. It would be exceedingly difficult, if not impossible, to develop any statistical information on this phenomenon. Nevertheless, the oral history of Elizabeth Burch of Fort Wayne, Indiana, offers poignant testimony of a child left behind:

> I was born [December 20, 1926 in Chester, Georgia] out of wedlock to Arlena Burch and John Halt. My mother went north and that's where they—all of it began in a little town called Albion, Michigan and she went back south to have me. . . . Aunt Clyde, that's my mother's sister, she was the baby and that was a little town called Albion, Michigan. That's where I was conceived at. That's where my mother went when she left Georgia. My mother decided well she go back up north. She married just to get away from home to go back north and this guy was working as a sharecropper and he had made enough money that year that he was willing to marry my mother and take her back up north. . . . So they left me with Miss Burch—Miss Mattie Elizabeth Burch, namesake which was my grandmother and that's where I grew up at and years passed and years went through I was just on the farm with my grandparents.[20]

The difficulty of putting aside enough money to send for their children placed a tremendous strain on many a domestic salary. It took Sara Brooks almost fifteen years to reconstitute her family, to retrieve her three sons left behind in Orchard, Alabama. With obvious pride in her accomplishment, Brooks explained, "The first one to come was Jerome. . . . Then Miles had to come because my father didn't wanna keep him down there no more because he wouldn't mind him. . . . Then Benjamin was the last to come." Brooks summed up her success, "So I come up to Cleveland with Vivian [her daughter], and after I came up, the rest of my kids came up here. I was glad—I was *very* glad because I had wanted 'em with me all the time, but I just wasn't able to support 'em, and then I didn't have no place for them, either, when I left and come to Cleveland cause I came here to my brother."[21]

Arguably, inasmuch as so many midwestern black women were absentee mothers—that is, their children remained in the South—their actual acculturation into an urban lifestyle became a long, drawn out, and often incomplete, process. On the other hand, as historians Peter Gottlieb and Jacqueline Jones persuasively maintain, black women served as critical links in the "migration chain."[22] They proved most instrumental in convincing family members and friends to move north. This concept of women as "links in a migration chain" begs elaboration. I suspect that it is precisely because women left children behind in the care of parents and other relatives that they contributed so much to the endurance and tenacity of the migration chain. Their attachment to the South was more than sentimental or cultural. They had left part of themselves behind.

Parental obligations encouraged many black women migrants to return south for periodic visits. Burch recalled that "my mother would come maybe once a year—maybe Christmas to visit" her in Georgia from Fort Wayne.[23] Still other midwestern women returned perhaps to participate in community celebrations and family reunions and to attend religious revivals. Of course, such periodic excursions southward also permitted display of new clothes and other accoutrements of success. Before she made the journey to Cleveland, Sara Brooks admitted delight in her sister-in-law's return visits. "I noticed she had some nice-lookin' little clothes when she come back to Orchard to visit. She had little nice dresses and brassieres and things, which I didn't have. . . . I didn't even have a brassiere, and she'd lend me hers and I'd wear it to church."[24] Indeed, Brooks's recollections raise a complex question—to what extent and how does the woman's relation to the South change over the course of the migrant's life? When do migrants move from being southerners in the North to southern northerners?

Unable or unwilling to sever ties to or abandon irrevocably the South, black women's assimilation to urban life remained fragmented and incomplete. It was the very incompleteness of the assimilation, however, that facilitated the "southernization" of the Midwest. Vestiges of southern black culture were transplanted and continuously renewed and reinforced by these women in motion. The resiliency of this cultural transference is reflected in food preferences and preparation styles, reliance on folk remedies and superstitions, religious practices, speech patterns, games, family structures and social networks, and music, most notably, the blues.[25] The southernization of urban midwestern culture was but one likely consequence of the migration chain women forged. In short, although unattached black women migrants may have traveled the initial distance to Chicago, Cleveland, Detroit, or Cincinnati, in one trip, as long as offspring, relatives, and friends remained in the South, psychological and emotional relocation was much more convoluted and, perhaps, more complicated than heretofore assumed.

Discussions of marital status and family obligations—specifically, whether the women migrants had children remaining in the South—are indirectly, perhaps, related to a more controversial topic of current interest to historians of nineteenth-century black migration and urbanization. In his study of violence and crime in post-Civil War Philadelphia, Roger Lane suggests that there was a marked decline in black birthrates in the city near the turn of the century. He attributes the decline in part to the rising incidence of syphilis, which left many black women infertile. He notes that "in Philadelphia in 1890 the black-white ratio was .815 to 1,000, meaning that black women had nearly 20 percent fewer children than whites, a figure that in 1900 dropped to .716 to 1,000, or nearly 30 percent fewer." Lane concludes, "All told, perhaps a quarter of Philadelphia's black women who reached the end of their childbearing years had at some time had exposure to the diseases and habits associated with prostitution. This figure would account almost precisely for the difference between black and white fertility in the city."[26]

Without reliance on the kinds of statistical data Lane employs in his analysis, the oral histories and autobiographies of midwestern black women migrants suggest an alternate explanation, though often overlooked in discussions of black birth decline. Sara Brooks, mother of five children, was still in her childbearing years when she embraced celibacy. She declared, "See, after Vivian was born I didn't have no boyfriend or nothin', and I went to Mobile, I didn't still have no boyfriend in a long time. Vivian was nine years old when Eric come. . . . But after Eric came along, I didn't have no boyfriend. I didn't want one because what I wanted, I worked for it, and that was that home." Brooks had realized her dream in 1957 with the purchase of her home and the reuniting of all of her children under one roof.[27]

For women, ignorant of effective birth control or unable to afford the cost of raising additional children alone, sexual abstinence was a rational choice. It should be pointed out that often deeply held religious convictions, disillusionment with black men, a history of unhappy and abusive marriages, adherence to Victorian ideals of morality, a desire to refute prevalent sexual stereotypes and negative images of black women as a whole, or even an earlier unplanned pregnancy may have informed many a decision to practice sexual abstinence among adult black women. Only latent acceptance of the myths concerning the alleged unbridled passions and animalistic sexuality of black women prevent serious consideration of the reality and extent of self-determined celibacy. Meanwhile, until we know more about the internal lives of black women, the suggestion of abstinence or celibacy as a factor limiting births should not be dismissed.

The fact that women who migrated north produced fewer children than their southern counterparts warrants further investigation. It is not enough to argue that prostitution, venereal disease, and infanticide account for declining black births in urban settings. Many other factors, in addition to abstinence, offer fruitful and suggestive lines of inquiry. Some scholars have asserted that children in urban as opposed to rural settings had rather insignificant economic roles and therefore their labor was not as important to family survival.[28]

As black women became more economically sufficient, better educated, and more involved in self-improvement efforts, including participation in the flourishing black women's club movement, they would have had more access to birth control information.[29] As the institutional infrastructure of black women's clubs, sororities, church groups, and charity organizations took hold within black communities, they gave rise to those values and attitudes traditionally associated with the middling classes. To black middle-class aspirants, the social stigma of having many children would have, perhaps, inhibited reproduction. Furthermore, over time, the gradually evolving demographic imbalance in the sex ratio meant that increasing numbers of black women in urban midwestern communities would never marry. The point is simply this, that not dating, marrying, or having children may very well have been a decision—a deliberate choice, for whatever reason—that black women made. On August 23, 1921, Sarah D. Tyree wrote tellingly about her own decision not to date. Tyree had a certificate from the Illinois College of Chiropracty, but was, at the time, taking care of aged parents and her sister's children in Indianapolis. She confided to her sister living in Muskegon, Michigan:

> I have learned to stay at home lots. I firmly believe in a womanly independence. Believe that a woman should be allowed to go and come where and when she pleases alone if she wants to, and so long as she knows who is right, she should not have to worry about what others think. It is not every woman who can turn for herself as I can, and the majority of women who have learned early to depend upon their male factors do not believe that their sister-woman can get on alone. So she becomes dangerously suspicious, and damagingly tongue-wagging. I have become conscious of the fact that because I am not married, I am watched with much interest. So I try to avoid the appearance of evil, for the sake of the weaker fellow. I do not therefore go out unaccompanied at night. There are some young men I would like to go out with occasionally if it could be understood that it was for the occasion and not for life that we go. I don't care to be bothered at any time with a fellow who has been so cheap and all to himself for 5 or 6 years I have all patients [*sic*] to wait for the proper one to play for my hand.[30]

Moreover, social scientists Joseph A. McFalls, Jr. and George S. Masnick persuasively argue that blacks were much more involved in birth control than previously assumed. They contend, "The three propositions usually advanced to support the view that birth control had little, if any, effect on black fertility from 1880 to 1940—that blacks used 'ineffective' methods, that blacks did not practice birth control 'effectively,' and that blacks used birth control too late in their reproductive careers to have had much of an effect on their fertility—simply have no empirical or even a priori foundation. There is no reason now to believe that birth control had little impact on black fertility during this period."[31] Not to be overlooked are the often chronic health problems overworked, undernourished, and inadequately housed poor black women undoubtedly experienced, especially during the Depression. In discussing the morbidity and mortality rates of blacks in Chicago, Tuttle observes that

> Chicago's medical authorities boasted of the city's low death rate, pointing to statistics which indicated that it was the lowest of any city in the world with a population of over one million. Their statistics told another story as well, however, and it was that Chicago's blacks had a death rate which was twice that of whites. The stillbirth rate was also twice as high; and the death rate from tuberculosis and syphilis was six times as high; and from pneumonia and nephritis it was well over three times as high. . . . The death rate for the entire city was indeed commendable, but the statistics indicated that the death rate for Chicago's blacks was comparable to that of Bombay, India.[32]

One more observation about the declining birth rate among northern black women should be made. Here it is important to note the dichotomy between black women who worked in middle- and working-class occupations. Middle-class working women, regardless of color, had fewer children than those employed in blue-collar jobs. The professional and semiprofessional occupations most accessible to black women during the years between the world wars included teaching, nursing, and social work, on the one hand, and hairdressing or dressmaking, on the other. In some of the smaller midwestern communities and towns, married women teachers, race notwithstanding, lost their jobs, especially if the marriage became public knowledge or the wife pregnant. At least one black woman schoolteacher in Lafayette, Indiana, confided that she never married, though she had been asked, because in the 1930s and 1940s to have done so would have cost her the position. The pressure on the small cadre of professional black women not to have children was considerable. The more educated they were, the greater the sense of being responsible, somehow, for the advance of the race and of black womanhood. They held these expectations of themselves and found them reinforced by the demands of the black community and its institutions. Under conditions and pressures such as these, it would be erroneous to argue that this is the same thing as voluntary celibacy. Nevertheless, the autonomy, so hard earned and enjoyed to varying degrees by both professional women and personal service workers, offered meaningful alternatives to the uncertainties of marriage and the demands of child rearing. The very economic diversity —whether real or imagined—that had attracted black women to the urban Midwest also held the promise of freedom to fashion socially useful and independent lives beyond family boundaries.

None of this is to be taken as a categorical denial of the existence of rampant prostitution and other criminal activity in urban midwestern ghettos. Too many autobiographies and other testimony document the place and the economic functions of prostitution in urban society to be denied. Indeed, Jane Edna Hunter's major contribution to improving black women's lives in Cleveland—the establishment of the Phillis Wheatley boarding homes—stemmed from her commitment to provide training, refuge, and employment for young migrating women who were frequently enticed or tricked into prostitution as a means of survival. She remarked on her own awakening, "The few months on Central Avenue made me sharply aware of the great temptations that beset a young woman in a large city. At home on the plantation, I knew that some girls had been seduced. The families had felt the disgrace keenly—the fallen ones had been wept and prayed over. . . . Until my arrival in Cleveland I was ignorant of the wholesale organized traffic in black flesh."[33]

Young, naive country girls were not the only ones vulnerable to the lure of seduction and prostitution. Middle-aged black women also engaged in sex for pay, but for them it was a rational economic decision. Sara Brooks did not disguise her contempt for women who bartered their bodies. She declared, while commenting on her own struggle to pay the mortgage on her house, "Some women would'a had a man to come and live in the house and had an outside boyfriend too, in order to get the house paid for and the bills." She scornfully added, "They meet a man and if he promises 'em four or five dollars to go to bed, they's grab it. That's called sellin' your own body, and I wasn't raised like that."[34]

Prostitution was not the only danger awaiting single migrating black women. Police in many midwestern towns seemed quick to investigate not only black men but also black women who appeared suspicious. Historian James E. DeVries records several encounters between black women and the police in Monroe, Michigan. "In January 1903, Gertie Hall was arrested after acting in a very nervous manner on the inter-urban trip from Toledo. An investigation by Monroe police revealed that Hall was wanted for larceny in Toledo, and she was soon escorted to that city." In another incident four years later involving fifteen-year-old Ahora Ward, also from Toledo, DeVries notes that she was "picked up and taken to jail. . . . As it turned out, she had been whipped by her mother and was running away from home when taken into custody."[35]

There exists a scholarly consensus about the origins and the destinations of the overwhelming majority of black migrants throughout the period between 1915 and 1945. Before turning to a discussion of the economic impetus, or the pull factors, for black women's migrations, I would like to interject another rarely explored "push" factor, that is, the desire for freedom from sexual exploitation, especially rape by white men, and to escape from domestic abuse within their own families. A full exploration of this theme requires the use of a plethora of sources including oral testimonials, autobiographies, biographies, novels, and court records. The letters, diaries, and oral histories collected by the Black Women in the Middle West Project and deposited in the Indiana Historical Society contain descriptions of domestic violence that fed the intraregioral movement of black women who had migrated from southern states. Elizabeth Burch explained why she left Fort Wayne for Detroit: "And my mother—

and my stepfather—would have problems. He would hit my mother and so, you know, beat upon my mother but he never did beat up on me. My mother would say—'Well you just don't put your hands on her. You better not, hear.' " To avoid these scenes, Burch moved to Detroit but later returned to Fort Wayne.[36]

Similarly, Jane Pauline Fowlkes, sister of the above-mentioned Sarah Tyree, was granted a divorce from her husband, Jess Clay Fowlkes, in Muskegon, Michigan, in 1923 and returned to her family and sister in Indianapolis. Granted the decree because her husband was found "guilty of several acts of extreme cruelty," Fowlkes retained custody of all three children.[37]

Although Sara Brooks's experiences are hardly representative, they are nevertheless suggestive of the internal and personal reasons black women may have had for leaving the South. Brooks vividly described the events that led her to leave her husband for the third and final time. When she ran away from home the last time, she did not stop running until she reached Cleveland almost a decade later. "When he hit me," she said, "I jumped outa the bed, and when I jumped outa the bed, I just ran. . . . I didn't have a gown to put on—I had on a slip and had on a short-sleeved sweater. I left the kids right there with him and I went all the way to his father's house that night, barefeeted, with that on, on the twenty-fifth day of December. That was in the dark. It was two miles or more and it was rainin'. . . . I walked, and I didn't go back."[38]

For whatever reasons Sara Brooks, Melinda, Jane Edna Hunter, and others wound up in the various midwestern cities, they expected to work and to work hard, for work was part of the definition of what it meant to be a black woman in America, regardless of region. The abundant economic opportunities, or pull factors, especially in automobile plants and, during the war, in the defense industries, had been powerful inducements for black male migrants. The dislocation of blacks in southern agriculture, the ravages of the boll weevil, floods, and the seasonal and marginal nature of the work relegated to them in the South were powerful push factors. Taken together, these factors help us to understand why 5 percent of the total southern black population left the South between 1916 and 1921.[39]

Black women shared with black men a desire for economic improvement and security. They too were attracted to midwestern cities, specifically those with a greater diversity of women's jobs. The female occupational structure of Chicago, for example, held the promise of more opportunity for black women than did the much more heavy-industry-dependent Pittsburgh.[40] Black men, however, were not as constrained. To be sure, the majority of neither group expected to secure white-collar jobs or managerial positions. None were so naive as to believe that genuine equality of opportunity actually existed in the North or the Midwest, but occasionally black women migrants did anticipate that more awaited them in Cleveland and Chicago than an apron and domestic servitude in the kitchens of white families, segregated hotels, and restaurants. Most were disappointed. Author Mary Helen Washington recalled the disappointment and frustration experienced by her female relatives when they migrated to Cleveland:

> In the 1920s my mother and five aunts migrated to Cleveland, Ohio from Indianapolis and, in spite of their many talents, they found every door except the kitchen door closed to them. My youngest aunt was trained as a bookkeeper and was so good at her work that her white employer at Guardian Savings of Indianapolis allowed her to work at the branch in a black area. The Cleveland Trust Company was not so liberal, however, so in Cleveland she went to work in what is known in the black community as private family.[41]

Scholars concur that although black women secured employment in low-level jobs in light industry, especially during the World War I years when overseas immigration came to a standstill, this window of opportunity quickly closed with the end of hostilities. Florette Henri calculates that "immigration dropped from 1,218,480 in 1914, to 326,700 in 1915, to under 300,000 in 1916 and 1917, and finally to 110,618 in 1918." This drop and the draft made it possible for black women to squeeze into "occupations not heretofore considered within the range of their possible activities," concluded a Department of Labor survey in 1918. Thus, the percentage of black domestics declined between 1910 and 1920, from 78.4 percent to 63.8 percent in Chicago, and from 81.1 percent to 77.8 percent in Cleveland.[42]

The study of migrations from the perspective of black women permits a close examination of the intersection of gender, class, and race dynamics in the development of a stratified workforce in midwestern cities. During the war years, a greater number of black women migrants found work in midwestern hotels as cooks, waitresses, and maids, as ironers in the new steam laundries, as labelers and stampers in Sears Roebuck and Montgomery Wards mail order houses, as common laborers in garment and lampshade factories, and in food processing and meat packing plants. But even in these places, the limited gains were short lived and easily erased. As soon as the war ended and business leveled off, for example, both Sears and Wards immediately fired all the black women.[43] In 1900, black women constituted 4 percent of the labor force in commercial laundries; by 1920 this figure had climbed to 6 percent. As late as 1930, a little over 3,000 black women, or 15 percent of the black female labor force in Chicago, were unskilled and semiskilled factory operatives. Thus, over 80 percent of all employed black women continued to work as personal servants and domestics. Historian Allan H. Spear points out that "negro women were particularly limited in their search for desirable positions. Clerical work was practically closed to them and only a few could qualify as school teachers. Negro domestics often received less than white women for the same work and they could rarely rise to the position of head servant in large households."[44]

In Milwaukee, especially during the Depression decades, black women were, as historian Joe W. Trotter observes, "basically excluded from this narrow industrial footing; 60.4 percent of their numbers labored in domestic service as compared to only 18.6 percent of all females."[45] To be sure, this was down from the 73.0 percent of black women who had worked as domestics in Milwaukee in 1900.[46] A decline of 13 percent over a forty-year period—regardless of from what angle it is viewed—is hardly cause for celebration.

Many reasons account for the limited economic gains of black women as compared to black men in midwestern industries. One of the major barriers impeding a better economic showing was the hostility and racism of white women. The ceiling on black women's job opportunities was secured tight by the opposition of white women. White females objected to sharing the settings, including hospitals, schools, department stores, and offices. Now 90, Sarah Glover migrated with her family to Grand Rapids, Michigan, from Alabama in 1922, where they had jobs working in the coal mines. Although she would in later years become the first practical nurse in the city, during her first seventeen years as a maid at Blodgett Hospital, she scrubbed the floors. After completing her housekeeping chores she would voluntarily help the nurses. She reminisced, "The nurses used to call me 'Miss Sunshine' because I would cheer up the patients. I'd come over and say you look good today or crack a joke. That used to get most of them smiling again." In spite of her good work record, excellent human relations skills, and eagerness, hospital officials deemed it a violation of racial rules and thus rejected Glover's appeal to become a nurse's aid.[47]

Historians Susan M. Hartmann and Karen Tucker Anderson convincingly demonstrate that although white women enjoyed expanded employment opportunities, black women continued to be the last hired and first fired throughout the Depression and World War II years. Employers seeking to avert threatened walkouts, slowdowns, and violence caved in to white women's objections to working beside or, most particularly, sharing restroom and toilet facilities with black women.[48] To be sure, many employers, as was the case with the Blodgett Hospital in Grand Rapids, harbored the same racist assumptions and beliefs in black inferiority, but camouflaged them behind white women's objections.

The black media was not easily fooled by racist subterfuges and remained keenly attuned to all excuses that rationalized the denial of job opportunities to black women. In its official organ, *Opportunity*, National Urban League officials catalogued the thinly veiled justifications white employers offered when discriminating against women:

> "There must be some mistake"; "No applications have heretofore been made by colored"; "You are smart for taking the courses, but we do not employ coloreds"; "We have not yet installed separate but equal toilet facilities"; "A sufficient number of colored women have not been

> trained to start a separate shift"; "The training center from which you come does not satisfy plant requirements"; "Your qualifications are too high for the kind of job offered"; "We cannot put a Negro in our front office"; "We will write you but my wife needs a maid"; "We have our percentage of Negroes."[49]

Trotter did, however, discover instances when the interests of white women occasionally promoted industrial opportunities for black women. "The white women of the United Steelworkers of America Local 1527, at the Chain Belt Company, resisted the firm's proposal for a ten-hour day and a six-day week by encouraging the employment of black women."[50] In a classic understatement, historian William Harris hesitantly asserts, "Black women apparently experienced more discrimination than black men in breaking into nonservice jobs."[51]

In their study of labor unions in Detroit, August Meier and Elliott Rudwick reveal that more than white women's hostility accounts for the employment discrimination and the job segregation black women encountered in the automobile industries. Throughout the World War II era, the Ford Motor Company hired only a token number of black women. According to Meier and Rudwick, "Black civic leaders and trade unionists fought a sustained and energetic battle to open Detroit war production to black women, but because government manpower officials gave discrimination against Negro females low priority, the gains were negligible when compared with those achieved by the city's black male workers." By March 1943, for example, the Willow Run Ford Plant employed 25,000 women, but less than 200 were black. Apparently, Ford was not alone or atypical in these anti-black-women hiring practices. Both Packard and Hudson employed a mere half dozen each at this time. Most of those employed in the plants, as was to be expected, worked in various service capacities—matrons, janitors, and stock handlers. Meier and Rudwick point out that "as late as the summer of 1943 a government report termed the pool of 25,000 available black women the city's 'largest neglected source of labor.' "[52]

Much more work needs to be done on the migration of black women. As difficult as the task may prove, historians must begin to probe deep into the internal world and lives of these women, who not only were Detroit's largest neglected source of labor but also remain the largest neglected, and still most obscure, component of Afro-American history. It is not enough to study black women simply because they are neglected and historically invisible. Rather, it is incumbent that we examine and interpret their experiences, for what this new information yields may very well bring us closer to a comprehensive and more accurate understanding of all of American history from colonial times to the present.

Notes

1. D. J. Steans, *Backward Glance: A Memoir* (Smithtown, N.Y.: Exposition Press, 1983), 1.

2. Thomas C. Cox, *Blacks in Topeka, Kansas, 1865-1915* (Baton Rouge: Louisiana State University Press. 1982); Nell Irvin Painter, *Exodusters: Black Migration to Kansas After Reconstruction* (New York: W. W. Norton, 1977); Quintard Taylor, "The Emergence of Black Communities in the Pacific Northwest, 1865-1910," *Journal of Negro History* 64 (1979), 342-45; Janice L. Ruff, Michael R. Dahlin, and Daniel Scott Smith, "Rural Push and Urban Pull: Work and Family Experience of Older Black Women in Southern Cities, 1880-1910," *Journal of Social History* 16 (Summer 1983), 39-48; James O. Wheeler and Stanley D. Brunn, "Negro Migration Into Southwestern Michigan," *Geographical Review* 58 (April 1968), 214-30. For a description of the development of eight all-black towns in northern communities of more than 1,000 population, see Harold M. Rose, "The All-Negro Town: Its Evolution and Function," *Geographical Review* 55 (1965), 362-81; Edwin S. Redkey, *Black Exodus: Black Nationalism and Back-to-Africa Movements, 1890-1969* (New Haven: Yale University Press, 1969), 150-94; Wilson Jeremiah Moses, *The Golden Age of Black Nationalism 1850-1925* (New York: Oxford University Press, 1978), 83-102.

3. Allan H. Spear, *Black Chicago: The Making of a Negro Ghetto, 1890-1920* (Chicago: University of Chicago Press, 1967); Peter Gottlieb, *Making Their Own Way: Southern Blacks' Migration to Pittsburgh, 1916-1930* (Urbana: University of Illinois Press, 1987); Kenneth Kusmer, *A Ghetto Takes Shape: Black Cleveland, 1870-1930* (Urbana: University of Illinois Press, 1976); Joe W. Trotter, *Black Milwaukee: The Making of an Industrial Proletariat, 1915-45* (Urbana: University of Illinois Press, 1985); Darrel E. Bigham, *We Ask Only a Fair Trial: A History of the Black Community of Evansville, Indiana* (Bloomington: Indiana University Press, 1987); Florette Henri, *Black Migration: Movement North, 1900-1920* (New York: Anchor, 1975); Richard W. Thomas, "From Peasant to Proletarian: The Formation and Organization of the Black Industrial Working Class in Detroit, 1915-1945" (Ph.D. diss., University of Michigan, 1976).

4. Trotter, *Black Milwaukee*, 25; Thomas, "From Peasant to Proletarian," 6-7; Henri, *Black Migration*, 52, 69. For a general overview of the historiography of black urbanization, see Kenneth Kusmer, "The Black Urban Experience in American History," in *The State of Afro-American History: Past, Present, and Future*, Darlene Clark Hine, ed. (Baton Rouge: Louisiana State University Press, 1986), 91-122. In an important study

of women in Chicago, historian Joanne J. Meyerowitz comments on the different migratory patterns of black and white women: "Black women followed different paths of migration to Chicago. In 1880 and in 1910, the largest group of black women adrift in Chicago, almost half, came from the Upper South states of Kentucky, Tennessee, and Missouri. A smaller group of migrants listed birthplaces elsewhere in the South. In 1880, one-fourth of all black women adrift came from the states of the Deep and Atlantic Coastal South; in 1910, almost one-third. During and after World War I, the stream of migrants from Mississippi, Alabama, Georgia, and other parts of the Deep South swelled to a flood": Joanne J. Meyerowitz, *Women Adrift: Independent Wage Earners in Chicago, 1880-1930* (Chicago: University of Chicago Press, 1988), 10.

5. Henri, *Black Migration*, 69; August Meier and Elliott Rudwick, *Black Detroit and the Rise of the UAW* (New York: Oxford University Press, 1979), 5-7: Spear, *Black Chicago*, 129-30.

6. Gottlieb, *Making Their Own Way;* James Borchert, *Alley Life in Washington: Family, Community, Religion, and Folklife in the City, 1850-1970* (Urbana: University of Illinois Press, 1980), 237. Borchert stresses throughout his study "the strong continuities not only between slave and alley culture, but also between alley culture and both rural and urban black cultures of the third quarter of the twentieth century" (pp. 237-278). Also see Kusmer, "The Black Urban Experience," 13; Dianne M. Pinderhughes, *Race and Ethnicity in Chicago Politics: A Reexamination of Pluralist Theory* (Urbana: University of Illinois Press, 1987); St. Clair Drake and Horace R. Cayton, *Black Metropolis: A Study of Negro Life in a Northern City* (New York: Harcourt, Brace & World, 1945; rev. edition, 1970).

7. Elizabeth Clark-Lewis, "'This Work Had a End': African-American Domestic Workers in Washington, D.C., 1910-1940," in *"To Toil the Livelong Day": America's Women at Work, 1780-1980,* Carol Groneman and Mary Beth Norton, eds. (Ithaca: Cornell University Press, 1987), 196-212, especially 211.

8. Clark-Lewis, "'This Work Had a End,' " 198-99; David M. Katzman, *Seven Days a Week: Women and Domestic Service in Industrializing America* (New York: Oxford University Press, 1979), 219-21.

9. Darlene Clark Hine, "Rape and the Inner Lives of Black Women in the Middle West: Preliminary Thoughts on the Culture of Dissemblance," *Signs: Journal of Women and Culture in Society* 14 (Summer 1989), 912-20.

10. H. F. Kletzing and William F. Crogman, *Progress of a Race* (1987; reprint, New York: Negro University Press, 1969), 193.

11. Lynda F. Dickson, "Toward a Broader Angle of Vision in Uncovering Women's History: Black Women's Clubs Revisited," *Frontiers* 9, 2 (1987), 62-68, especially 67.

12. Ibid., 67.

13. See Moses, *The Golden Age of Black Nationalism,* chap. 5, "Black Bourgeois Feminism Versus Peasant Values: Origins and Purposes of the National Federation of Afro-American Women," 103-31; Darlene Clark Hine, *When the Truth Is Told: Black Women's Culture and Community in Indiana, 1875-1950* (Indianapolis: National Council of Negro Women, Indianapolis Section, 1981), 49-78.

14. Lawrence W. Levine, *Black Culture and Black Consciousness: Afro-American Folk Thought From Slavery to Freedom* (New York: Oxford University Press, 1977), 274.

15. Gottlieb, *Making Their Own Way*, 46-49, 52; Jacqueline Jones, *Labor of Love, Labor of Sorrow: Black Women, Work, and Family From Slavery to the Present* (New York: Basic Books, 1985), 159-60.

16. Carol Tanis, "A Study in Self-Improvement," *Grand Rapids Magazine* 42 (January 1987), 41-44 (quote on 43). The complete records and minute books of the Grand Rapids Study Club are located in the Grand Rapids Public Library, Grand Rapids,

Michigan. They are among the most thorough and extensive records, spanning the years between the 1920s and the early 1980s, of a midwestern regional black women's club I have found.

17. Sara Brooks, *You May Plow Here: The Narrative of Sara Brooks*, Thordis Simonsen, ed. (New York: Touchstone Edition, Simon and Schuster, 1987), 195-96.

18. Steans, *Backward Glance*, 17.

19. Jane Edna Hunter, *A Nickel and a Prayer* (Cleveland: Elli Kani, 1940), 65-66. Hunter's papers are located at the Western Reserve Historical Society, Cleveland, Ohio.

20. The Maddy Bruce Story, May 18, 1984, transcript, Deborah Starks Collection, book 1, folder 4, Oral Histories, Black Women in the Middle West (BWMW) Project, Fort Wayne, Indiana (Indiana Historical Society, Indianapolis, Indiana). There is considerable confusion surrounding the spelling of the name in the transcript. Sometimes her name is spelled Burch, which is the way she spelled it in the text of the oral history. The listing of the oral history, however, is under Bruce. For the sake of consistency in the narrative, I refer to her as Burch.

21. Brooks, *You May Plow Here*, 211-14, 216-17.

22. Gottlieb, *Making Their Own Way*, 49-50; Jones, *Labor of Love, Labor of Sorrow*, 156-60. Also see Earl Lewis, "Afro-American Adaptive Strategies: The Visiting Habits of Kith and Kin Among Black Norfolkians During the First Great Migration," *Journal of Family History* 12 (1987), 407-20.

23. The Maddy Bruce Story.

24. Brooks, *You May Plow Here*, 195.

25. LeRoi Jones, *Blues People: The Negro Experience in White America and the Music That Developed From It* (New York: William Morrow, 1963), 105-07; Sandra R. Leib, *Mother of the Blues: A Story of Ma Rainey* (Amherst: University of Massachusetts Press, 1981), 21-22, 78-79; Daphne Duval Harrison, *Black Pearls: Blues Queens of the 1920s* (New Brunswick: Rutgers University Press, 1988), 18-21.

26. Roger Lane, *The Roots of Violence in Black Philadelphia, 1860-1900* (Cambridge, Mass.: Harvard University Press, 1986), 130, 158-59.

27. Brooks, *You May Plow Here*, 206, 109. For a provocative discussion of earlier black women who also practice abstinence, see Rennie Simson, "The Afro-American Female: The Historical Context of the Construction of Sexual Identity," in *The Powers of Desire: The Politics of Sexuality*, Anne Suitow, Sharon Thompson, and Christine Stansall, eds. (New York: Monthly Review Press, 1983), 229-35. Simson's observations warrant quoting at length: "[Harriet] Jacob's attempt to maintain control over her life is also shown in her pattern of living after her escape to freedom in the North. She mentioned no sexual attachments and relied on herself for financial support. [Elizabeth] Keckley too learned self-reliance. A brief marriage with a Mr. Keckley ended in divorce as she found him 'a burden instead of a helpmate.' No children issued from this marriage as Keckley did not wish to bring any more slaves into the world and thus fulfill her function as a breeder. When her marriage was terminated she said of her husband, 'Let charity draw around him the mantle of silence.' Keckley never mentioned another sexual relationship and, like Jacobs, she remained self-supporting for the rest of her life" (p. 232). For a discussion along the same vein, see Darlene Clark Hine, "Female Slave Resistance: The Economics of Sex," *Western Journal of Black Studies* 3 (Summer 1979), 123-27. For additional insight into incidences of domestic violence in the aftermath of emancipation, see Ira Berlin, Steven F. Miller, and Leslie F. Rowland, "Afro-American Families in the Transition From Slavery to Freedom," *Radical History Review* 42 (November 1988), 89-121, especially 99-100.

28. Stewart E. Tolnay, "Family Economy and the Black American Fertility Transition," *Journal of Family History* 11, 3 (1986), 272-77.

29. The Minute Book of the 1935 meetings of the Grand Rapids Study Club notes that among other issues, one topic earmarked for discussion was birth control. On January 10, 1935, the Study Club met for a discussion: "Public Institutions—Prisons, Asylums, Hospitals, etc." The question that focused the discussion was, "Who Belongs in Prison—Habitual Drunkard? Prostitutes? Homosexual? Non-supporter?" box 1 (Grand Rapids Public Library, Grand Rapids, Michigan), Gerda Lerner, "Early Community Work of Black Club Women," *Journal of Negro History* 59 (1974), 158-67. For a probing examination of black club women's work and institution building in one midwestern city, see Earline Rae Ferguson, "The Woman's Improvement Club of Indianapolis: Black Women Pioneers in Tuberculosis Work, 1903-1938," *Indiana Magazine of History* 84 (September 1988), 237-61.

30. Sarah Darthulin Tyree to Jennie P. Fowlkes, August 23, 1921. Frances Patterson Papers, box 1, folder 2, BWMW Project (Indiana Historical Society, Indianapolis, Indiana).

31. Joseph A. McFalls, Jr. and George S. Masnick, "Birth Control and the Fertility of the U.S. Black Population, 1880 to 1980," *Journal of Family History* 6 (Spring 1981), 103.

32. William M. Tuttle, Jr., *Race Riot: Chicago in the Red Summer of 1919* (New York: Atheneum, 1982), 164.

33. Hunter, *A Nickel and a Prayer*, 68. Also see for a judicious discussion of white and black prostitution, Ruth Rosen, *The Lost Sisterhood: Prostitution in America, 1900-1918* (Baltimore: Johns Hopkins University Press, 1982). See Thomas Connelly, *The Response to Prostitution in the Progressive Era* (Chapel Hill: University of North Carolina Press, 1980), 48-66, for a discussion of the relations between prostitution and European immigration.

34. Brooks, *You May Plow Here*, 219.

35. James E. DeVries, *Race and Kinship in a Midwestern Town: The Black Experience in Monroe, Michigan, 1900-1915* (Urbana: University of Illinois Press, 1984), 90-91.

36. The Maddy Bruce Story, May 18, 1984.

37. Divorce Decree: Jesse Clay Fowlkes vs. Jane Pauline Fowlkes, April 12, 1923. Frances Patterson Papers, box 2, folder 3 (Indiana Historical Society, Indianapolis, Indiana). J. C. Fowlkes was ordered to pay $7.50 per week for the support of the children.

38. Brooks, *You May Plow Here*, 219. For an insightful historical analysis of the meaning of wife beating and battered women's resistance, see Linda Gordon, *Heroes of Their Own Lives: The Politics and History of Family Violence, Boston 1880-1960* (New York: Viking, 1988), 250-88.

39. Gerda Lerner, *Black Women in White America: A Documentary History* (New York: Vintage, 1973), 238-39; Jones, *Labor of Love*, 161-64; Gottlieb, *Making Their Own Way*, 107-9.

40. Spear, *Black Chicago*, 29, 34, 155; Henri, *Black Migration*, 142, 168.

41. Mary Helen Washington, *Invented Lives: Narratives of Black Women, 1860-1960* (New York: Anchor, 1987), xxii.

42. Henri, *Black Migrations*, 52.

43. Spear, *Black Chicago*, 151-55; Henri, *Black Migration*, 143-44.

44. Spear, *Black Chicago*, 34.

45. Trotter, *Black Milwaukee*, 14, 47, 81, 171, 203.

46. Ibid., 174.

47. Tanis, "A Study in Self-Improvement," 42.

48. Susan M. Hartmann, "Women's Organizations During World War II: The Interaction of Class, Race and Feminism," in *Women's Being, Women's Place: Female Identity and Vocation in American History*, Mary Kelley, ed. (Boston: G. K. Hall, 1979); Karen Tucker Anderson, "Last Hired, First Fired: Black Women Workers During World War II," *Journal of American History* 64 (June 1982), 96-97.

49. George E. DeMar, "Negro Women Are American Workers, Too," *Opportunity* 21 (April 1943), 41-43, 77. For a description of the stratified workforce in Milwaukee, see Trotter, *Black Milwaukee*, 159, 171. Trotter notes that "where black females worked in close proximity to whites, the work was stratified along racial lines. At the Schroeder Hotel, for example, black women operated the freight elevator, scrubbed the floors, and generally performed the most disagreeable maid's duties. Conversely, white women worked the passenger elevator, filled all clerical positions, and carried out light maid's duties" (p. 159).

50. Trotter, *Black Milwaukee*, 174.

51. William H. Harris, *The Harder We Run: Black Workers Since the Civil War* (New York: Oxford University Press, 1982), 64.

52. Meier and Rudwick, *Black Detroit*, 136, 153-54. Trotter also notes that in spite of vigorous efforts to extend the benefits of the FEPC to black women, the Fair Employment Practices Committee focused on traditionally white female-dominated industries. Yet, he notes, the complaints of black women of racial discrimination in heavy industries like Allis Chalmers, Nordberg, and Harnishchfeger were "frequently dismissed by the FEPC due to insufficient evidence, although some of their charges were as potently documented as those of black men," Trotter, *Black Milwaukee*, 171.

Making the Second Ghetto in Metropolitan Miami, 1940-1960

Raymond A. Mohl

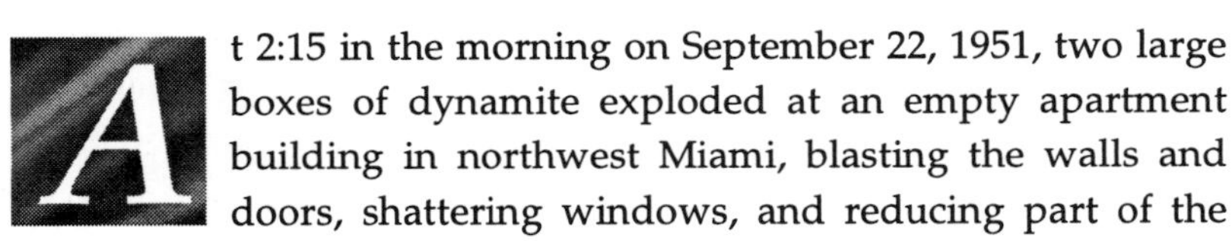

At 2:15 in the morning on September 22, 1951, two large boxes of dynamite exploded at an empty apartment building in northwest Miami, blasting the walls and doors, shattering windows, and reducing part of the building to rubble. No one was injured, but the explosion rocked the neighborhood for blocks in every direction. Two more massive dynamite bombings in November and December 1951 damaged other sections of the apartment complex. Few in the community were surprised at these events, and everyone who lived there knew exactly what it was all about.[1]

The dynamited building was part of a larger apartment complex owned by two Miami real estate developers who were in the process of converting their rental units from white occupancy to black. The racial confrontation at the Carver Village complex brought together several powerful combustible elements—elements that typified the racial transition of neighborhoods in postwar Miami and many other cities. Overcrowded black residential areas were bursting at the seams; African American families aggressively sought more and better housing and to escape the inner-city ghetto. Working-class whites resisted the intrusion of black people into white neighborhoods on the fringes and frontiers of the ghetto. White neighborhood associations, some linked to the Ku Klux Klan, were willing to use bombs and violence to maintain the color line, and African Americans were willing to resist the intimidators. Slumlords, blockbusters, and real estate developers were busily at work facilitating the process of racial transition, and profiting handsomely as well. Racial zoning had been outlawed in Florida in 1946, and restrictive real estate covenants had been banned by the U.S. Supreme Court in 1948. Nevertheless, through a variety of public policy decisions, government at every level continued to play a significant role in shaping the racial characteristics of residential areas.[2] In "Deep South" Miami in 1951, these varied elements of urban residential change combined to produce an explosive mix.

Residential segregation by race has been a persistent characteristic of urban life in twentieth-century America. Some recent historical research, beginning with Arnold R. Hirsch's book, *Making the Second Ghetto: Race and Housing in Chicago, 1940-1960* (1983), has begun to focus on questions of race and housing during the years between 1940 and 1960. During this period, dramatic demographic changes were reshaping postwar urban America. A new Great Migration of African Americans from the South was underway—over five million African Americans moved north and west between 1940 and 1970; hundreds of thousands of additional rural southern migrants settled in such southern cities as Atlanta, Memphis, Houston, and Miami. At the same time, and especially after 1945, millions of white Americans began moving to the sprawling new suburbs. A massive racial/spatial reorganization of the American metropolis took place after World War II. In Chicago, Detroit, Cleveland, Boston, Philadelphia, Washington, Atlanta, Los Angeles, and many other cities, the pressure of rising black population triggered the racial transition of neighborhoods. African Americans pushed out the boundaries of the ghetto, moving into the neighborhoods left behind by whites departing for the suburban fringe. As blacks moved in, for-sale signs sprouted, often overnight, and remaining white residents eventually joined the suburban trek. The conjunction of these two demographic shifts in the postwar era produced the newer, second ghetto.[3]

The process of residential transition was not an easy one. Indeed, the frontiers of neighborhood change were marked by years of protest, demonstration, and violence in many postwar cities. At the deepest level, racial/spatial transitions lay behind the Detroit race riots of 1943. In Philadelphia, racial incidents over housing were commonplace, with more than two hundred such incidents taking place in the first six months of 1955 alone. In postwar Chicago, fierce battles erupted over public housing location and neighborhood transitions. The wave of antiblack violence in changing neighborhoods was often stirred up by white homeowner groups and "protective" or "improvement" associations, occasionally by white hate groups such as the Columbians in Atlanta, America Plus in Los Angeles, and the Ku Klux Klan in many cities. It is also quite evident, although more research is needed on this point, that the real estate industry was deeply involved in shaping the urban land market—initially in maintaining the boundaries of the first ghetto, and then in managing the growth and expansion of the second ghetto.[4]

This essay represents a preliminary effort to provide a framework for understanding the dynamics of second-ghetto development in metropolitan Miami. Three shaping forces producing urban change help to frame the analysis: first, the impact of public policy decisions; second, the African American push for new housing and subsequent patterns of white violence; and third, the role of real estate interests in breaking the color line. The interaction of these powerful forces brought a rapid racial and spatial transformation to the Miami metropolitan area.

The Impact of Public Policy

In the first three decades of the twentieth century, residential segregation seemed permanently etched on Miami's urban landscape. Early efforts to define territorial boundaries between white and black residential areas occasionally led to racial conflict and violence, as in 1920 when some African Americans crossed over into "white territory." But by 1930, most of Miami's African American population of about 29,000 was clustered residentially in a confined, inner-city area known at the time as "Colored Town," later called Overtown. By the 1940s, this area had grown to 105 city blocks built up mostly with shotgun shacks and slum housing. A second, much smaller black neighborhood, mostly of black Bahamian immigrants, had developed in a section of Coconut Grove, a few miles south of Miami's business district. Still smaller black farming communities sprang up early in the twentieth century along the Florida East Coast Railway tracks as they stretched south of Miami toward Homestead. African Americans were concentrated in Overtown and a few other segregated districts because local policies of racial zoning left few other areas open to black settlement.[5]

Given the nature of race relations in south Florida in the 1920s and 1930s, black residential patterns in those years generally were dictated by public policy decisions. In the 1930s, for instance, Miami's white business leaders were interested in pushing out the boundaries of the relatively confined downtown business district into the nearby black community. New Deal public housing programs during the Depression decade provided the first such opportunity. A New Deal public housing project for blacks named Liberty Square was completed in 1937 on undeveloped land five miles northwest of the central business district. The city's white civic elite conceived of this project as the nucleus of a new black community that might siphon off the population of Overtown and permit downtown business expansion. The availability of federal housing funds mobilized the civic elite, who seized this opportunity to push the blacks out of the downtown area.[6]

Other plans were proposed in the 1930s to achieve the same goal. In 1936, for example, the Dade County Planning Board proposed a "negro resettlement plan." The idea was to cooperate with the city of Miami "in removing [the] entire Central Negro town" to three "model negro towns" located on the distant agricultural fringes west of Miami. A year later, in a speech to the Miami Realty Board, former

Coral Gables developer George Merrick proposed "a complete slum clearance effectively removing every negro family from the present city limits." The idea of black removal from the central district died hard. In 1945, Miami civic leaders were still discussing "the creation of a new negro village that would be a model for the entire United States." Although never implemented, slum clearance plans in 1946 called for the removal of African Americans from Miami's central area to a distant new housing development west of Liberty City. In the late 1940s and early 1950s, federal housing officials pursued similar plans for new public and private housing for African Americans on the metropolitan fringe. As late as 1961, the *Miami Herald* was reporting on new plans to eliminate Overtown to facilitate downtown business expansion.[7]

None of these proposals was ever officially implemented, mostly because of a lack of funding and local opposition to federal intrusion, but some racial transitions in residential neighborhoods began to occur by the 1940s. New Deal housing agencies such as the Home Owners Loan Corporation and the Federal Housing Administration contributed to changing racial patterns. Through their appraisal policies, both agencies "redlined" Miami's black community and nearby white residential neighborhoods and even completely undeveloped areas. These early appraisal decisions, plotted out on the "residential security maps" of the Home Owners Loan Corporation as early as 1936, dictated the course of future neighborhood development. These redlining decisions, for which the local real estate industry provided much of the appraisal data, essentially designated the northwest area of Miami and Dade County for future black settlement.[8]

Until the early 1950s, through its policy of racial zoning, the Dade County Planning Board sought to control the gradual expansion of black settlement, mostly within the northwest area that had already been redlined.[9] But local real estate people—slumlords, developers, and speculators—with an interest in the profit potential in black housing had much to do with the specific pace and direction of residential change. The rapid growth of Dade County's African American population between 1940 and 1960 added a special urgency to the need for new black housing. The black population in the Miami metropolitan area grew from 49,518 in 1940 to 64,947 in 1950, a growth rate of 31.2 percent, and to 137,299 by 1960, a decennial growth rate of 111.4 percent. Hemmed in for decades by race and custom, Miami's African American population now sought to push out of the inner-city ghetto into new housing areas. The combined impact of these influences—of black population growth, the quest for better housing, real estate speculation, and redlining—began to produce racial change in Miami neighborhoods by the late 1940s.[10]

By mid-century, three new areas of substantial black residence had taken shape in the Miami metropolitan area. First, the Liberty Square housing project had become the center of a new and rapidly growing African American community known as Liberty City, as new housing for blacks sprouted on the largely undeveloped land around the New Deal project and as some nearby white housing turned over. Second, a slightly smaller built-up area called Brownsville, a district of perhaps 100 blocks and a bit closer to downtown Miami, became mostly black by the late 1940s. Originally known as "Brown's Sub," Brownsville was home to a few scattered black farmers by 1920, but by 1940 the neighborhood was built up with small houses owned by working-class whites. Intense real estate activity in Brownsville in the 1940s produced Miami's first true second-ghetto racial turnover. Finally, in the postwar era, in white, working-class Opa-locka on the more distant fringes of northwest Miami, white builders began construction of several large apartment complexes and single-family housing developments for black residents. Between 1946 and 1951, almost 2,000 new units of black housing went up on undeveloped land in and near Opa-locka. By the early 1950s, then, a corridor of black residential development was fanning out to the northwest of the central city.[11]

More changes came in the 1950s. The construction of the Miami expressway system in the late 1950s and early 1960s had a dramatic impact on Dade County's urban land market and on racial settlement patterns. The federal interstate highway program provided Miami's white business and political leaders with a new opportunity to recapture inner-city land for urban redevelopment. It also supplied

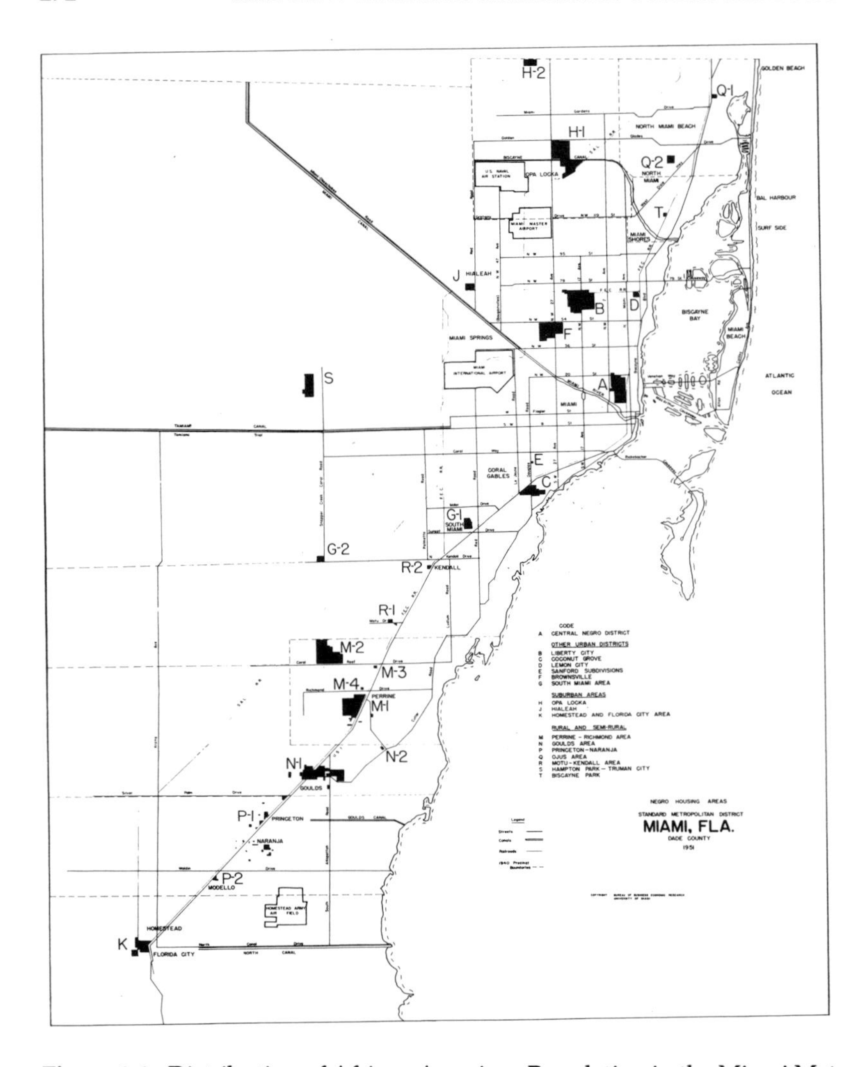

Figure 9.1. Distribution of African American Population in the Miami Metropolitan Area in 1951, Showing the Beginnings of Second-Ghetto Development in Brownsville, Liberty City, and Opa-Locka

SOURCE: Reinhold P. Wolff and David K. Gillogly, *Negro Housing in the Miami Area* (Coral Gables, 1951).

a new rationale for pushing Overtown's blacks to more distant residential areas on the northwestern fringe of the metropolitan area. As early as 1956, the Florida State Road Department, in conjunction with local officials and business people, routed Interstate 95 directly through Overtown and into downtown Miami. Alternative plans using an abandoned railroad corridor for the downtown expressway were rejected, as the planners stated, to provide "ample room for the future expansion of the central business district in a westerly direction." When the downtown leg of the Interstate 95 expressway was completed in the mid-1960s, it ripped through the center of Overtown, wiping out extensive housing as well as the black business district. One massive expressway interchange alone took up twenty square blocks of densely settled land and destroyed the housing of about 10,000 people. Some 40,000 African Americans made Overtown home before the interstate came, but less than 8,000 now remain in an urban wasteland dominated by the expressway. By the end of the expressway building era, little remained of Overtown to recall its days as a thriving center of African American community life.[12]

The building of Miami's expressway and consequent housing destruction in Overtown speeded the growth of Miami's second ghetto in the late 1950s and after. As early as 1957, as the expressway plans became public, the Greater Miami Urban League, the *Miami Times* (the city's black newsweekly), and even the *Miami Herald* expressed concern about where displaced blacks would find new housing. Federal policy had purposely separated highway building from housing, and that continued to be the case until the early 1960s. Thus, for much of the expressway era in Miami, there was no official linkage between highway building and urban renewal activities, on one hand, and public housing construction, relocation assistance, or mortgage assistance, on the other hand. But Overtown's thousands of dislocated residents did find new housing, and, as we shall see, they did have some help from segments of the local real estate industry. Mostly, they found new housing in the northwest corridor of black residence that had already begun to emerge in the late 1940s. As the *Miami Herald* noted in 1957 in an article titled "What About the Negroes Uprooted by the Expressway?" even without the assistance of any government agency, blacks were already moving into "former white areas along the fringes" of the ghetto. Liberty City began pushing out its boundaries in all directions, as did now mostly black Brownsville. Ultimately, by the end of the 1950s, Brownsville had merged with Liberty City, creating one sprawling second-ghetto area. Further north, white flight in the early 1960s from Opa-locka and from Carol City, another white suburban development, opened up still more housing for African Americans.[13]

Thus, by the time the Miami expressway system tore through central Overtown in the early 1960s, the process of second-ghetto growth in northwest Miami had been underway for almost two decades. Redlining and other public policy decisions paved the way. The location of the Liberty Square public housing project in the 1930s virtually dictated the growth of black residential neighborhoods in the surrounding area. No public housing was built in Miami between 1942 and the late 1950s, mainly because of local political hostility to federal programs. However, new housing projects after 1958 were located in the path of the expanding corridor of black housing (a pattern common elsewhere, as well), thus speeding the process of racial transition.[14]

The pace of residential change was given a new impetus in the mid-1950s by the issue of school desegregation. For instance, the Orchard Villa subdivision just to the south of Liberty City was transformed from a primarily white community to a primarily black one in less than one year after the Dade County School Board decided in February 1959 to admit four black pupils to the neighborhood elementary school as a "pilot" project. The Orchard Villa School opened with 235 white children in September 1958; a year later, after protest meetings, demonstrations, and rapid white flight, the Orchard Villa School began the new school year with only eight white students. By 1961, the changing neighborhood pattern had pushed beyond Orchard Villa to Floral Park, another white, working-class subdivision, where the same process of racial turnover was repeated.[15]

Another dimension of public decision making—federal immigration policy—affected Miami's housing patterns as well. The arrival, beginning in 1959, of several hundred thousand Cuban exiles, who began carving out their own areas of residential space in metropolitan Miami—west and southwest of the CBD and in Hialeah to the far northwest—also shaped the local housing market, effectively limiting the housing choices of African Americans displaced from Overtown by redevelopment activities.[16] The cumulative impact of public policy decisions on Miami's second-ghetto development over several decades was indeed quite powerful. According to several sociological studies, of more than 100 large American cities, Miami had the highest degree of residential segregation by race between 1940 and 1960.[17]

The Black Push for Housing and White Resistance

In 1958, *Ebony* magazine published an article by journalist Carl T. Rowan titled "Why Negroes Move to White Neighborhoods." Rowan began with a discussion of an August 1957 incident in Levittown, Pennsylvania, in which an African American family was harassed and mobbed after moving into the new community of 60,000 suburbanites. Rowan noted that housing segregation, perhaps more than anything else, symbolized racial injustice in the United States. But African Americans were struggling to break out of the ghetto, Rowan contended; like most Americans, they wanted better housing and schools for their children, and less crime, congestion, and squalor. These were rational choices made by people pursuing the American dream of home ownership in decent, safe communities. Rowan's article typified a substantial journalistic and housing literature on the racial/spatial transformations of the 1950s. It also emphasized the degree to which African Americans pushed the limits of housing choice in postwar America. Human agency, reflected in tens of thousands of individual and family decisions to move into white neighborhoods, had much to do with the making of the second ghetto.[18]

Perhaps even more striking, African American agency on housing choice was exercised within the context of several decades of white harassment, intimidation, and violence. Whites resisted black intrusion into new neighborhoods, but blacks challenged the system individually by moving to white residential areas and collectively

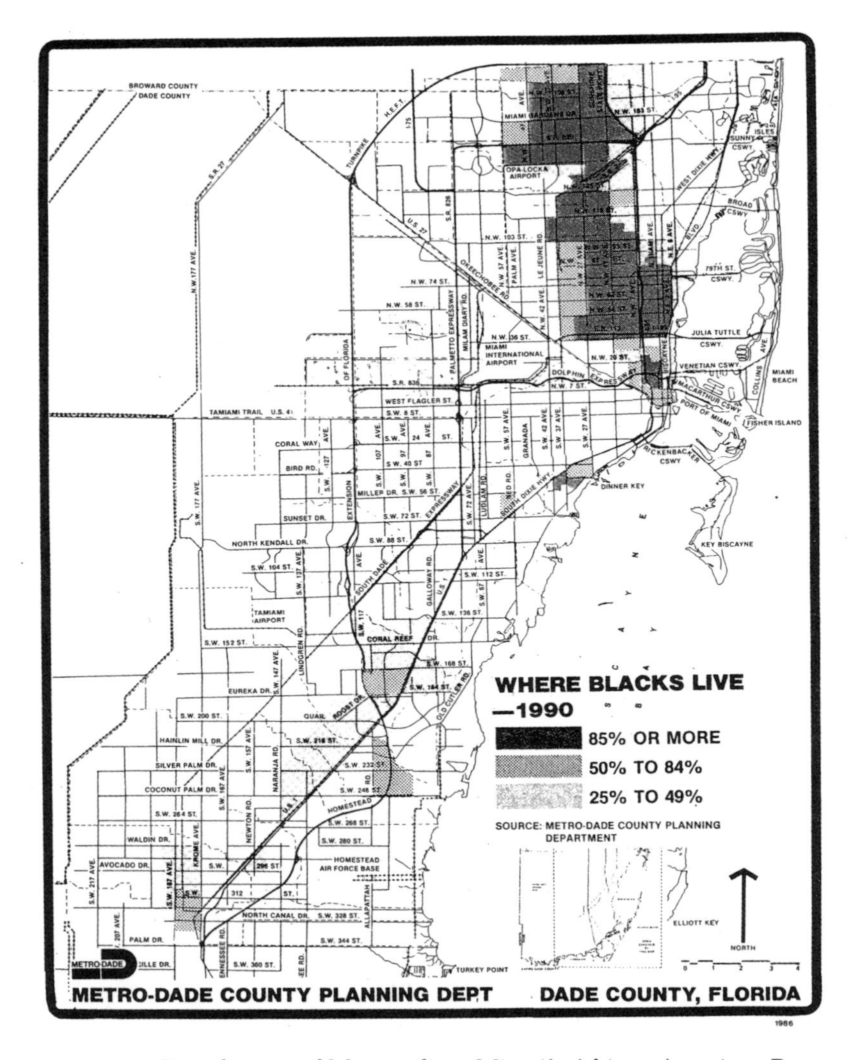

Figure 9.2. Distribution of Metropolitan Miami's African American Population in 1990, Showing the Heavy Concentration of Blacks in the Northwest —The Culmination of Fifty Years of Second-Ghetto Development
SOURCE: Metro-Dade County Planning Department.

through their churches and such groups as the National Urban League and the NAACP. This pattern of challenge and resistance, marked by violence and conflict, played an important role in shaping the new contours of Miami's postwar second ghetto.

Turf wars on the racial frontiers of Miami neighborhoods were intense between 1945 and the late 1950s. Punctuated by white protest marches, Klan bombings, cross burnings, police harassment, and other forms of racial violence, the process of residential turnover was not an easy one. The Ku Klux Klan had been rejuvenated in postwar Miami. Although it had relatively few members, Miami's John B. Gordon Klan No. 5 played an active role in second-ghetto neighborhood violence. In August 1945, for instance, two African American families crossed the so-called red line and purchased homes in a white residential area near Brownsville. The blacks were harassed by county health, zoning, and police officials and were eventually jailed for zoning violations. Once the blacks were released on bond, the Ku Klux Klan burned a fiery, ten-foot cross in their neighborhood—a warning, a black reporter wrote, to keep African Americans from pushing out the boundaries of their residential neighborhoods. Three months later, the Klan struck again, burning five crosses in the same neighborhood while armed whites paraded through the area in cars and trucks. It was a classic racial confrontation, as whites resisted the black push for new housing in white areas. In November 1947, the Klan burned not only a cross but the homes of two black families who had moved across the unofficial boundary line separating the races. According to John A. Diaz, a black Miami reporter, these incidents marked the initiation of a forceful and focused Klan campaign to keep blacks out of white neighborhoods. The campaign of intimidation and violence continued in 1948, as fiery Klan crosses illuminated night skies in Miami's black districts—mostly in Brownsville—on numerous occasions in May, October, and November of that year. There is only one good explanation for the persistence of the Klan's vicious terror campaign—obviously unintimidated, African Americans continued moving into white Brownsville in the search for better housing.[19]

Second-ghetto tensions flared dangerously once more in July 1951. John Bouvier and Malcolm Wiseheart, owners of the white-occupied Knight Manor apartment complex on the eastern fringes of Liberty City, moved whites out of part of the complex, renamed it Carver Village, and opened it for black rentals. A newly organized Dade

County Property Owners Association charged in an appeal to Governor Fuller Warren that "by vacating white tenants and replacing them with colored tenants," the owners were disturbing the tranquillity of "a long established white neighborhood." A local Citizens Action League sprang up "to protect our Southern way of life." Others complained that "the negro area is to be extended into what has always been a white area."[20]

The Klan injected itself into the Carver Village controversy, distributing hate literature, posting warning signs, and organizing armed white motorcades through the area. Random shootings occurred, injuring at least one African American man. Leaving little to the imagination, the Klan torched giant wooden Ks at four locations around Carver Village (cross burning violated a 1951 Florida law, thus the resort to burning Ks). Blacks began moving into Carver Village in August 1951, followed by the dynamite bombings of September, November, and December, mentioned at the beginning of this essay. Jewish synagogues and schools and a Catholic church in Miami were also bombed during this period, suggesting the coordinated work of white hate groups such as the Ku Klux Klan. Reporting on the incident for *The Nation,* investigative journalist Stetson Kennedy noted that the Klan had "long used terror to keep Negroes inside the ghettos assigned to them." Later investigation, in fact, revealed links between the Klan and the leadership of the Dade County Property Owners Association. White supremacists were willing to use violence and dynamite to maintain the color line and contain Miami's second ghetto. However, there is little evidence that blacks were intimidated, and before long both Carver Village and the adjoining Knight Manor apartment complex were black occupied.[21]

Racial skirmishing continued on Miami's residential frontiers throughout the 1950s. African Americans carved out extensive new areas of housing as Liberty City and Brownsville pushed out their boundaries. Typical, perhaps, was the 1957 incident involving Frank Legree, the first black Miamian to purchase a house in the Orchard Villa subdivision. Soon after moving into this white neighborhood separating Liberty City from Brownsville, Legree began receiving hate mail and harassing phone calls, threatening that his house

Figure 9.3. Dynamited Apartment Building at Carver Village, Undergoing Transition to Black Occupancy, 1951
SOURCE: Florida State Archives, Tallahasse, used by permission.

would be dynamited. A newly organized white supremacist group, the Seaboard White Citizens Council, headed by a notorious former Klansman, organized daily pickets outside Legree's house. The Dade County Property Owners Association, still active seven years after the Carver Village bombings, sought legal means to force Legree out of the neighborhood. Local residents picketed the Orchard Villa Elementary School to prevent Legree's seven-year-old son from attending. Finally, one night in February 1957, police arrested four men planting a seven-foot, kerosene-soaked cross in Legree's front lawn. The four ultimately received short jail terms, but sporadic harassment and picketing of Legree's house continued for months. Meanwhile, the Miami NAACP, despite death threats against its leaders, rallied to Legree's support and attacked the race-baiters in the white neighborhood associations. Legree's persistence paved the way for other black homeowners in Orchard Villa and nearby Floral Park, also stimulating rapid white flight from these new second-ghetto areas.[22]

African Americans in Miami became the targets of harassment, intimidation, and violence throughout the period of second-ghetto development, but they were not passive victims and rarely seemed intimidated. From the 1930s, Miami's African American community had confronted Jim Crow, often aggressively. In 1939, for instance, the Negro Citizens Service League, which later became the National Urban League branch, conducted a voter registration campaign for a nonpartisan municipal election. The night before the election, the Klan held a massive downtown rally, then paraded through Overtown, burned twenty-five crosses, and warned blacks to stay home on election day. Blacks responded with an unprecedented voter turnout.[23]

Similarly, the flurry of Klan activity in Brownsville in the late 1940s did not seem to deter blacks from seeking better housing, nor did the dynamite bombings at Carver Village. A typical response in Miami's African American community may have been reflected in this headline in one of the city's black papers, the *Miami Tropical Dispatch,* in February 1949: "KKK Burns Three Crosses—So What?" The paper went on to editorialize that "the day has passed when the mere burning of a cross somewhere in their section or the parading of hooded cowards can wreak fear" among African Americans. The paper also noted that the burned crosses were so pathetic and poorly constructed that they were not worth photographing for publication. There is no evidence from Miami suggesting that African Americans were forced out of new neighborhoods by white racists, as happened in Detroit, Chicago, and other cities. Throughout this period in Miami, the NAACP, the Urban League, the Civil Rights Congress, the Congress of Racial Equality, many activist black preachers, and the black press worked for integration, racial change, civil rights, and better housing. NAACP lawyers represented blacks who were breaking the residential color line. Human agency—the black push for new housing—was driving the second-ghetto process throughout the postwar years.[24]

Miami's African American press, in fact, took hard-hitting stands against residential segregation and racism. The well-established *Miami Times,* founded in 1924, consistently spoke out for equal rights, as did smaller, short-lived black papers such as the *Miami Whip* and the *Miami Tropical Dispatch.* Self-described as "an aggressive and progressive Negro weekly," the *Tropical Dispatch* in 1947 published an editorial titled "The Northwest Territory"—a sarcastic commentary on official Miami efforts to concentrate blacks in the northwest section of Dade County. The editorial opposed forcible efforts to relocate Overtown's blacks to the northwest fringes, but also recalled that the famous Northwest Ordinance of 1787 had outlawed slavery: "If Miami's Northwest Territory is to provide such unstinted liberties for the hemmed-in Negroes of the city, then on with the plan, for there will be many too glad to welcome the opportunity extended." The editor of the *Tropical Dispatch,* a man named Daniel Francis, lived a few blocks from Carver Village. When that apartment complex was blown up in the early morning hours of September 22, 1951, Francis grabbed his shotgun and rushed over to Carver Village to help. This was not the response of an intimidated black man.[25] The growth of Miami's second ghetto, then, took shape within the context of substantial racial conflict—conflict involving white harassment, intimidation, and violence but also involving a forceful black response and a persistent black push for more and better housing.

The Impact of the Real Estate Industry

Dynamite bombings, cross burnings, shootings, and other forms of racial/spatial conflict grabbed a big share of the newspaper headlines in Miami and elsewhere. Behind the headlines, however, and behind the process of residential change was the real estate industry —the hidden hand that shaped the urban land market and often managed the growth and development of the second ghetto. It should come as no surprise to students of urban history and race relations to learn that slumlords, blockbusters, and real estate developers had much to do with changing residential patterns in the modern American city. What is surprising is that so little historical research has been devoted to this subject. Part of the problem may lie in the fact that real estate and property-transfer records are difficult to use—tedious might be a better word—or they are purposely confusing, with slumlords, blockbusters, and investors in inner-city real estate often using dummy corporations and holding companies to protect their identities.

In Dade County, however, a preliminary survey of official property-transfer records has revealed a rich trove of accessible data to trace the role of the real estate industry in the making of Miami's second ghetto. These records include abstract books, a direct-seller index, a reverse-buyer index, a corporation index, corporation record books, deed books, plat books, the township-section-range map, and aerial maps. The abstract books are particularly useful, because they record transfer of property—buyer and seller—within platted subdivisions. The seller and buyer indexes also provide windows to observe the real estate market in action. Although research in these records is still at an early stage, enough has been revealed to outline the activities of several key real estate players who shaped Miami's second ghetto. Much of what follows comes from an initial examination of those records, as well as from more traditional sources.[26]

Miami's first second-ghetto developer was Floyd W. Davis, a Miami area land speculator and builder. By the early 1930s, Davis had acquired several large tracts of undeveloped land in Miami's northwest section, most of it outside Miami's city limits in unincorporated Dade County. On a small piece of this land, Davis laid out and built a tiny subdivision accommodating a few hundred people by the mid-1930s. Named Liberty City by Davis, this small development paved the way for future second-ghetto expansion in northwest Miami. Davis was deeply involved in the advance planning for federally supported public housing in the area. Along with Miami attorney and civic leader John Gramling, Davis was a partner in the Southern Housing Corporation, a limited-dividend corporation organized in 1933 to seek federal funds for black housing on Davis's land in northwest Miami. While Davis remained in the background, Gramling served as the front man, gathering support from the civic leadership in Miami and corresponding with federal housing officials in Washington. That project failed for lack of funds, but Gramling persisted in a subsequent push for black public housing in Miami. Federal approval for the Liberty Square public housing project came in 1935. Davis sold forty acres of his land to the U.S. government for the project. As the owner of much of the surrounding land, Davis stood to profit enormously from the future expansion of black housing in the area.[27]

The connection between Davis and Gramling was not widely known at the time, but it helps to explain Gramling's active role in promoting public housing for blacks, and especially his insistence on the northwest Miami location for the project. Gramling appeared to federal housing officials as a disinterested, even altruistic, civic leader, but as Davis's personal attorney, it is obvious that he had a personal stake in housing decisions for African Americans in Miami. Acting together, Davis and Gramling set the stage for Miami's initial second-ghetto development. The Liberty Square project served as the nucleus for what eventually became the sprawling African American community known today as Liberty City.

Davis and Gramling remained involved in black housing into the 1940s. In 1941, they formed the New Myami Development Corporation and purchased 500 acres of undeveloped land on the western fringes of Miami with the intention of developing housing for African Americans. With racial zoning still in effect, Davis and Gramling appealed to the Dade County Zoning Board to rezone their tract for black housing. Other developers who owned adjacent tracts and whites who lived nearby were outraged, and petitions of protest flooded the zoning board. The New Myami developers had a curious coalition of supporters, however. The Dade County Knights of the Ku Klux Klan supported the plan for new black housing on the city's western fringe, primarily because such housing would "eliminate the many smaller negro communities that now exist in the center of the white districts." Even more striking was the support Davis and Gramling gained from Miami's African Americans. Dozens of petitions to the zoning board, embellished with hundreds of signatures, attested to the wide appeal of new housing in the African American community. Ultimately, the zoning board rejected the New Myami plan, and the company's tract eventually was absorbed by the Miami airport, which was undergoing expansion during the war years. The failure of the New Myami project launched metropolitan Miami on a more traditional path of second-ghetto development—the racial turnover of existing neighborhoods.[28]

If Floyd Davis was Miami's first second-ghetto developer, Wesley E. Garrison was the city's first major "blockbuster." A wealthy, white real estate speculator and builder, and also politically active in Dade County's small Republican Party, Garrison essentially managed the racial transition of Brownsville during the 1940s. Beginning in 1938, through two separate companies—Garrison Investment Corporation and Garrison Home Builders, Inc.—Garrison aggressively bought houses and empty lots in white, working-class Brownsville, then subdivided, built, rented, and sold property to African Americans. Between 1943 and 1945, for instance, Dade County property-transfer records show that Garrison Investment Corporation purchased eight tracts of land in Brownsville, while Garrison Home Builders bought thirty-five properties. He continued buying, subdividing, building, and selling through the rest of the decade. Garrison also provided mortgages for black buyers, facilitating the process of home buying in new neighborhoods, because African Americans had difficulty getting access to the local mortgage market. Garrison ran up against Dade County's practice of racial zoning, as well as against Ku Klux Klan cross burnings and intimidation, but he remained undeterred. In fact, he posted bond for several African Americans arrested in 1945 and 1946 for buying his homes in white-only Brownsville, and then financed a legal challenge to the county's racial zoning ordinance. A Florida circuit judge upheld the right of African Americans to purchase homes anywhere, and, after the county appealed, the Florida Supreme Court ultimately declared Dade County's racial zoning ordinance unconstitutional. The legal ban on racial zoning speeded the neighborhood transition that Garrison had initiated in Brownsville.[29]

In the late 1940s and early 1950s, several other real estate developers and builders became active in the urban land market in transitional neighborhoods. Two of these men—John A. Bouvier and Malcolm B. Wiseheart—were well-known Miami slumlords with a wide range of investments in Overtown, Coconut Grove, Brownsville, and Liberty City. Bouvier was also a member of the Dade County Zoning Board during this period and was thus in a position to influence policy decisions on housing issues. Through several real estate development companies, notably South Kingsway Corporation and Fiftieth Street Heights, Inc., Bouvier and Wiseheart were actively buying property in transitional areas for black housing. Full-page ads in the *Miami Herald* in 1949 announced their "Kingsway" project in Coconut Grove—"Better Homes for Better Negroes," read the ad. Located on the fringes of existing black development, the Kingsway project pushed out the boundaries of the ghetto. Similarly, Fiftieth Street Heights, Inc. ultimately built twenty large, two-story apartment buildings on the white edges of Brownsville.[30]

Bouvier and Wiseheart were also the owners of the Knight Manor apartment complex in the white Edison Center area near Liberty City. In 1951, they renamed part of the complex Carver Village and rented apartments to blacks, with the explosive consequences mentioned earlier. Neighborhood whites believed, apparently with good reason, that the movement of blacks into Knight Manor was the opening wedge for designating the entire area for black residency, with the landowners and realtors profiting immensely. In fact, property-transfer records indicate that Bouvier and Wiseheart had purchased a large amount of vacant land in the Edison Center neighborhood. They were well positioned to profit from any racial transition of the area. If Wesley Garrison was managing the racial transformation of Brownsville, Bouvier and Wiseheart were instrumental in facilitating the territorial expansion of Liberty City.[31]

A somewhat different form of second-ghetto expansion was facilitated by Luther L. Brooks, who represented the interests of Overtown slumlords for more than thirty years. Brooks managed and later owned the Bonded Collection Agency, which by 1960 represented more than one thousand white and black landlords who collectively owned more than 10,000 rental units, mostly in Overtown. Managing a fleet of radio-dispatched cars, Luther Brooks was the best-known

ATTENTION CITIZENS

You Are Cordially Invited To Inspect Beautiful "KINGSWAY"

SUNDAY – NOVEMBER 13–2 to 5 P.M.

3500 SOUTH DOUGLAS ROAD

THE ANSWER TO SLUM CLEARANCE

"Better Homes for Better Negroes"

- ★ Triple "A" Construction
- ★ Hurricane and Fire Proof
- ★ Terrazzo Floors
- ★ Modular Wall Units
- ★ Jalousie Doors
- ★ Sunken Garbage Cans
- ★ Insulated Roofs
- ★ All Modern Conveniences

Come and see for yourself

what has been done for Negroes who want good homes – what has been done for Coconut Grove! This development was made possible because of the good judgment and fearlessness of your Miami Planning Board and four of your Miami Commissioners.

These homes constitute the answer, the only answer, to the obstructionists who tried by every means possible to block this housing.

KINGSWAY, INC.

John A. Bouvier Jr., Pres. Malcolm B. Wiseheart, Sec'y

Figure 9.4. Newspaper Ad for South Kingsway Black Housing, Developed by Bouvier and Wiseheart

SOURCE: *Miami Herald*, November 13, 1949.

white man in black Miami, as well as a power broker of sorts in white Miami. Brooks began collecting rents for Miami slumlords in the 1930s, so he was well aware of efforts to relocate blacks from the Overtown area. Defending their investment in Overtown rental housing, Brooks and the slumlords bitterly fought against the new Liberty Square public housing project in the mid-1930s. In the late 1940s and early 1950s, Brooks led the real estate interests in their fight against urban renewal and public housing, and he initially opposed the Overtown route of Interstate 95—all these projects destroyed downtown slum housing and threatened the profits of the slumlords. According to a *Miami Herald* investigation of slum housing in the early 1950s, those profits were very high—averaging 23 percent annually on invested capital. Thus, the interests of these downtown slumlords were quite different from those of such second-ghetto builders as Davis, Garrison, Bouvier, and Wiseheart.[32]

The construction of the Miami expressway system, with its massive destruction of downtown slum housing, ultimately led Brooks and the slumlords into the second ghetto. Once they realized that the expressway route was irrevocable, they also recognized that thousands of Overtown blacks would need new housing somewhere else and that the slum owners could transfer their activities to Brownsville, Liberty City, and nearby white areas of transition. By 1960, Brooks and the Bonded Collection Agency had decisively embraced the second ghetto. For years, the Overtown slumlords had resisted the encroachments of the central business district. In 1961, however, in a long *Herald* article about him, Brooks publicly announced his support for the expansion of the business district into Overtown, and the consequent relocation of the entire African American community to the northwest area around Liberty City. Brooks and the slumlords had come to recognize the profit potential of the second ghetto. As the *Miami Herald*'s urban affairs reporter wrote of Brooks in 1963, he "helped break the boundaries of the old Negro ghettos," moving black families "into border areas, then pushing the borders." In the absence of any official relocation efforts until the mid-1960s, the Bonded Collection Agency conducted an unofficial relocation program of its own, providing the moving trucks as well as the relocation housing in second-ghetto areas. Through his role as a rental agent for black housing, Luther Brooks emerged as an important shaper of Miami's second ghetto.[33]

Brooks, Garrison, and the other realtors and developers who built the second ghetto were thought of as renegades by Miami's mainstream real estate community. After all, they were violating long-accepted canons of the real estate profession. In the 1920s, the National Association of Real Estate Boards (NAREB) drafted a code of ethics that made the sale of homes in white neighborhoods to African Americans a breach of professional standards. The Florida Real Estate Commission replicated the NAREB code on race, as reflected in guidelines published in the late 1930s for Florida realtors. The NAREB deleted the racial provision of its ethics code in 1950, but the policy continued to be observed informally throughout the real estate industry. Real estate people took it for granted that the movement of blacks into white neighborhoods would undermine property values, a position also asserted in the appraisal standards of the Home Owners Loan Corporation, the Federal Housing Administration, and the Veterans Administration. The National Association of Home Builders pursued similarly discriminatory practices, advocating home building for blacks only in segregated neighborhoods. These discriminatory codes and practices prevailed among mainstream real estate people in Miami as well, and they persisted well into the 1950s.[34]

Miami's second-ghetto builders may have appeared as renegades to their real estate colleagues at the Miami Realty Board, but their reputations were secure in the African American community. Luther Brooks did not always get along with Miami's black leaders, but generally he was held in high esteem among his Overtown tenants. Wesley Garrison was often praised effusively by the *Miami Times* for his commitment to black housing and his willingness to break the color line. Bouvier and Wiseheart were commended for their development projects. The builders of the second ghetto, after all, were opening up substantial areas of new housing for African Americans at a time of rapid population growth and when overcrowding in Overtown had reached dangerous and unhealthy levels. The term

Figure 9.5. Typical Newspaper Ad for New Apartment Construction in Liberty City, Although Curiously, a White Woman Is Shown in the Swimming Pool
SOURCE: *Miami Times*, June 3, 1950.

blockbuster generally carried certain pejorative connotations, but we must be sensitive to the fact that African Americans may have had an entirely different perspective on the process of blockbusting and second-ghetto growth. For Miami's African Americans, Wesley Garrison was a true community builder.[35]

Yet altruism had little to do with the building of Miami's second ghetto. The profit motive was a powerful incentive, which explains why some realtors and developers got into the business of black housing. Building the second ghetto was cost-effective and highly profitable. First, the property was cheap, whether it was undeveloped land on the fringes of built-up settlement or existing housing in neighborhoods affected by white flight. Second, for the builders, construction costs were relatively inexpensive. Building codes were nonexistent or rarely enforced in black areas. Thus, builders of black housing could get away with building small houses with poor quality materials, without garages or carports, and with poor plumbing and shabby construction. Apartment builders crammed an excessive number of units on small lots, built mostly one-bedroom units for families with children, and provided no recreation facilities or parking space. Subdivisions plotted by second-ghetto developers generally had poor drainage, no sewers, narrow and often unpaved streets, no sidewalks, no street lighting, and no parks or recreation facilities. Often, these developments were distant from public schools and lacked adequate public transportation to get people to work in the downtown district. Third, FHA-insured mortgage money was often available, so that many developers needed little initial capital. Finally, there was little competition from the real estate mainstream, and there was a big demand for black housing.[36]

The dominant belief in the nation's real estate industry held, as one appraisal expert put it in 1948, that "neighborhoods change, but never for the better." Real estate analysts uniformly ticked off a laundry list of explanations for the decline of residential neighborhoods. Ubiquitously present and high up on that list was the euphemistic notation: "infiltrations of unharmonious racial groups," real estate code for the movement of African Americans into new residential areas. But some real estate people either rejected the

SPECIAL

ANNOUNCEMENT

To

COLORED CITIZENS

Get out of those slums

Do You Own Your Lot?

Then call the

D. & H.

CONSTRUCTION CO. Inc.

Phone 48-1119

For a limited time only, the D. and H. Construction Co., Inc. will build a two bedroom CBS home on your lot in Liberty City, Coconut Grove, or Brown's Sub for the small down payment of one thousand dollars balance payable like rent.

We have built many houses in Miami and Dade County and a telephone call to our office will bring a representative out to show you sample houses. Now is your opportunity to 'GET OUT OF THOSE SLUMS' Do not confuse our homes with "project homes." Every D. and H. Home is individually built and a separte set of blue prints is made by our force of experienced architects and you, as owner, are consulted as to your desires.

STILL AVAILABLE—One beautiful 52 x 100 lot in Brown's Sub.

Call 48-1119
For Further Details

Our Prices Range from $6,950.00 up

Figure 9.6. Newspaper Ad for New Black Housing, Encouraging Migration to Second-Ghetto Neighborhoods

SOURCE: *Miami Times*, April 15, 1950.

"infiltration theory" or sought to profit from it. In fact, a considerable real estate literature at the time argued the case that African Americans made reliable mortgage customers, and that residential property values rose after blacks moved into transitional neighborhoods.[37]

Clearly, Miami's second-ghetto developers treated as myth the widely held idea that African American buyers and renters were poor economic risks. The postwar Miami service and tourist economy, in fact, provided steady employment opportunities for black workers, so a higher proportion of black families had the financial resources to purchase homes than in many other urban areas. As the Greater Miami Urban League noted in 1954, "Enlightened sources view the non-white population as a potential housing market which has not been fully explored." Similarly, a race relations adviser for the Housing and Home Finance Agency noted in a 1948 Miami field report that African Americans were "able, willing, and anxious" for decent housing, "thus providing a substantial market for both rental units and privately owned homes." In a variety of ways, then, Miami's second-ghetto realtors and builders found incentive and profit in their work.[38]

Miami's second ghetto, then, grew as a consequence of black housing demands, public policy decisions, and energetic real estate activity. This process of racial change was not unique to the Miami area. During the decades that spanned midcentury, the racial landscape of urban America was undergoing dramatic transformation. We need more detailed studies of this process of racial/spatial transition. We need to pay closer attention to the key players in the process—the African Americans who aggressively sought new housing, as well as the real estate people, the slumlords, blockbusters, and developers—who collectively were the real shapers and builders of the postwar second ghetto. And we need to recognize that for the African American community in the pre-civil rights era, the second ghetto, because it opened up new and better housing, seemed to promise positive change. The tragedy, of course, is that those promises were never completely fulfilled.

Notes

1. *Miami Daily News*, September 22, 1951; Teresa Lenox, "The Carver Village Controversy," *Tequesta: The Journal of the Historical Association of Southern Florida* 50 (1990), 39-51.

2. On racial zoning, see Christopher Silver, "The Racial Origins of Zoning: Southern Cities From 1910-40," *Planning Perspectives* 6 (1991), 189-205. On restrictive covenants, see Robert C. Weaver, "Race Restrictive Housing Covenants," *Journal of Land and Public Utility Economics* 20 (August 1944), 183-193; Herman H. Long and Charles S. Johnson, *People vs. Property: Race Restrictive Covenants in Housing* (Nashville, 1947); Tom C. Clark and Philip B. Perlman, *Prejudice and Property: An Historic Brief Against Racial Covenants* (Washington, D.C., 1948); Oscar I. Stern, "The End of the Restrictive Covenant," *Appraisal Journal* 16 (October 1948), 434-42; Clement E. Vose, *Caucasians Only: The Supreme Court, the NAACP, and the Restrictive Covenant Cases* (Berkeley, 1959); Jack Greenberg, *Race Relations and American Law* (New York, 1959), 275-312; and, more generally, Davis McEntire, *Residence and Race: Final and Comprehensive Report to the Commission on Race and Housing* (Berkeley, 1960), and Arnold R. Hirsch, "With or Without Jim Crow: Residential Segregation in the United States," in Arnold R. Hirsch and Raymond A. Mohl, eds., *Urban Policy in Twentieth-Century America* (New Brunswick, 1993), 65-99. For evidence that local governments in metropolitan Miami continued to practice racial zoning well into the 1950s, see Frank S. Horne to Warren R. Cochrane, July 11, 1952, Records of the Housing and Home Finance Agency (hereafter referred to as HHFA Records), Record Group 207, box 750, National Archives; Hubert M. Jackson to A. R. Hanson, Miami Field Trip Report, August 15, 1952, ibid.

3. For studies on second-ghetto development, see Arnold R. Hirsch, *Making the Second Ghetto: Race and Housing in Chicago, 1940-1960* (Cambridge, 1983); Thomas J. Sugrue, "The Structures of Urban Poverty: The Reorganization of Space and Work in Three Periods of American History," in Michael B. Katz, ed., *The "Underclass" Debate: Views From History* (Princeton, 1993), 87-117; Thomas J. Sugrue, "Crabgrass-Roots Politics: Race, Rights, and the Reaction Against Liberalism in the Urban North, 1940-1964," *Journal of American History* 82 (September 1995), 551-578; W. Edward Orser, *Blockbusting in Baltimore: The Edmondson Village Story* (Lexington, 1994); Henry Louis Taylor, Jr., "Social Transformation Theory, African Americans and the Rise of Buffalo's Post-Industrial City," *Buffalo Law Review* 39 (Spring 1991), 569-606; several of the essays in Henry Louis Taylor, Jr., *Race and the City: Work, Community, and Protest in Cincinnati, 1820-1970* (Urbana, 1993); Hillel Levine and Lawrence Harmon, *The Death of an American Jewish Community* (New York, 1992); Donald R. Deskins, Jr., *Residential Mobility of Negroes in Detroit, 1837-1965* (Ann Arbor, 1972). For a contemporary statistical study, see Housing and Home Finance Agency, *Housing of the Nonwhite Population, 1940 to 1950* (Washington, D.C., 1952).

4. Walter White and Thurgood Marshall, *What Caused the Detroit Riot?* (New York, 1943); Dominic J. Capeci, Jr., *Race Relations in Wartime Detroit: The Sojourner Truth Housing Controversy of 1942* (Philadelphia, 1984); John F. Bauman, *Public Housing, Race, and Renewal: Urban Planning in Philadelphia, 1920-1975* (Philadelphia, 1987), 161-62; Hirsch, *Making the Second Ghetto*. See also Charles Abrams, *Forbidden Neighbors: A Study of Prejudice in Housing* (New York, 1955), 81-119, 169-90; Robert E. Forman, *Black Ghettos, White Ghettos, and Slums* (Englewood Cliffs, 1971). On the Columbians in Atlanta, see Stetson Kennedy, *The Klan Unmasked* (Boca Raton, 1990), 120-60. For an insightful and influential contemporary dramatization of racially

changing neighborhoods, see Lorraine Hansberry's play, with later film versions, *A Raisin in the Sun* (New York, 1959).

5. *Miami Daily Metropolis,* June 30, July 2, July 8, 1920; *Miami Herald,* July 2, 1920; Paul S. George, "Colored Town: Miami's Black Community, 1896-1930," *Florida Historical Quarterly* 56 (April 1978), 432-77; G. M. Hopkins, *Plat Book of Greater Miami, Fla. and Suburbs* (Philadelphia, 1925); G. M. Hopkins, *Plat Book of Greater Miami, Fla. and Suburbs* (Philadelphia, 1936); "Statement of Edward T. Graham, for the Negro Service Council, Miami, Fla.," *Study and Investigation of Housing,* Hearings Before the Joint Committee on Housing, Miami, Fla., October 27, 1947, 80th Congress, 1st Session (Washington, D.C., 1948), 1019-21. For the early Bahamian immigration, see Raymond A. Mohl, "Black Immigrants: Bahamians in Early Twentieth-Century Miami," *Florida Historical Quarterly* 65 (January 1987), 271-97.

6. Raymond A. Mohl, "Trouble in Paradise: Race and Housing in Miami During the New Deal Era," *Prologue: Journal of the National Archives* 19 (Spring 1987), 7-21; Paul S. George and Thomas K. Peterson, "Liberty Square, 1933-1937: The Origins and Evolution of a Public Housing Project," *Tequesta: The Journal of the Historical Association of Southern Florida* 49 (1988), 53-68.

7. Dade County Planning Board Minutes, August 27, 1936, in George E. Merrick Papers, box 2, Historical Association of Southern Florida, Miami; Dade County Planning Council, "Negro Resettlement Plan," 1937, in National Urban League Papers, Part 1, Series 6, box 56, Library of Congress, Washington, D.C.; George E. Merrick, *Planning the Greater Miami for Tomorrow* (Miami, 1937), 11; *Miami Herald,* April 5, 1945, May 28, 1961; *Pittsburgh Courier,* July 20, 1946; A. L. Thompson to Robert C. Redman, Miami Field Trip Reports, July 12, 1948, November 1, 1949, HHFA Records, Record Group 207, box 750; Frank S. Horne to Julius A. Thomas, December 27, 1949, ibid.

8. On redlining in Miami, see "Security Area Map, Miami, Florida," and "Analysis of Realty Area Map of Miami, Florida," both 1936, Records of the Home Owners Loan Corporation, Record Group 195, National Archives, Washington, D.C. (hereafter cited as HOLC Records); "Security Area Map, Miami, Florida," and "Security Area Descriptions: Metropolitan Miami, Florida," both 1938, HOLC Records, Record Group 195; Mohl, "Trouble in Paradise," 7-21.

9. On racial zoning and "controlled" expansion of Miami's black residential areas, see *Miami Herald,* November 1, 1946, May 29, 1947, December 2, 1952; Dade County Planning Board, *Biennial Report, 1949-1951* (Miami, 1951); Warren M. Banner, *An Appraisal of Progress, 1943-1953* (New York, 1953), 20-24; James W. Morrison, *The Negro in Greater Miami* (Miami, 1962), 6-12.

10. On black population growth, see Raymond A. Mohl, "The Settlement of Blacks in South Florida," in Thomas D. Boswell, ed., *South Florida: The Winds of Change* (Miami, 1991), 112-39. For studies of black housing needs in the 1940s, see Miami Housing Authority, *Report of Low Rent Housing Needs: Miami, Florida and Vicinity* (Miami, 1942); Miami Planning Board, *Dwelling Conditions in the Two Principal Blighted Areas: Miami, Florida* (Miami, 1949); Reinhold P. Wolff, *Greater Miami Population and Housing Survey: Real Estate Division* (Coral Gables, 1949); "Negro Housing in Dade County, *Miami Residential Research* 4 (May 1951), 1-8.

11. Reinhold P. Wolff and David K. Gillogly, *Negro Housing in the Miami Area: Effects of the Postwar Housing Boom* (Coral Gables, 1951); Dade County Planning Board, *Survey of Negro Areas* (Miami, 1951), 8-14, 22-26, 34-36; Elizabeth L. Virrick, "New Housing for Negroes in Dade County," in Nathan Glazer and Davis McEntire, eds., *Studies in Housing and Minority Groups* (Berkeley, 1960), 135-43; Harold M. Rose, "Metropolitan Miami's Changing Negro Population, 1950-1960," *Economic Geography*

40 (July 1964), 221-38; Pat Morrissey, ed., *Miami's Neighborhoods* (Miami, 1982), 19-21, 73-75, 106-8.

12. Miami Planning and Zoning Board, *The Miami Long Range Plan: Report on Tentative Plan for Trafficways* (Miami, 1955); Wilbur Smith and Associates, *A Major Highway Plan for Metropolitan Dade County, Florida, Prepared for State Road Department and Dade County Commission* (New Haven, 1956), 33-44; Raymond A. Mohl, "Race and Space in the Modern City: Interstate-95 and the Black Community in Miami," in Hirsch and Mohl, eds., *Urban Policy in Twentieth-Century America*, 100-58.

13. *Miami Times*, March 16, 1957; Greater Miami Urban League, "Statement on Expressway and Housing," typescript, 1957, National Urban League Papers, Part 1, Series 1, box 107; J. E. Preston to LeRoy Collins, April 26, 1957, LeRoy Collins Papers, box 126, Florida State Archives, Tallahassee; *Miami Herald*, March 4, 1957; Paul C. Watt, "Relocation of Persons Displaced by Highway Construction," Metro-Dade County Manager's Office, Administrative Report, February 13, 1959, mimeo, in Elizabeth L. Virrick Papers, Historical Association of Southern Florida, Miami; Metro-Dade County Planning Department, *Mobility Patterns in Metropolitan Dade County, 1964-1969* (Miami, 1970). On Carol City, see Clyde C. Wooten et al., *Psycho-Social Dynamics in Miami* (Coral Gables, 1969), 531-54; *Miami Herald*, April 2, 1971.

14. Haley Sofge, "Public Housing in Miami," *Florida Planning and Development* 19 (March 1968), 1-4; Mohl, "Trouble in Paradise," 7-21; *Miami Times*, August 20, 1965, September 23, 1966.

15. Dade County School Board, "Orchard Villa Survey," November 13, 1958, typescript, Dade County Public School Collection, Historical Association of Southern Florida, Miami; Shirley Zoloth, Clipping Scrapbook, 1957-1959, in author's possession (Zoloth, a founding member of the Miami chapter of the Congress of Racial Equality in 1958, was deeply interested in school integration during this period); *New York Times*, August 16, 1959; *Miami Herald*, May 29, May 30, 1961. School desegregation in Miami during this period can also be followed in *Southern School News*, 1958-1959, a monthly publication that tracked the progress of integration in the South. The racial change of neighborhoods in relation to school attendance in the 1950s can be traced in Dade County School Board, "Studies Requested in Board of Public Instruction Meeting on August 17, 1955," August 1, 1956, mimeo, in LeRoy Collins Papers, box 139, University of South Florida, Tampa.

16. On Cuban residential patterns, see Morton D. Winsberg, "Housing Segregation of a Predominantly Middle-Class Population: Residential Patterns Developed by the Cuban Immigration Into Miami, 1950-74," *American Journal of Economics and Sociology* 38 (October 1979), 403-18; B. E. Aguirre et al., "The Residential Patterning of Latin American and Other Ethnic Populations in Metropolitan Miami," *Latin American Research Review* 15, 2 (1980), 35-63; and on black-Cuban conflict generally, Raymond A. Mohl, "On the Edge: Blacks and Hispanics in Metropolitan Miami Since 1959," *Florida Historical Quarterly* 69 (July 1990), 37-56.

17. For studies of Miami's "index of residential segregation," see Donald O. Cowgill, "Trends in Residential Segregation of Non-Whites in American Cities," *American Sociological Review* 21 (February 1956), 43-47; Karl E. Taeuber and Alma F. Taeuber, *Negroes in Cities: Residential Segregation and Neighborhood Change* (Chicago, 1965), 39-41; Annemette Sorenson et al., "Indexes of Racial Residential Segregation for 109 Cities in the United States, 1940-1970," *Sociological Focus* 8 (1975), 125-42.

18. Carl T. Rowan, "Why Negroes Move to White Neighborhoods," *Ebony* (August 1958), 17-23. For typical discussions of the subject, see Robert C. Weaver, "Northern Ways," *Survey Graphic* 36 (January 1947), 43-47; George B. Nesbitt, "Relocating

Negroes From Urban Slum Clearance Sites," *Land Economics* 25 (August 1949), 275-88; Charles Abrams, "The Segregation Threat in Housing," *Commentary* 7 (February 1949), 123-31; Charles Abrams, "The New 'Gresham's Law of Neighborhoods': Fact or Fiction," *Appraisal Journal* 19 (July 1951), 324-37; Thomas F. Farrell, "Object Lesson in Race Relations," *New York Times Magazine* (February 12, 1950), 16, 36-37; "Discrimination in Housing: A Debate," *New York Times Magazine* (July 21, 1951), 13, 52-54; "Migratory Population Changes by Race," *Real Estate Analyst* 23 (June 30, 1954), 260-61; Hannah Lees, "Not Wanted: Negro Neighbors," *Atlantic Monthly* 197 (January 1956), 59-63; Paul F. Coe, "The Nonwhite Population Surge to Our Cities," *Land Economics* 35 (August 1959), 195-210. For book-length discussions, see Charles S. Johnson, *Patterns of Negro Segregation* (London, 1944); Robert C. Weaver, *The Negro Ghetto* (New York, 1948); Abrams, *Forbidden Neighbors.*

19. *Pittsburgh Courier,* August 11, November 17, 1945, February 23, 1946, November 15, November 22, 1947; *Atlanta Daily World,* November 11, December 1, 1945; "Lynchings and Mob Violence," 1947, typescript, Papers of the Civil Rights Congress, microfilm edition, Part 2, Reel 17; "Pattern of Violence," *New South* 4 (March 1949), 3-5. The *Pittsburgh Courier,* which had a Florida edition, was widely read in the Miami area. The paper employed a black correspondent, John A. Diaz, as its Florida editor. Diaz regularly reported on black Miami for the *Courier* in the 1940s and early 1950s. No issues of the *Miami Times,* the city's black weekly, survive prior to 1948, which makes the *Pittsburgh Courier* reportage especially important.

20. Ira D. Hawthorne to Fuller Warren, August 28, 1951, September 25, 1951, Fuller Warren Papers, box 22, Florida State Archives; Ira D. Hawthorne to Miami City Commission and Dade County Commission, September 10, 1951, Warren Papers, box 22; Lorine S. Reder to Fuller Warren, August 8, 1951, Warren Papers, box 21.

21. *Miami Herald,* July 14, 1951; *Miami Times,* September 29, December 1, 1951; Stetson Kennedy, "Miami: Anteroom to Fascism," *The Nation* 173 (December 22, 1951), 546-47; Joe Alex Morris, "The Truth About the Florida Race Troubles," *Saturday Evening Post,* June 21, 1952, 24-25, 50, 55-58; William S. Fairfield, "Florida: Dynamite Law Replaces Lynch Law," *The Reporter* 7 (August 5, 1952), 31-34, 41; Abrams, *Forbidden Neighbors,* 120-36; Lenox, "The Carver Village Controversy," 39-51.

22. *Miami Times,* March 2, March 9, March 23, 1951, February 28, 1991; *Miami Herald,* February 5, 1957; *Miami News,* February 25, 1957, September 21, 1958; *Southern School News,* 3 (March 1957), 16; "The Daring Plot Against Miami Negroes," *Jet* 11 (March 28, 1957), 12-15.

23. On the 1939 election, see *Miami Herald,* May 1, May 2, May 3, May 12, 1939; "Miami Klan Tries to Scare Negro Vote," *Life* (May 15, 1939), 27; Ralph J. Bunche, *The Political Status of the Negro in the Age of FDR* (Chicago, 1973), 199-200, 307; Hugh D. Price, *The Negro and Southern Politics: A Chapter of Florida History* (New York, 1957), 23.

24. For the quotation about the Klan, see *Miami Tropical Dispatch,* February 19, 1949. For the civil rights movement in Miami during this period, see Raymond A. Mohl, "The Pattern of Race Relations in Miami Since the 1920s," in David Colburn and Jane Landers, eds., *The African-American Heritage of Florida* (Gainesville, 1995), 326-65; Raymond A. Mohl, " 'South of the South'? Jews, Blacks, and the Civil Rights Movement in Miami, 1945-1960" (unpublished paper).

25. *Miami Tropical Dispatch,* June 7, 1947; Lenox, "The Carver Village Controversy," 42. For the aggressiveness of the *Miami Whip,* see Leon L. Lewis to Fuller Warren, August 2, 1950, Warren Papers, box 21, Florida State Archives. Lewis was editor of the *Miami Whip.*

26. All of the records mentioned above can be researched in the Official Records Library, Office of Dade County Clerk, 44 West Flagler Street, Miami.

27. Reverse-Buyer Index, 1933-1935, 1936-1937, Dade County Official Records Library; Hopkins, *Plat Book of Greater Miami* (1925); Floyd W. Davis to A. R. Clas, July 28, 1935, August 13, 1935, Records of the Public Housing Administration (hereafter cited as PHA Records), Record Group 196, box 301, National Archives; Mohl, "Trouble in Paradise," 8-14.

28. Floyd W. Davis to Dade County Zoning Commission, December 16, December 22, 1941, Dade County Building and Zoning Department Records, Metro-Dade County Government Center, Miami; Davis to Dade County Commission, January 28, 1942, ibid.; Louise Pencke (executive secretary of the Miami Board of Realtors) to I. T. Blount, July 26, 1941, ibid.; Pencke to N. P. Lowrey, January 26, 1942, ibid.; George H. Swanson to Dade County Zoning Commission, August 4, 1941, ibid.; Dade County Klans No. 24 and No. 26 to Dade County Commission, January 21, 1942, ibid.; and dozens of petitions to the Dade County Zoning Commission supporting and opposing the New Myami project, ibid.; *Miami Herald*, July 24, 1941.

29. Reverse-Buyer Index, 1943-1945, 1946-1948, Dade County Official Records Library. The legal challenge to racial zoning, financed by Garrison, can be followed in *Pittsburgh Courier*, August 11, November 17, December 1, 1945, February 16, February 23, March 2, March 30, April 20, May 11, 1946. On the exclusion of African Americans from the Miami mortgage market, see Thompson to Redman, November 1, 1949, HHFA Records, Record Group 207, box 750; "Statement of Edward T. Graham," 1019-21.

30. For Fiftieth Street Heights activity in Brownsville, see Deed Books, 1946-1948, Book No. 2893, pp. 407, 409, 411, 413, Dade County Official Records Library; Dade County Planning Board, *Survey of Negro Areas*, 34. For a Fiftieth Street Heights apartment project, see also Wolff and Gillogly, *Negro Housing in the Miami Area*, 8. Reverse-buyer indexes, 1938-1951, Dade County Official Records Library, demonstrate that Bouvier and Wiseheart had a voracious appetite for property of all kinds throughout the Miami area. For newspaper reportage on Bouvier and Wiseheart real estate activity in black neighborhoods, see *Miami News*, March 23, 1947, November 3, November 28, November 31, 1948; *Miami Herald*, January 5, February 3, February 4, 1949. For the Kingsway project advertisement, see *Miami Herald*, November 13, 1949.

31. Bouvier and Wiseheart purchased a large tract of land near Knight Manor in 1947. See Deed Books, 1946-1948, Book No. 2886, p. 429, Dade County Official Records Library.

32. On Brooks, see *Miami Herald*, April 14, 1963. For slumlord opposition to public housing in the 1930s, see Floyd W. Davis to A. R. Clas, July 28, August 13, 1935, PHA Records, Record Group 196, box 301; John C. Gramling to A. R. Clas, June 24, July 12, 1935, ibid., box 301; Gramling to Clas, June 29, 1935, ibid., box 298. For the 1950s, see Harry Simonhoff, "Low Rent Housing and Negro Segregation," *Jewish Floridian*, March 31, 1950; *Miami Herald*, January 22, May 15, June 25, 1950, March 17-21, 1958; Dade County Commission, Minutes, February 28, 1950, microfilm edition, Florida Room, Miami-Dade Public Library, Miami; Joseph Barnett to F. C. Turner, December 23, 1957, U.S. Department of Transportation Records, Record Group 30, box 70, National Archives; Elizabeth L. Virrick to Arthur Field, March 14, 1958, Elizabeth L. Virrick Papers, Historical Association of Southern Florida; Virrick to Marion Massen, August 31, 1958, ibid; and Elizabeth L. Virrick Clipping Scrapbooks, Virrick Papers, ibid., covering the late 1940s and 1950s. Elizabeth L. Virrick was a housing reformer and social activist from the late 1940s until her death in 1990.

33. Juanita Greene, "He'd Shift Negro District, Build a New Downtown," *Miami Herald*, May 28, 1961; Juanita Greene, "Luther Brooks: Behind-the-Scenes 'Mover,' "

Miami Herald, April 14, 1963; Mohl, "Race and Space in the Modern City," 129-34. For Brooks's obituary, see *Miami Herald,* December 31, 1988.

34. Hirsch, "With or Without Jim Crow," 75; Greenberg, *Race Relations and American Law,* 300-1; W. H. Poe and L. E. Broome, *Principles and Practices of the Real Estate Business* (Orlando, 1937), 122-23; McIntire, *Residence and Race,* 175-98; National Association of Home Builders, *Home Builders Manual for Land Development* (Washington, D.C., 1953), 17. See also Abrams, *Forbidden Neighbors,* 150-68; C. Louis Knight, "Blighted Areas and Their Effects Upon Urban Land Utilization," *Annals of the American Academy of Political and Social Science* 148 (March 1930), 133-38; Rose Helper, *Racial Policies and Practices of Real Estate Brokers* (Minneapolis, 1969). For analysis of race, housing, and neighborhood change presented in typical real estate texts of the era, see Arthur A. May, *The Valuation of Residential Real Estate* (New York, 1942), 73-75, 106-8; Arthur M. Weimer and Homer Hoyt, *Principles of Urban Real Estate* (New York, 1948), 120-41; Henry E. Hoagland, *Real Estate Principles* (3rd ed.; New York, 1955), 64-65, 236-37; Paul F. Wendt, *Real Estate Appraisal: A Critical Analysis of Theory and Practice* (New York, 1956), 132-34.

35. *Miami Herald,* April 14, 1963; *Miami Times,* March 15, 1950, September 29, 1951, November 22, 1952, September 5, 1953, March 23, 1957; *Miami Tropical Dispatch,* February 12, 1949. The interests of the Miami Realty Board are reflected in Stuart B. McIver, *The Greatest Sale on Earth: The Story of the Miami Board of Realtors, 1920-1980* (Miami, 1980).

36. Dade County Commission, Minutes, April 3, 1945, microfilm edition, Florida Room, Miami-Dade Public Library; Wolff and Gillogly, *Negro Housing in the Miami Area,* 3-11; Dade County Planning Board, *Survey of Negro Areas;* Elizabeth L. Virrick, "Our Slums: Expensive Incubator" (Miami Foundation for Civic Education), *The Citizen's Business* 1 (September 1953), 1-2; Elizabeth Virrick, "Negro Housing in Dade County," April 1956, typescript, in Virrick Papers; *Pittsburgh Courier,* August 30, 1947.

37. George L. Schmutz, "Sidelights on Appraisal Methods," *Review of the Society of Residential Appraisers* 14 (June 1948), 18; George A. Phillips, "Racial Infiltration," ibid., 16 (February 1950), 7-9. For favorable discussion within the real estate and housing industry about African Americans as home buyers, see Elsie Smith Parker, "Both Sides of the Color Line," *Appraisal Journal* 11 (January 1943), 27-34; (July 1943), 231-49; "Realty for Negro," *Business Week,* August 25, 1945, 54-58; George W. Beehler, Jr., "Colored Occupancy Raises Values," *Review of the Society of Residential Appraisers* 11 (September 1945); Oscar I. Stern, "Long Range Effect of Colored Occupancy," ibid., 12 (January 1946), 4-6; Belden Morgan, "Values in Transition Areas," ibid., 18 (March 1952), 5-10; Margaret Kane, "Opportunities in a Neglected Market," *Insured Mortgage Portfolio* 13 (Fourth Quarter 1948), 6-8, 32-33; Margaret Kane, "A Wider Field for Mortgage Lending," ibid., 14 (Fourth Quarter 1949), 15-18; C. O. Stuart, "Private Enterprise Clears a Slum," ibid., 16 (Third Quarter 1951), 3-5; Roland M. Sawyer, "FHA Program Aids Minorities," ibid., 16 (Fourth Quarter 1951), 6-9; Stanley W. Kadow, "Observations on the Minority-Group Market," ibid., 18 (Summer 1954), 16-18; Luigi M. Laurenti, "Effects of Nonwhite Purchases on Market Prices of Residences," *Appraisal Journal* 20 (July 1952); Catherine Bauer, "The Home Builders Take a 'New Look' at Slums—and Raise Some Questions," *Journal of Housing* 10 (November 1953), 371-73, 389; Norris Vitchek, "Confessions of a Block-Buster," *Saturday Evening Post* 235 (July 14, 1962), 15-19; Chester Rapkin and William G. Grigsby, *The Demand for Housing in Racially Mixed Areas* (Berkeley, 1960), 81-105.

38. H. Daniel Lang, *Food, Clothing and Shelter: An Analysis of the Housing Market of the Negro Group in Dade County* (Miami, 1954), 35; Thompson to Redman, July 12, 1948, HHFA Records, Record Group 207, box 750.

10

African Americans in the City

The Industrial Era, 1900-1950

Joe W. Trotter

ver the past three decades, African American urban history emerged as a new scholarly field within American urban and Afro-American history. Black urban history did not unfold within a vacuum, however. It built on its late-nineteenth- and early-twentieth-century roots in the discipline of sociology. From sociology, urban history adopted the community study format and explored virtually every aspect of black urban life—its spatial, class, and particularly, racial dimensions. Although the field benefited from its sociological heritage, it also transcended that heritage by emphasizing dimensions of change over time. It demonstrated how one era in the black urban experience gave way to another and set the stage for a new round of social transformations.[1]

Author's Note: The author wishes to thank the University of Illinois Press and Indiana University Press for use of materials previously published in his book *Black Milwaukee: The Making of an Industrial Proletariat, 1915-45* (Urbana, 1985) and his edited volume *The Great Migration in Historical Perspective: New Dimensions of Race, Class, and Gender* (Bloomington, 1991).

As African American urban history fully emerged during the 1960s and 1970s, urban community studies not only proliferated but proved remarkably resilient. They accommodated a growing array of chronological units, geographical areas, topics, theoretical frameworks, and methodologies.[2] In recent years, for example, black urban history expanded from its original focus on northern cities to a focus on the urban South and West; from the period before 1930 to the Depression and World War II; from ideas, thoughts, and responses of black elites to increasing emphasis on those of black workers; from preoccupation with black men to a rising interest in the history of black men, women, and gender issues; and from extensive interest in the conjunction of social, economic, and political issues to a growing focus on the intersection of these with the cultural dimensions of black urban life.[3] Accordingly, this essay examines how the field developed over the past thirty years. It shows how the field evolved out of the discipline of sociology, entered the historical discipline, and incorporated a growing variety of regions, topics, and conceptual orientations. Finally, this essay suggests that despite existing blind spots and areas that require additional research, the field matured and is now strong enough to support a historical synthesis of black urban life during the first half of the twentieth century.

The Sociological Heritage

Black urban history is deeply rooted in the discipline of sociology. Its sociological genesis is tied to the publication of W. E. B. Du Bois's *The Philadelphia Negro: A Social Study* (1899). Working under the auspices of the sociology department at the University of Pennsylvania, Du Bois studied Philadelphia's black community, particularly its Seventh Ward, for over fifteen months. He constructed an elaborate questionnaire, consulted a variety of contemporary and historical sources, and examined virtually every aspect of life among the city's black residents. The study covered patterns of population growth and spatial distribution, education, and literacy; economic status, social conditions, and institutional life; politics, social classes, and especially, race relations, which penetrated and shaped his concern with other facets of black life in the city. Like white social reformers in the United States and Europe during the late nineteenth century, Du Bois adopted the tenets of the expanding social "sciences." He believed that the compilation and analysis of statistical data and "facts" would lead to the solution of social problems, including the problems of poverty, crime, disease, and especially, the constraints of racism and prejudice that hampered the lives of urban blacks. While analyzing numerous dimensions of black life in the city, as Du Bois put it, the final "design" of the study was "to lay before the public such a body of information as may be a safe guide for all efforts toward the solution of the many Negro problems of a great American city."[4]

The Philadelphia Negro helped to launch the field of American and African American urban sociology. Yet it exhibited important limitations. Although Du Bois offered a brief historical context for his study, he offered little guidance on how the black urban experience changed over time, how one era gave way to another, and how such changes set the stage for a new round of social transformations. Moreover, for all of his attention to the internal differentiation of black life by region of birth and social class, he tended to view southern blacks as devoid of cultural resources that enabled them to cope with life in the city; he also gave insufficient attention to class as a process, whereby blacks participated in a larger pattern of working-class formation at the same time that they became increasingly segregated in one area of the urban landscape. Subsequent studies would build on his splendid beginning, but also exhibit his central shortcomings.

Black urban life was the subject of only a few studies before World War I, but gained increasing scholarly attention with the onset of the Great Migration. Reinforcing Du Bois's emphasis on race relations were a growing number of studies that focused on the magnitude, causes, and consequences of black urban migration. Although focusing explicitly on patterns of southern black migration to northern cities, the migration studies adopted the essential components of the race relations model. During World War I and its early aftermath, a variety of studies highlighted the interrelationship between black migration and changes in patterns of race relations. Although most of these studies focused on black migration to a variety of cities, the most comprehensive of them—Charles S. Johnson's *The Negro in Chicago: A Study of Race Relations and a Race Riot* (1922)—offered the most explicit illustration of race relations scholarship during the war and early postwar years. Conducted for the Chicago Commission on Race Relations, *The Negro in Chicago* examined all facets of the black urban experience and illuminated the forces that led to the eruption of racial violence. Like Du Bois, however, Johnson's study used history as a backdrop to understanding race relations, offered little insight into shifting class relations within the black urban community, and gave insufficient attention to the role of southern black culture in shaping the black urban experience. Although *The Negro in Chicago* added another city to knowledge of black urban life, coming as it did in the wake of the bloody Chicago Race Riot of 1919, it helped to link scholarship on black urban life even more closely to the race relations paradigm.[5]

In the meantime, sociologists at the University of Chicago elaborated on the race relations model and set the stage for a new decade of scholarship. In 1913, the sociologist Robert E. Park left Tuskegee Institute and joined the Chicago school of sociologists at the University of Chicago. In the midst of the increasing migration of blacks into urban America, Park elaborated on his theory of the race relations cycle. Using insights from his research on West Coast Asians and Hawaiians, as well as observations on black life in the rural South, Park concluded that race relations evolved through four major phases: contact, competition, accommodation, and assimilation. Although he initially focused on black life in the South, he increasingly turned to the implications of southern black migration to the urban North. By the late 1930s, Park had moved from a racial caste conceptualization of blacks within the race relations cycle to the increasing view of blacks as a racial minority. The Great Migration, he argued, diminished the distances between blacks and whites at the different class levels and set in motion the transformation of black Americans from a "caste to that of a racial minority." Consequently, Park set the experiences of blacks within a broader framework that offered comparisons with the history of European immigrants. Although he recognized the complications that attended race relations in practice, he detracted from the historical aspects of his theory by arguing that the process universally repeated itself and was "apparently progressive and irreversible."[6]

Ernest Burgess, Roderick McKenzie, and E. Franklin Frazier elaborated on Park's ideas of the race relations cycle and developed a complex class and spatial analysis of black urban life. Frazier became the most energetic student of Park's ideas, and, based on field research in Chicago and later Harlem, he soon published *The Negro Family in Chicago* (1932), his Ph.D. dissertation at the University of Chicago, and "Negro Harlem: An Ecological Study" (1937). Although he used black families as the primary focus of his study, he examined a broad range of social, economic, and political dimensions of black life; he showed how the black population arranged itself by social class across the urban landscape, making it difficult to sustain the notion of a uniformly depressed black ghetto; and, like Du Bois, he used his research to illuminate patterns of black-white interactions.[7]

Frazier also emphasized the disorienting effects of urbanization on black families and their communities. He underplayed the value of southern black culture in helping blacks to reestablish their lives in an urban context. Nonetheless, Frazier highlighted how racial barriers aggravated race relations and prevented blacks from fully participating in American society. To his credit, he also argued that the "world of the city" was not merely "a destroyer," but also "a builder, of traditions"—if only whites would permit black life to take an unhampered course in the larger political economy.[8] Still, like his earlier counterparts, Frazier used history to set the general context for his discussion of the late 1920s and early 1930s, but showed little interest in systematically documenting the ways that black urban family and community life emerged out of earlier patterns and set the stage for later changes. Indeed, although he offered important insights into the development of social classes within black Chicago, he was primarily interested in the spatial manifestations of these classes, rather than in the historical development of classes and how they shaped the black urban experience. Although his research on Harlem expanded the number of cities under investigation, his choice of cities reinforced knowledge of black life in the larger urban centers, but left open the need for careful studies of black life in other cities, where black populations also dramatically increased in the wake of the Great Migration. Moreover, by building on Park's race relations cycle theory, Frazier's work raised but did not resolve the issue of blacks as another minority group. Social anthropologists would soon challenge this proposition and push scholarship on black urban life in a different direction.

As Frazier, Park, and other members of the Chicago school forged the race relations cycle theory of black urban life, others followed the lead of social anthropologist W. Lloyd Warner. Building on his work among the Murgin ethnic group in Australia, *A Black Civilization* (1937), the *Yankee City* series on New England towns (1941), and, with Burleigh and Mary Gardner, the study of southern race relations in *Deep South* (1941), Warner challenged the field's growing emphasis on blacks as merely another ethnic minority.[9] Accordingly, Warner and his associates adopted the "caste and class" model of race relations, emphasizing the persistence of racial separation in social, economic, cultural, and political life of town and country in *Deep South.* Although blacks and whites developed internal class systems, they argued that blacks occupied a "lower caste" and found it exceedingly difficult to challenge both the legal and extralegal barriers that they faced. Thus, for Warner and others, the barriers facing blacks offered compelling evidence that they could not be construed as another minority group.

Cognizant of the limitations of their work on black life in the deep South, Warner and his associates quickly turned to research on black life in a northern industrial city. Warner's ideas soon gained expression in the research and writings of St. Clair Drake and Horace R. Cayton. In 1945, Drake and Cayton published their classic two-volume study *Black Metropolis: A Study of Negro Life in a Northern City.* Drake received his training in social anthropology at Dillard University and at the University of Chicago; he had also worked for two years on the Deep South project. For his part, Cayton had obtained his training under Robert E. Park at the University of Chicago. Working collaboratively with W. Lloyd Warner, Drake and Cayton brought the disciplines of sociology and social anthropology together in a single study. *Black Metropolis* exceeded previous studies of black urban life in important ways. It expanded on the race relations cycle model, but included an exceedingly detailed analysis of black cultural life (using the community's own self-designation as "Bronzeville"), along with the usual emphasis on black social, economic, political, and institutional life. It offered the richest and most complete discussion of black life within the lower, middle, and upper classes of the African American community.[10]

Unlike other sociological studies of the period, *Black Metropolis* also reflected the growing militance of the black community itself. Drake and Cayton viewed the issue of racial violence and friction from a different angle than their earlier counterparts. For example, as they put it,

> It is conceivable that the Negro question—given the moral flabbiness of America—is incapable of solution. Perhaps not all social problems are soluble. Indeed it is only in America that one finds the imperative to assume that all social problems *can* be solved without conflict. To feel that a social problem cannot be solved peacefully is considered almost immoral. Americans are required to appear cheerful and optimistic about a solution, regardless of evidence to the contrary.[11]

Despite important differences from other race relations studies, Drake and Cayton reinforced scholarship on one of the nation's largest cities and continued to emphasize patterns of black-white interactions. While adding new cultural data and avoiding Warner's use of "caste," for example, they adopted the notion of "color line" to discuss the nature of race relations, giving such relations a prominent position in their conceptualization. Drake and Cayton also adopted the term *ghetto* as the most powerful visual evidence of the color line in "Bronzeville." As such, their study served to reinforce Warner's model of caste and class. As Warner put it, "This evidence strongly supports the hypothesis that, while there is a noticeable difference between *Deep South* and *Black Metropolis* . . . the *type* of status relations controlling Negroes and whites remains the same and continues to keep the Negro in an inferior and restricted position. . . . In short, there is still a status system of the caste type."[12]

In the early postwar years, Robert Weaver built on earlier sociological notions of the "ghetto" and offered a pointed historical argument. In his book *The Negro Ghetto* (1948), Weaver argued that the black ghetto was a relatively recent development. According to Weaver, the pattern of black housing segregation in northern cities was a product of the massive migration of blacks to northern cities during World War I. As we will see, the first generation of black urban historians would take sharp issue with Weaver's chronology and push the origins of the northern black ghetto back toward the turn of the twentieth century. Even before Weaver elaborated on Drake and Cayton's notion of the ghetto as the most significant visual image of the color line, however, the Swedish economist Gunnar Myrdal had adopted the larger caste-class idea in his massive study *An American Dilemma: The Negro Problem and Modern Democracy* (1944).[13] Based on the premise that racial discrimination contradicted white America's fundamental ideals of social democracy and human equality, Myrdal's ideas gained widespread acceptance during the postwar years. In 1954, for example, the U.S. Supreme Court used his ideas to support its decision to desegregate public schools. Yet it was Park's race relations cycle theory that regained a footing by the late 1950s. Under the growing impact of the Cold War and under the expanding civil rights movement, Park's notion of blacks as another immigrant group appeared even more persuasive to some analysts.[14]

Between 1900 and 1950, a variety of studies advanced knowledge of black life in cities and established the foundation for the emergence of black urban history as a field. These studies would bequeath to scholars of black urban history important theoretical and methodological approaches, particularly the urban community study. Studies of black urban community life provided insights into a broad range of demographic, social, economic, political, and cultural dimensions of black life in cities. Spatial, class, and especially, race relations gained significant attention. Yet black urban sociology offered little guidance on how the black urban experience changed over time, how one era supplanted another, and how such changes set in motion forces that would shape subsequent developments. Moreover, for all their sensitivity to the internal differentiation of black life by social class and region of origin, such studies viewed class in largely static rather than dynamic terms, that is, as a process of class formation; and, more important, such studies underplayed the very positive impact of lower-class southern black culture on black life in cities. Finally, black urban sociology made its contributions on the basis of a few major cities of the North and Midwest, giving little attention to black life in the urban South, West, and even smaller cities of the urban North. Nonetheless, in the postwar years, historians would build on this sociological heritage, emphasize dimensions of change over time, and carve out a place for black urban life in the discipline of history.

Emergence and Development of Black Urban History as a Field

Despite the dramatic resurgence of black migration to urban America during and following World War II, historians gave little attention to black urban life during the 1950s. Under the impact of the modern civil rights movement, they turned their attention to the origins of Jim Crow, the institution of slavery, and the early emancipation years. When they did turn to research on black urban life, as Kenneth Kusmer notes, they tended to turn as often to black life in antebellum cities as to black life in the twentieth century. Nonetheless, in his book *The Newcomers: Negroes and Puerto Ricans in a Changing Metropolis* (1959), historian Oscar Handlin helped to reestablish the image of urban blacks as another immigrant group and framed questions that the first generation of black urban historians would ask—that is, were African Americans simply the latest immigrant group to arrive in American cities? Was it merely a matter of time before they would gain a foothold on the urban escalator, experience upward mobility, and join European immigrants in a multiethnic city?[15]

Black urban history fully emerged during the 1960s and 1970s. Responding to the rapid growth of nearly all black communities in the urban North, the first generation of black urban historians focused almost exclusively on African American life in northern cities. Scholars built on the preceding sociological studies, adopted the urban community study format, and used the "ghetto" as the primary conceptual and theoretical framework for understanding the black urban experience. In other words, historians incorporated the chief strengths of the earlier sociological tradition, placed the canons of historical scholarship up front, and provided new and much needed temporal perspectives. As such, they documented the origins and development of the ghetto, locating its beginnings in the events of the late nineteenth and early twentieth centuries. They also dispelled the notion that blacks were simply another immigrant group—by documenting the critical role that white racial hostility and prejudice played in the growth and development of black urban communities. As I have written elsewhere, the ghetto synthesis—emphasizing the external and internal forces that gave rise to nearly all black communities in northern cities—fully emerged in a series of studies published between 1963 and 1980: Gilbert Osofsky's *Harlem: The Making of a Ghetto, 1890-1930* (1963); Allan Spear, *Black Chicago: The Making of a Negro Ghetto, 1890-1920* (1967); David Katzman, *Before the Ghetto: Black Detroit in the Nineteenth Century* (1973); Kenneth L. Kusmer, *A Ghetto Takes Shape: Black Cleveland, 1870-1930* (1976); and Thomas Philpott, *The Slum and the Ghetto* (1978).[16]

During the 1980s and early 1990s, however, historians expressed growing dissatisfaction with certain features of African American urban history. Despite its numerous strengths, the ghetto synthesis tended to overplay the significance of residential segregation, treated important facets of black life in highly pathological terms, and gave inadequate attention to the importance of class formation within the black urban community. In his *Alley Life in Washington: Family, Community, Religion, and Folklife in the City, 1850-1970* (1980), James Borchert emphasized the ways that black migrants' primary group and folk culture enabled them to survive their difficult and harsh encounter with the urban environment. As such, he also helped to facilitate the emergence of new perspectives on African American urban life. Studies by Earl Lewis, *In Their Own Interests: Race, Class, and Power in Twentieth-Century Norfolk, Virginia* (1991); Darlene Clark Hine, "Black Migration to the Urban Midwest: The Gender Dimension, 1915-1945" (1991); James Grossman, *Land of Hope: Chicago, Black Southerners, and the Great Migration* (1989); Peter Gottlieb, *Making Their Own Way: Southern Blacks Migration to Pittsburgh, 1916-30* (1987); Dennis Dickerson, *Out of the Crucible: Black Steelworkers in Western Pennsylvania, 1875-1980* (1986); and Joe W. Trotter, *Black Milwaukee: The Making of an Industrial Proletariat, 1915-45* (1985), all pinpointed certain limitations of the ghetto model and sought to overcome them.[17] "Ghettoization," these studies suggest, "was indeed a significant development, but the ghetto dwellers were also Afro-Americans from agricultural, urban domestic, personal service, and common laborer jobs moving into factory employment."[18]

Recent scholarship exhibits increasing sensitivity to the dynamics of migration, working-class formation, the role of black women, and cultural issues. Until recently, for example, scholars of black population movement emphasized how blacks were *pushed* out of the rural South by intolerable socioeconomic and political conditions, on one hand, and *pulled* into the urban North by the labor demands of wartime production, restrictions on European immigration, better race relations, and access to citizenship rights, on the other. Recent studies challenge this static image of black migration. Although recognizing the discriminatory social and economic conditions that blacks faced in both rural and urban economies, the new scholarship demonstrates how African Americans used their kin and friendship networks, pooled their resources, shared information, and played a pivotal role in organizing their own movement into the expanding cities of the North and South. In short, these studies suggest that southern black culture played a dynamic and very positive role in the development of the black proletariat and the growth of black urban communities. As historian Peter Gottlieb put it, black migration to the urban North represented a process of African American "self-transformation." Similarly, James Grossman documents the Great Migration as a "grassroots" social movement. Moreover, as

Clark Hine notes, these processes were quite gendered and involved the agency of black women no less than black men.[19]

During the 1970s and 1980s, black urban history incorporated a growing number of new geographical areas, cities, topics, temporal units, and theoretical and methodological approaches. Although initially expanding our understanding of the larger urban centers like Chicago and New York, black urban history brought under increasing scrutiny smaller black populations in the urban North, South, and West. Since the early 1970s, for example, the field added careful studies of black communities in Detroit, Cleveland, Milwaukee, and Evansville, as well as Washington, D.C., Louisville, and Norfolk. Topeka, San Francisco, Los Angeles, and Seattle also received increasing scholarly attention.[20] Black urban history not only accommodated a growing range of regions and cities of various types but also generated interest in a number of specialized topics. In varying degrees, studies of black urban education, women, family life, business, professions, and culture all expanded as potential subfields. Studies by Michael Homel, Vincent Franklin, and Judy Jolly Mohraz illuminated the history of black education in cities like Chicago, Philadelphia, and Indianapolis.[21] In his seminal study, *The Black Family in Slavery and Freedom* (1976), labor historian Herbert Gutman pushed historical research on the black family into the early-twentieth-century city. He argued that in southern and northern cities, the black urban family remained a viable two-headed structure despite the destructive heritage of slavery and the disruptions of black migration from the rural farms to urban America. Although studies of black women, like their white counterparts, unfolded in close conjunction with family studies and challenged Gutman's implicit assumption that female-headed families were necessarily weak, scholars of black women's history gradually pushed the history of black urban women in a variety of new directions. Studies by Darlene Clark Hine, Jacqueline Jones, Elsa Barkley Brown, Evelyn Brooks Higginbotham, Earl Lewis, and Dolores Janiewski, among others, all reflect new directions in the history of black women in cities.[22] Seminal studies of the black press, insurance companies, nurses, social workers, the Harlem Renaissance, the

Garvey Movement, and black folk and popular culture also slowly emerged during the period.[23]

Scholars also examined new dimensions of black politics, comparative black-white group experiences, and race relations. In her study *Race and Ethnicity in Chicago* (1987), Dianne Pinderhughes offers an alternative to the pluralist models of black urban political history and places black politics within a comparative framework. Focusing on the experiences of Afro-Americans, Italians, and Poles between 1910 and 1940, she systematically reconstructs the economic and historical context in which these groups existed and compares the "route and speed of entry" into political participation, as measured by voting, nomination, and election to public office, and access to municipal jobs, education, and consideration before the criminal justice system. Contrary to pluralist models, she argues that Chicago developed a "racial hierarchy," which repeatedly subordinated blacks to other groups.[24] Other historical studies of comparative group experiences reached similar conclusions, although they often underplayed important aspects of black community and cultural life.[25] Along with politics and comparative group experiences, studies of urban racial violence proliferated. Since publication of William Tuttle's *Race Riot: Chicago in the Red Summer of 1919* (1970) and Elliott Rudwick's *Race Riot at East St. Louis, July 2, 1917* (1964), for example, we now have studies of racial violence in a variety of southern and northern cities: Tulsa, Atlanta, Houston, New Orleans, Detroit, and Springfield. In her study of the Springfield race riot, historian Roberta Senechal analyzes the available historical and social science literature and offers the provocative conclusion that the Springfield conflict resulted from "factors less visible and less tangible than a measurable shortage of jobs, black residential expansion or labor strife."[26]

As black urban history expanded the geographical and topical dimensions of its work, it also broadened its chronological boundaries. Whereas studies of the 1960s and 1970s focused on the pre-Depression years, the 1980s and early 1990s brought increasing attention to the 1930s and 1940s. Recent studies illuminate developments in southern, northern, and western cities during the war and interwar years, giving close attention to the ways that developments of the

1920s gave way to events of the Depression and World War II, particularly the local manifestations of the New Deal, the Congress of Industrial Organizations, and the March on Washington Movement to end discrimination in defense industries.[27] On the other hand, studies by Mark Naison, Robin D. G. Kelley, Cheryl Lynn Greenberg, and Nancy Grant focus exclusively on the Depression years. Naison documents the special role of the Communist Party in Harlem, and Kelley illuminates the role of the party in the urban and rural South. Greenberg offers a much needed community study of black life in the nation's largest metropolis during the period. Framing her study with the Harlem Riots of 1935 and 1943, Greenberg documents the impact of racial discrimination and segregation on Harlem residents, their political responses, and white reactions to black political demands. She also shows how blacks, insecure during the 1920s, sank even deeper into unemployment, poverty, and suffering during the 1930s. Still, Greenberg concludes that the New Deal programs helped numerous African Americans and their families to survive hard times. Reinforcing research on black urban life during the Depression and World War II is Nancy L. Grant's study of southern blacks and New Deal programs, covering such cities as Chattanooga and Knoxville.[28]

African American urban history not only added new time periods, topics, and regional interests but also adopted new theoretical perspectives and methodologies. Whereas the ghetto model made the issue of increasing spatial segregation central to its mission, by the 1980s, writers exhibited a growing interest in strengthening various aspects of the model itself as well as with developing alternative approaches. Kenneth Kusmer, for example, aimed to strengthen the framework by introducing a comparative perspective that emphasized the differential timing, causes, and consequences of ghetto formation across regions. According to Kusmer, it was large northern cities like Chicago, New York, and Philadelphia that exhibited the most intense forms of racially segregated urban spaces, whereas southern and western cities had the lowest levels of ghettoization. Moreover, according to Kusmer, a midwestern city like Cleveland exhibited patterns of ghetto formation that represented a medium between extremes. In a groundbreaking essay nearly a decade later, Kusmer also deepened our understanding of the causal factors that promoted the rise of black ghettos, by adding what he called structural factors (e.g., certain urban technological changes as in the transition to electric streetcars during the late nineteenth century) alongside the usual emphasis on internal and external forces, that is, the interplay between white racism and black responses.[29] Although he situated his study within the framework of the ghetto model, historian James Borchert broke ranks with the prevailing ghetto conceptualization. He documented the lives of Washington, D.C.'s poorest black residents, the alley dwellers. He imaginatively reconstructed alley life from the usual primary and secondary sources, but added the novel use of photographic analysis to his repertoire of research tools. Moreover, he was intensely interested in the very positive ways that migrants retained a cohesive culture that enabled them to survive the rigors of life in the city. As he put it, "In contrast to the 'conventional' social scientific and historical theory that folk migrants undergo a period of disorder as a result of their exposure to urban life, migrants actually used their primary groups and folk experiences to create strategies which enabled them to survive the often harsh and difficult urban experience."[30]

Urban historian Earl Lewis opens the field to yet another theoretical formulation—what he calls "the relationship between power and culture in the industrial phase of American history." Lewis emphasizes how African Americans developed strategies for empowerment that focused on improvements at home, at work, or in some combination of the two. He argues that black urban history needs to account for the shifting relationship between the different strategies that urban blacks devised and "the culture" that they "constantly recreated, a culture that bound blacks to one another, even while it distinguished the working class from the elite."[31] In his new book on Detroit, a radically revised version of his 1976 Ph.D. dissertation, historian Richard Walter Thomas abandons his earlier proletarian framework and adopts what he calls a "community-building" approach to the black urban experience.[32] Finally, as discussed above, I advanced the notion of proletarianization or working-class formation as an alternative, but complementary, framework to ghetto formation. *Black Milwaukee* emphasized the expansion of the black working class and the emergence of new black elites, which produced inter- and intraracial tensions and conflicts that shaped black life in cities no less than the expanding urban ghetto itself.[33] These different conceptualizations and approaches suggest that black urban history not only emerged during the 1960s and 1970s, it also developed the necessary flexibility to grow beyond its initial conceptual and methodological boundaries. With the exception of David Katzman's study of Detroit,[34] most historians eschewed the earlier race relations and caste-class approaches to black urban life, but nonetheless adopted the urban community format. Thus, black urban history still owes much to its sociological past.

Areas for Future Research

Despite important strides in the development of African American urban history, as the 1990s got under way, important conceptual and substantive gaps remained. Black women and questions of gender received increasing attention, for example, but few studies conceptualize the black male's experience in explicitly gender terms. Moreover, the black family as a dynamic and changing institution is an uneven and incomplete dimension of black urban history. Similarly, few studies analyze the history of black education and literacy in smaller centers of black population and in the urban South and West. Indeed, historians give insufficient attention to education and literacy, even within established historical case studies of black life in large northeastern and midwestern cities.[35] The same can be said about black urban health and medical history.[36] In short, we need much better insights into gender issues, literacy, and the development of Afro-American medical knowledge and practices and how they changed as blacks moved into cities in increasing numbers.

Black business and professional people received the lion's share of attention in early studies of black urban history. Therefore, until recently we knew far more about them than we did about other segments of the black community. As suggested above, however, we are now revising our understanding of the role of black workers and women in shaping black urban life. Thus, a study of black elites that takes this important scholarship as a starting point should prove especially enlightening. Moreover, whereas specialized studies by Walter Weare and Alexa Benson Henderson deepen our knowledge of black business and professional people in the urban South, we now need similar studies of northern and western cities. New studies of black economics, culture, and politics are also very much needed. Despite extensive knowledge of black urban political behavior and beliefs, we have yet to explore what Robin Kelley calls "infrapolitics," that is, how informal modes of black political culture intersected with and shaped larger and more formalized patterns of urban institutional life, culture, and politics. In a groundbreaking paper on the subject, Kelley suggests how such a study would uncover a variety of "sequestered social spaces that tend to fall between the cracks of political history" and reintegrate these into discussions of the larger political economy. Such studies would also address more fully the impact of wars, military service, and the explosion of mass-consumerism on the black urban experience.[37] Finally, if black urban history is to remain a vibrant field of scholarship, it will need to explore a broad range of other issues and possible lines of research—studies of black urban life that take the neighborhood, region, and state as units of analyses; studies of black immigrants; comparative studies (of black life in different cities, of blacks and other nonwhite ethnic groups); analyses of substantially long periods of time, including the post-World War II era; and studies of the specific impact of blacks on the city-building process itself. Fortunately, new books by Quintard Taylor and Henry Louis Taylor illuminate the development of African American urban communities over more than a century of time and clarify the relationship between race, class, and the city-building process.[38]

Conclusion

Although there is much work yet to be done, black urban history has moved a long way from its sociological roots. It has also moved beyond its initial formulation in the ghetto synthesis. Following its inception as a historical specialty, it slowly and then rapidly incorporated new geographical areas, cities, topics, time periods, and theoretical and methodological approaches. Available scholarship suggests that the rise of black urban America was an exceedingly complex process, which varied tremendously over time, from city to city, from region to region, by class, by gender, and by a host of other variables. Yet its central theme seems quite clear—in the face of hostility from without and substantial fragmentation from within, African Americans created a new community in the urban environment. The complicated transition of agricultural workers to a new urban-industrial foundation stood at the center of this process. In short, over the past thirty years, black urban history matured, and despite existing shortcomings, provides the basis for a new and more dynamic black urban history.

Notes

1. Kenneth Kusmer, "The Black Urban Experience in American History," in Darlene Clark Hine, ed., *The State of Afro-American History: Past, Present, and Future* (Baton Rouge, 1986); Joe W. Trotter, "Afro-American Urban History: A Critique of the Literature," in Joe W. Trotter, *Black Milwaukee: The Making of an Industrial Proletariat, 1915-45* (Urbana, 1985), 264-82; John H. Stanfield, "The 'Negro Problem' Within and Beyond the Institutional Nexus of Pre-World War I Sociology," *Phylon* 43 (Fall 1982), 187-201; Stanfield, "Race Relations Research and Black Americans Between the Two World Wars," *Journal of Ethnic Studies* 11 (Fall 1983), 61-93; Walter A. Jackson, *Gunnar Myrdal and America's Conscience: Social Engineering and Racial Liberalism, 1938-1987* (Chapel Hill, 1990); James E. Blackwell and Morris Janowitz, eds., *Black Sociologists: Historical and Contemporary Perspectives* (Chicago, 1974).

2. Kusmer, "The Black Urban Experience in American History"; Trotter, "Afro-American Urban History: A Critique of the Literature"; Joe W. Trotter, *The Great Migration in Historical Perspective: New Dimensions of Race, Class, and Gender* (Bloomington, 1991).

3. Kusmer, "The Black Urban Experience in American History"; Trotter, "Afro-American Urban History: A Critique of the Literature"; Trotter, ed., *The Great Migration in Historical Perspective.*

4. W. E. B. Du Bois, *The Philadelphia Negro: A Social Study* (1899; reprinted New York, 1967), 1.

5. Charles S. Johnson (for Chicago Commission on Race Relations), *The Negro in Chicago: A Study of Race Relations and a Race Riot* (1922; reprinted New York, 1969).

6. Trotter, "Afro-American Urban History," 265-66.

7. Ibid., 266-68.

8. E. Franklin Frazier, *The Black Family in Chicago* (Chicago, 1932), 243.

9. W. Lloyd Warner, "A Methodological Note," in St. Clair Drake and Horace R. Cayton, eds., *Black Metropolis, Vols. 1 and 2* (New York, 1945), 769-82; Trotter, "Afro-American Urban History," 267-69.

10. Drake and Cayton, *Black Metropolis, Vols. 1 and 2.*

11. Drake and Cayton, *Black Metropolis,* 766.

12. Ibid., 781.

13. Trotter, "Afro-American Urban History," 269-70, 272-73.

14. Ibid.

15. Kusmer, "The Black Urban Experience," 96-97; Trotter, "Afro-American Urban History," 270-71; Oscar Handlin, *The Newcomers: Negroes and Puerto Ricans in a Changing Metropolis* (Garden City, 1959).

16. Trotter, "Afro-American Urban History," 271-75.

17. For a discussion of recent studies, see Trotter, "Black Migration in Historical Perspective: A Review of the Literature," in Trotter, ed., *The Great Migration in Historical Perspective,* 1-21.

18. Quote from Trotter, *Black Milwaukee,* xi; Trotter, "Black Migration in Historical Perspective," in Trotter, ed., *The Great Migration in Historical Perspective,* 1-21. Other studies also helped to carve out more space for black workers within African American urban history: Nell Irvin Painter, *The Narrative of Hosea Hudson: His Life as a Communist* (Cambridge, 1979); William H. Harris, *Keeping the Faith: A. Philip Randolph, Milton P. Webster, and the Brotherhood of Sleeping Car Porters, 1925-37* (Urbana, 1977); Robert J. Norrell, "Caste in Steel: Jim Crow Careers in Birmingham, Alabama," *Journal of American History* 73 (December 1986); Daniel Rosenberg, *New Orleans Dockworkers: Race, Labor, and Unionism, 1892-1923* (Albany, 1988); Eric Arnesen, *Waterfront Workers of New Orleans: Race, Class, and Politics, 1863-1923* (New York, 1991); and Robert Korstad and Nelson Lichtenstein, "Opportunities Found and Lost: Labor, Radicals, and the Early Civil Rights Movement," *Journal of American History* 75 (December 1988), 786-811.

19. Trotter, "Black Migration in Historical Perspective."

20. David Katzman, *Before the Ghetto: Black Detroit in the Nineteenth Century* (Urbana, 1973); Richard Walter Thomas, *Life for Us Is What We Make It: Building Black Community in Detroit, 1915-1945* (Bloomington, 1992); Dennis C. Dickerson, *Out of the Crucible: Black Steelworkers in Western Pennsylvania, 1875-1980* (Albany, 1986); Kenneth Kusmer, *A Ghetto Takes Shape: Black Cleveland, 1870-1930* (Urbana, 1976); Trotter, *Black Milwaukee;* Darrel E. Bigham, *We Ask Only a Fair Trial: A History of the Black Community of Evansville, Indiana* (Bloomington, 1987); James Borchert, *Alley Life in Washington: Family, Community, Religion, and Folklife in the City, 1850-1970* (Urbana, 1980); George Wright, *Life Behind a Veil: Blacks in Louisville, Kentucky, 1865-1930* (Baton Rouge, 1985); Earl Lewis, *In Their Own Interests: Race, Class, and Power in Twentieth-Century Norfolk, Virginia* (Berkeley, 1991); Ronald H. Bayor, "Roads to Racial Segregation: Atlanta in the Twentieth Century," *Journal of Urban History* 15 (November, 1988), 3-21; Thomas C. Cox, *Blacks in Topeka, Kansas, 1865-1919: A Social History* (Baton Rouge, 1982); Douglas H. Daniels, *Pioneer Urbanites: A Social and Cultural History of Black San Francisco* (Philadelphia, 1980); Quintard Taylor, "The Emergence of Black Communities in the Pacific Northwest, 1865-1910," *Journal of Negro History* 64 (Fall 1979), 342-45; Quintard Taylor, *The Forging of a Black Community: Seattle's Central District from 1870 Through the Civil Rights Era* (Seattle, 1994); Lawrence De Graaf, "The City of Black Angels: Emergence of the Los Angeles Ghetto, 1890-1930," *Pacific Historical Review* 39 (August 1970), 328-50; Albert Broussard, *Black San Francisco: The Struggle for Racial Equality in the West, 1900-1954* (Lawrence, 1993); Shirley Moore, *To Place Our Deeds: The Black Community in Richmond, California* (University of Illinois Press, Urbana, forthcoming).

21. Michael Homel, *Down From Equality: Black Chicagoans and the Public Schools, 1920-41* (Urbana, 1984); Vincent P. Franklin, *The Education of Black Philadelphia: The Social and Educational History of a Minority Community, 1900-1950* (Philadelphia, 1979); Judy Jolley Mohraz, *The Separate Problem: Case Studies of Black Education in the North, 1900-1930* (Westport, 1979); David Tyack, "Growing Up Black: Perspectives on the History of Education in Northern Ghettos," *History of Education Quarterly* 9 (1969), 287-97.

22. Herbert Gutman, *The Black Family in Slavery and Freedom, 1750-1925* (New York, 1976); Darlene Clark Hine, "Lifting the Veil, Shattering the Silence: Black Women's History From Slavery to Freedom," in Darlene Clark Hine, ed., *The State of Afro-American History;* Darlene Clark Hine, "Black Migration to the Urban Midwest: The Gender Dimension," in Trotter, ed., *The Great Migration in Historical Perspective;* Elsa Barkley Brown, "Womanist Consciousness: Maggie Lena Walker and the Independent Order of St. Luke," *Signs* 14 (Spring 1989), 610-33; Evelyn Brooks, "The Feminist Theology of the Black Baptist Church, 1880-1900," in Amy Swerdlow and Hanna Lessinger, eds., *Class, Race, and Sex: The Dynamics of Control* (Boston, 1983); Evelyn Brooks Higginbotham, *Righteous Discontent: The Women's Movement in the Black Baptist Church, 1880-1920* (Cambridge, 1993); Jacqueline Jones, *Labor of Love, Labor of Sorrow: Black Women, Work, and the Family, From Slavery to Freedom* (New York, 1985); Dolores Janiewski, *Sisterhood Denied: Race, Gender, and Class in a New South Community* (Philadelphia, 1985); Sharon Harley and Rosalyn Terborg-Penn, eds., *The Afro-American Woman: Struggles and Images* (Port Washington, 1978). Forthcoming studies by Tera Hunter on black household workers and Stephanie Shaw on black professional women will offer additional insights into the interplay of class, race, and gender in the lives of black women within the urban community: *Household Workers in the Making: Afro-American Women in Atlanta and the New South, 1861-1920* and *"What a Woman Ought to Be and to Do": Black Professional Women During the Jim Crow Era* (Chicago, forthcoming).

23. Robert Buni, *Robert L. Vann of the Pittsburgh Courier* (Pittsburgh, 1974); Emma Lou Thornbrough, *T. Thomas Fortune* (Chicago, 1972); Stephen R. Fox, *The Guardian of Boston: William Monroe Trotter* (New York, 1971); Henry Lewis Suggs, ed., *The Black Press in the South, 1865-1979* (Westport, 1983); Walter B. Weare, *Black Business in the New South: A Social History of the North Carolina Mutual Life Insurance Company* (1973; reprinted Durham, 1993); Alexa Benson Henderson, *Atlanta Life Insurance Company: Guardian of Black Economic Dignity* (Tuscaloosa, 1990); Darlene Clark Hine, *Black Women in White* (Bloomington, 1989); Cynthia Neverdon-Morton, *Afro-American Women of the South and the Advancement of the Race, 1895-1925* (Knoxville, 1989); Nathan Huggins, *The Harlem Renaissance* (London, 1971); David Levering Lewis, *When Harlem Was in Vogue* (New York, 1979); Lawrence W. Levine, *Black Culture and Black Consciousness: Afro-American Folk Thought From Slavery to Freedom* (New York, 1977); on the Garvey Movement, see notes 16, 17, and 20 above, as well as Tony Martin, *Race First: The Ideological and Organizational Struggles of Marcus Garvey and the Universal Negro Improvement Association* (Westport, 1976); and Judith Stein, *The World of Marcus Garvey: Race and Class in Modern Society* (Baton Rouge, 1986). See also August Meier, *Negro Thought in America, 1880-1915: Racial Ideologies in the Age of Booker T. Washington* (Ann Arbor, 1963).

24. Dianne Pinderhughes, *Race and Ethnicity in Chicago Politics: A Re-examination of Pluralist Theory* (Urbana, 1987).

25. John Bodnar, Roger Simon, and Michael Weber, *Lives of Their Own: Blacks, Italians, and Poles in Pittsburgh, 1900-1960* (Urbana, 1982); Stephan Thernstrom, *The Other Bostonians: Poverty and Progress in the American Metropolis, 1880-1970* (Cam-

bridge, 1973); Olivier Zunz, *The Changing Face of Inequality: Urbanization, Industrial Development, and Immigrants in Detroit, 1880-1920* (Chicago, 1982). See also Ivan H. Light, *Ethnic Enterprise in America: Business and Welfare Among Chinese, Japanese, and Blacks* (Berkeley, 1973) and Ira Katznelson, *Black Men, White Cities: Migration to Cities in the U.S., 1900-1930 and Britain, 1948-68* (London and New York, 1973).

26. William A. Tuttle, *Race Riot: Chicago in the Red Summer of 1919* (1970; reprinted New York, 1984); Elliott Rudwick, *Race Riot at East St. Louis, July 2, 1917* (Carbondale, 1964); William Ivy Hair, *Carnival of Fury: Robert Charles and the New Orleans Race Riot of 1900* (Baton Rouge, 1976); Scott Ellsworth, *Death in a Promised Land: The Tulsa Riot of 1921* (Baton Rouge, 1982); Robert V. Haynes, *A Night of Violence: The Houston Riot of 1917* (Baton Rouge, 1976); Dominic J. Capeci, Jr. and Martha Wilkerson, *Layered Violence: The Detroit Riots of 1943* (Jackson, 1991); Herbert Shapiro, *White Violence, Black Response: From Reconstruction to Montgomery* (Amherst, 1988); Roberta Senechal, *The Sociogenesis of a Riot: Springfield, Illinois, in 1908* (Urbana, 1990).

27. Lewis, *In Their Own Interests;* Dickerson, *Out of the Crucible;* Taylor, *Black Seattle;* Broussard, *Black San Franciscans;* Moore, *To Place Our Deeds.*

28. Mark Naison, *Communists in Harlem During the Depression* (Urbana, 1983); Robin D. G. Kelley, *Hammer and Hoe: Alabama Communists During the Great Depression* (Chapel Hill, 1990); Cheryl Lynn Greenberg, *Or Does It Explode: Black Harlem in the Great Depression* (New York, 1991); Nancy L. Grant, *TVA and Black Americans: For the Status Quo* (Philadelphia, 1990).

29. Kusmer, *A Ghetto Takes Shape* and "The Black Urban Experience."

30. Borchert, *Alley Life,* xii.

31. Lewis, *In Their Own Interests,* 4.

32. Thomas, *Black Detroit.*

33. Trotter, *Black Milwaukee,* xi-xiv, 264-82.

34. See Katzman, *Before the Ghetto,* 213-16.

35. See note 20 above.

36. An important exception is Edward H. Beardsley, *A History of Neglect: Health Care for Blacks and Mill Workers in the Twentieth-Century South* (Knoxville, 1987). See also the efforts of David McBride, " 'God is the Doctor': Medicine and the Black Working-Class in New York City, 1900-1950," (paper delivered at the American Historical Association Meeting, December 1990). Also relevant is Clark Hine, *Black Women in White.*

37. Robin D. G. Kelley, " 'We Are Not What We Seem': Rethinking Black Working-Class Opposition in the Jim Crow South," *Journal of American History* 80 (June 1993), 75-112.

38. Taylor, *The Forging of a Black Community;* Henry Louis Taylor, ed., *Race and the City: Work, Community, and Protest in Cincinnati, 1820-1970* (Urbana, 1993). For an effort to uncover the impact of African Americans along with working class and poor white on the city-building process, see Steven J. Hoffman, "Behind the Facade: The Constraining Influence of Race, Class and Power on Elites in the City-Building Process, Richmond, Virginia, 1870-1920" (Ph.D. diss., Carnegie Mellon University, 1993).

11

African Americans in the City Since World War II

From the Industrial to the Postindustrial Era

Kenneth L. Kusmer

he African American urban experience since World War II, a period marked by the transition of the United States from an industrial to a postindustrial order, is a study in continuity as well as change. Its origins and initial direction, its emerging problems and hopeful possibilities, lay to a considerable extent in the era of industrial dominance of the first four decades of the twentieth century. It was during that period that African Americans first migrated to northern metropolises in great numbers; that the formation of black ghettos initially took place; and that discrimination in public accommodations became widespread in northern communities and established in law and practice in the

Author's Note: Earlier versions of this essay were presented in April 1993 at the annual meeting of the Organization of American Historians and in April 1994 at the University of Iowa. The author would like to thank Ronald Bayor, Quintard Taylor, Randall Miller, James Borchert, Susan Borchert, and Susan Pizzano for reading and commenting on this article.

South. It was also during the decades between 1900 and 1940, however, that African Americans first had the opportunity to take part in industrial work; that the dynamic expansion of black communities in the North and West laid the groundwork for the beginnings of modern black urban politics; and that the formation and growth of distinctive black institutions, including social welfare, religious, and civil rights organizations, became a vital reality in many cities.[1]

Black migration to the cities slowed in the 1930s, but the need for industrial labor during World War II promoted a new exodus from the South. The World War II migration was as intense as that of World War I, and it led to a similarly hostile response from many whites. Numerous race riots broke out in the North, including a violent encounter in Detroit in 1943 that left twenty-five blacks and nine whites dead.[2] In spite of this, the wartime migration brought renewed hope to black communities that had been economically devastated by the Great Depression.[3] As industry began to revive in 1941-1942, African Americans were at first excluded or often restricted to menial jobs. Black protest and a vast expansion of industrial demand in 1942, however, finally reopened factory work to blacks in the urban North and West.[4] The shift from the discriminatory American Federation of Labor unions to the more racially inclusive Congress of Industrial Organizations assisted black workers in gaining a foothold, for the first time, in unionized industrial work.[5] Politically, as well, the era of the 1930s and World War II marked an important turning point, one that would continue into the postwar era. By the eve of the Great Depression, Republican party machines in northern cities had long taken the black vote for granted. Beginning in the mid-1930s, then accelerating during the 1940 election and afterward, urban blacks in the North and West shifted strongly to the Democratic Party, buoyed by the positive, if halting, improvement in treatment accorded African Americans by the Franklin D. Roosevelt administration. In the process, black voters became an important part of the New Deal coalition that would dominate the postwar era, and gradually a new group of black Democratic political leaders emerged in cities across the North.[6] Finally, important decisions regarding federal involvement in the cities, especially in the area of housing policy, first enunciated in the 1930s, would continue to affect African American urbanites for decades afterward. Although blacks received more than their share of much sought after public housing constructed under the New Deal, such housing was usually segregated. The Home Owners Loan Corporation, set up in 1933, helped many thousands of Americans save their homes from foreclosure. It also, however, initiated the policy of "redlining" by local real estate interests, establishing a practice that would make it difficult for inner-city blacks to obtain mortgages and contributing to the deterioration of largely black neighborhoods in many cities.[7]

A Historical Framework

Although many important trends of black urban development were established during the interwar years, new forces of economic and social change that began during the 1930s and 1940s and accelerated during subsequent decades would create a much different context for African Americans living in cities. The impact of the automobile, of growing federal involvement in the economy, and of the shift from industrial to postindustrial occupations would be profound. By the 1970s, the centralized, industrial metropolises of the East and Midwest were in decline as they struggled with the loss of basic industries, the shift to a service economy, and the exodus of population (and tax base) to expanding suburbs or distant Sunbelt cities. It was into this new socioeconomic environment that black Americans completed a process of rural to urban migration that had been over a century in the making.

Although a few historians have recently begun to produce works dealing specifically with the postwar era of black urban development, most of the current discussion of the status of African Americans in cities during the last half of the twentieth century has been framed by sociologists, economists, geographers, and anthropologists. Too often, however, the historical perspective of social-scientific studies of urban racial conditions is limited in scope or overly reliant on a heavily statistical analysis of social change. Interdisciplinary knowledge in this area has yielded important insights, but what is most needed now is a historical framework that will provide a grounding for such knowledge. The last half century of African American urban history falls logically into three phases: (1) 1942 to the early 1960s, (2) the middle and late 1960s, and (3) 1970 to the present. Each phase has its distinctive features and problems, but there are some aspects that are best understood by considering the entire fifty-year period as a whole.

The basis of post-World War II black urban development was the massive migration of blacks from the rural South to the cities of the North (and, to a lesser extent, the West and South), a migration that continued largely unabated until the late 1960s. This movement of people, which was actually far greater than the "Great Migration" of the World War I era, has been little studied by historians.[8] It is clear that two fundamental causes of this second migration were the mechanization of agriculture in the South, which drove black sharecroppers off the land, and the opportunity for better work in the urban North.[9] We need to know much more about who these migrants were, their motivations and expectations.[10] What is clear, however, is that their impact on the cities that received them was enormous. As late as 1940, African Americans made up less than 10 percent of the total population of most northern cities. Three decades later, they comprised a third of the residents of Chicago and Philadelphia and about 40 percent of the population of Cleveland, St. Louis, and Detroit; Newark, Gary, and Washington, D.C. had by then already become black-majority cities. During this period, Los Angeles also experienced substantial black in-migration. The situation in the urban South was somewhat different. There, the black percentage of the urban population had been substantially higher than that of northern cities in 1940 but changed little during the next thirty years. Important exceptions were Atlanta and New Orleans, where African Americans made up, respectively, 51 and 45 percent of the total population by 1970.[11]

The influx of black migrants coincided with a reshaping of the metropolis, as increasing numbers of city dwellers as well as businesses relocated to the rapidly expanding suburban areas, a movement of people and capital that was accelerated by federally financed urban renewal and highway construction in the 1950s and 1960s.[12] Although this metropolitan expansion could have provided an opportunity to reverse the trend toward residential segregation of the races, in fact, the exact opposite resulted. The influx of migrants sometimes led to an expansion of the ghettos formed originally in major northern cities after World War I, but more often, as Arnold Hirsch has shown, it laid the groundwork for the creation of new, much larger "second" (or third) ghettos as well—such as the west-side ghetto of Chicago, the Avondale section of Cincinnati, north and west Philadelphia, or the Hough Avenue district of Cleveland.[13]

Although some degree of segregation of blacks in inner-city areas might have been expected as a result of their relative poverty, the high level of ghettoization that existed by the 1960s must primarily be understood as a product of a long-standing pattern of white hostility to residential integration, combined with new structural factors that reinforced the trend of increasing segregation. The beginning of the exodus of the white middle class from central cities in the late 1940s and early 1950s opened up central-city housing stock for the expanding black population, but less impersonal factors were responsible for channeling African Americans into specific areas. At the neighborhood level, restrictive covenants were sporadically used to restrict blacks prior to their being declared unconstitutional in 1948. Throughout the 1940s and 1950s, however, the more informal practices of real estate dealers and white homeowners' organizations operated far more effectively to limit black housing options.[14] In 1957, in San Francisco, even baseball star Willie Mays found it impossible to buy a home in an exclusive, all-white section of that city.[15] In addition, during the late 1940s and 1950s, violence against blacks attempting to move into previously all-white neighborhoods became common. Except for Hirsch, few scholars have given much attention to this phenomenon, but physical attacks on black homeowners (which sometimes escalated into small-scale riots) were important in shaping the contours of the new postwar ghettos.[16]

Beyond the neighborhoods, the policies of both local elites and the federal government also contributed to an intensification of residential segregation. During the years of heaviest black migration to the cities (1945 to the mid-1960s), urban renewal and highway construction often decimated older black neighborhoods, forcing relocation in rapidly ghettoizing areas, or in some cases creating physical barriers that confined African Americans to certain areas.[17] It was also during this period, as Kenneth Jackson and others have demonstrated, that the redlining policy of major banks and insurance companies accelerated the decline of many inner-city neighborhoods. At the same time, suburbs effectively slowed housing integration by blocking construction of subsidized housing for the poor, while federal support for segregated public housing in the inner cities reinforced the growing concentration of urban black populations in ghettos.[18]

The black ghettos of the early twentieth century had developed mostly in rapidly industrializing northern cities. Because of their unique urban form (less centralized) and economic function (commercial or light industry rather than heavy industry), most southern cities had considerably lower levels of residential segregation than northern metropolises prior to World War II. There was some tendency toward clustering of the black population in distinct areas, but large, clearly defined ghettos were much less common in the South than in the North. A similar pattern existed in rapidly growing metropolises of the Midwest (such as Milwaukee and Minneapolis) and the West (Los Angeles, San Francisco, Seattle), although in contrast to the South most of these cities had relatively small black communities.[19] A distinctive feature of the post-World War II trend of urban racial concentration was that it affected cities in *all* regions. In Houston, Atlanta, and Los Angeles, no less than in Chicago, Detroit, and Philadelphia, the forces of economic transformation, racism, and governmental policy creating the postindustrial city also channeled the black populations of American cities increasingly into well-defined ghettos. By the end of the 1960s, state Douglas Massey and Nancy Denton in summarizing this trend, "The average black city dweller lived in a neighborhood where the vast majority of his or her neighbors were also black."[20]

If patterns of residential segregation in northern and southern cities converged after 1945, the opposite was true of separation of the races in public accommodations. In the 1950s and early 1960s, whites in many southern communities, goaded by demagogic politicians, put up intense resistance to any change in the system of racial apartheid that had long been a defining feature of race relations in the region.[21] In marked contrast, in the North (as well as in some border cities like Washington, D.C.) overt discrimination against blacks in public accommodations declined during the postwar era. The movement to integrate hotels, restaurants, and theaters in the North is a chapter in the career of Jim Crow that has yet to be written.[22] What little research has been done indicates that there was wide variation in the level of white opposition to integration of these public accommodations, but resistance was never as fierce as it was in the South. The struggle in the North, often spearheaded by local chapters of the NAACP, began during World War II and continued, in some cases, for two decades. Integration of Pittsburgh's public swimming pools came about only after a "protracted campaign" in the 1950s. In a number of Indiana cities, as Emma Lou Thornbrough has shown, resistance to Jim Crow facilities did not end until well after the passage of a state civil rights law in 1963.[23] In Cincinnati it was not until 1961, after a decade of struggle by civil rights groups, that a local amusement park was finally integrated. In most northern communities, however, the battle for equal treatment in public accommodations of this sort was over by the beginning of the 1960s.[24]

In politics, too, the 1940s and 1950s was a period of relative improvement, but in this regard northern black communities made much more rapid strides than did those in the South. Voter registration drives increased the number of black voters in southern cities after World War II, and they were able to elect black councilmen in Richmond and Nashville in the late 1940s and a white liberal to the Montgomery city council in 1953. Even where it did not determine elections, the increase in black voting led to some improvements in treatment of blacks by city governments. In Atlanta, for example, the first African American police officers were hired in 1948. The limitations of black political influence over such matters, however, were illustrated by the fact that as late as 1959 African Americans still made up only 6.1 percent of the force. In addition, there were severe restrictions on black police who, until 1962, patrolled only in black neighborhoods and were prohibited from arresting whites.[25]

Unlike their southern counterparts, African Americans in the North and West had long had the right to participate in the political system. With few exceptions, however, their numbers and organizational strength were too small to exert much power prior to World War II. The growth and increasing concentration of the black populations of northern cities after 1942, however, made possible a substantial increase in the number of black elected officials, now almost always Democrats. Black representation in city councils grew steadily, at least in large northern cities. In 1944, Harlem's Adam Clayton Powell, Jr. joined Chicago's William J. Dawson as the second African American member of Congress, and by 1964 there were six blacks in Congress, five from the North and one from Los Angeles.[26] Despite this trend, however, the new black political leaders of the postwar era played a largely subsidiary role within local political systems dominated by whites, "allowed to be council members, or even in a few cases state legislators [but] denied powerful citywide or countywide offices." The patron-client politics identified by Martin Kilson as symptomatic of black urban politics earlier in the century continued until the mid-1960s, when the size of the black voter base and emergence of a group of younger, more militant black leaders began to upset this arrangement.[27] In Chicago—by far the most studied city of the post-World War II period—black votes were of vital importance to the reigning Democratic Party coalition in the 1950s and early 1960s. Yet the loyalty of the city's African American voters did not earn them a fair share of political patronage, and Mayor Richard J. Daley consistently ignored the demands of reformers and civil rights groups for more black police and better schools and services in the black neighborhoods.[28]

The general deterioration of the schools in inner-city neighborhoods in northern cities was particularly vexing to blacks. Schools in many northern communities, although not segregated by law as in the South, had nevertheless exhibited a pattern of de facto segregation as early as the 1920s, largely as a result of ghettoization of the black population.[29] This pattern continued and intensified after 1945, accompanied by underfunding of schools in the inner city and an unstated policy, in many instances, of drawing school district boundaries to artificially promote segregation. The result, by the 1960s, was a high degree of isolation of minorities in the public schools of the North, and a growing frustration among African Americans over the limitations on their political power to effect changes in schools or other important aspects of urban government.[30]

Scattered evidence indicates that variations on the Chicago model of a growing, but still highly circumscribed, black political base was common in large cities outside the South during the two decades following World War II. The experience of some cities, however, shows that it was not universal. The black population of Boston remained too small (the city attracted few southern migrants) and that of Pittsburgh too geographically divided to permit the development of a successful ward-based politics. At-large council elections in both cities contributed to black electoral weakness. In Los Angeles, also, despite substantial in-migration and higher levels of economic progress than elsewhere, African Americans were virtually excluded from political influence prior to the early 1960s as a result of gerrymandering that fragmented both the black and Latino vote. A study of ten other cities (including San Francisco) revealed similar circumstances elsewhere in California.[31]

If blacks in the urban North had limited political power over housing, schools, and other city services, they had even less control over declining economic opportunities that, as early as the mid-1950s, were beginning to affect African American communities of northern industrial centers. The focus, in the work of "underclass" theorists and some other scholars, on the negative impact of structural changes in the economy on the black poor since 1970 ignores both the earlier stage of that development as well as the external factor of racism in hiring practices that often accompanied it.[32] As early as the late 1930s, the exodus of jobs from central cities to the suburban ring was evident to some observers.[33] World War II, however, gave a temporary boost to the industrial economy of northern cities, at the same time that it laid the groundwork for the shift of the defense industry to new areas in the South and West.[34]

The relatively strong economy of the late 1940s and 1950s masked the beginning of deindustrialization in the North. During this period, a number of structural changes began to affect the economy of northern cities: the relocation of plants to suburban or rural areas; automation (often as a means of labor control as much as an economizing measure); a decline of defense spending in the older industrial centers; and the increasing use of overtime by employers to avoid hiring additional workers with full benefits.[35] As Thomas Sugrue has shown in his case study of Detroit, these broad structural changes had a greater impact on black workers than whites because they often were reinforced by discriminatory hiring policies of employers or exclusionary practices of unions. In the 1940s and early 1950s, African Americans had a genuine chance to obtain jobs in the auto industry, but they had little opportunity for advancement to skilled or white-collar jobs. Even at the entry level, hiring practices were not centrally controlled, and some plant managers discriminated against blacks. Discrimination was far more widespread in smaller industries, and some skilled trades, notably in construction, remained virtually all white as late as the mid-1960s because of union policies.[36]

Not "impersonal" structural change alone, but the combination of discrimination with structural change, limited black employment opportunities. Compounding this problem was the fact that job loss occurred at the same time the black population was increasing as a result of migration from the South. Young migrants, says Sugrue, "found that the entry-level operative jobs that had been open to their fathers or older siblings in the 1940s and early 1950s were gone." In Birmingham, one of the few unionized southern industrial centers, black steelworkers still experienced a loss of income relative to whites between 1950 and 1970 due to a decline of industrial work combined with long-standing patterns of discrimination in promotions. In Detroit and elsewhere, there was a rise in the black unemployment rate to double that of whites as early as the mid-1950s. With some fluctuations, it has remained at that level ever since.[37] As Mark

Stern reminds us, black workers in Detroit, Birmingham, and other industrial centers who had good, unionized jobs in autos or steel were much better off than many black workers in the 1950s, who too frequently "suffered from irregular work, low wages, and poverty" even at the high point of postwar American economic prosperity.[38] By 1960, however, the number of higher-paying jobs available to blacks in the North was not keeping pace with the size of the population.

In the early 1960s, with national attention focused on the conflict over civil rights in the South, few whites were prepared to acknowledge that racial conditions in the northern cities were deteriorating, or that black patience with poor housing, inadequate schools, and high unemployment was reaching the breaking point. In 1964, riots in Philadelphia, New York, and Rochester gave some indication of the trouble to come, but it was the massive racial disturbance that convulsed the Watts ghetto of Los Angeles for four days in August 1965 that sent the first clear message that "politics as usual" in the ghettos was no longer working. The riot began, as many later riots would, with an incident involving white police arresting blacks in the ghetto. By the time the National Guard had restored order, 34 people had died, 1,000 had been injured, and 4,000 arrested; property loss was estimated at $35 million.[39] Initially, however, even the Watts riot did not lead to greater public awareness of urban racial conditions. The official report of the McCone Commission that investigated the riot concluded that it was caused by a small group of poorly educated, delinquent troublemakers. This "riffraff" theory bore a striking resemblance to the "outside agitator" theory often used by southern white segregationists to explain civil rights protests in the early 1960s. It reassured whites that the African Americans who rioted in Watts were not representative of their community. In effect, the report delegitimized the grievances of blacks living in the urban North.[40]

The events of the next three years discredited the McCone report. Each summer brought dozens of racial disturbances, culminating in the violent riots that took place in Detroit and Newark in 1967 that left 66 people dead and over 1,000 injured. The riots of the 1960s marked a definite change in the historical pattern of interracial violence. Unlike the riots of the World War I era, when white mobs assaulted blacks, the ghetto riots of the 1960s took place in black neighborhoods. They usually began as a result of a perceived incident of police brutality and for the most part involved black attacks on white property.[41] This pattern reflected the growing geographic isolation of blacks in the postindustrial metropolis, in which rapidly expanding suburbs remained predominantly white while the inner cities housed an increasing number of African Americans in ghettos that were much larger than those of the World War I era. This isolation made those white businesses still remaining in the ghetto easy targets for blacks who felt increasingly alienated. In summarizing information about participants in the ghetto riots of the 1960s, the Kerner Commission specifically rejected the theory that the typical rioter was "a hoodlum, habitual criminal, or riffraff; nor was he a recent migrant [or] a member of the uneducated underclass." Instead, the commission found the typical rioter to be a young man who had grown up in the city where the riot had taken place, was somewhat better educated than the average African American, but was "almost invariably underemployed or employed at a menial job." The typical rioter had a strong sense of racial pride and, though well informed politically, was "highly distrustful of the political system and political leaders."[42]

Social scientists who have studied the riots of the 1960s have disagreed with some aspects of this profile, but it would be a mistake to focus too much on the question of who rioted versus who did not. The causes of the riots were clearly rooted in socioeconomic conditions that had evolved over a period of at least two decades, conditions that broadly affected blacks living in inner-city areas. Although no comprehensive historical study of the riots exists, the question of why riots occurred in some cities rather than others has received considerable attention. The most persuasive of these studies compare the circumstances of blacks and whites in cities where riots occurred, rather than focusing on the conditions in black communities alone. One such study, which extended the analysis from central cities to larger metropolitan areas, determined that riots were most likely to take place in cities where rapid economic growth was occurring in new suburban areas, rather than in cities where ghetto conditions were most deplorable.[43] This general structural approach points us in the right direction because it places the riots in a context of broader changes in the postindustrial order. Such demographic studies are largely lacking in historical content, however, and most deal only superficially with the background of race relations and black community development in specific cities.

A deeper understanding of the historical significance of the riots must also address the wide difference in perceptions of the civil disorders by blacks and whites living in the same cities. Although it received little attention at the time, the difference between black and white perceptions of the Watts disturbance was so great that many blacks did not even use the word *riot* to describe it. Many African Americans, whether they took part in the civil disturbance or not, perceived it as a *revolt* whose purpose was to make whites aware of conditions in the ghetto. An innovative study of racial content in the Los Angeles press by Paula Johnson and her colleagues helps explain why this was so. At the same time the spatial isolation of blacks was increasing during the postwar era, African Americans were also becoming "psychologically invisible" to whites as a result of declining news coverage in the mainstream press.[44] An analysis of electronic media would probably reinforce this conclusion.[45] A few days after the Watts uprising, a group of young blacks in Los Angeles told Martin Luther King, Jr. that they felt the riot was a victory "because we made them [white people] pay attention to us."[46] The widening gap in socioeconomic conditions, coupled with increasing invisibility of these problems to whites, helps to explain such views. Although few recognized it at the time, the rapidly growing African American communities of northern cities had largely been excluded from the fruits of urban renewal, suburbanization, and "growth politics" promoted by white political and business leaders after World War II.[47] Whatever their specific causes, the riots of the 1960s were one result of this increasing, and unacknowledged, inequality —a trend that coincided with the crisis of the declining industrial city.[48]

The postriot era saw a sharp decline in black migration from the South, and beginning in the 1970s, there was even some return migration to that region or migration to the West, as the prosperity of some parts of the Sunbelt attracted Americans of all races.[49] By the 1980s, the black populations of many northern cities had stabilized or even declined in some cases. Despite increased black suburbanization after 1970, however, the exodus of whites still exceeded that of blacks. As a result, the percentage of central-city populations that were black continued to increase—although at a much slower rate than during the 1945-1970 period.[50]

The civil rights laws of the 1960s eliminated segregated public accommodations, forced the integration of schools in the South, and ostensibly prohibited discrimination in housing. The functional effect of these laws, however, has been ambiguous at best. The legal desegregation of schools did not, in most cases, end de facto segregation in northern cities and often provided only a temporary respite from such conditions in the South. In the North, the Supreme Court ruled in 1973 that mandated busing of students between Detroit and its predominantly white suburbs was unconstitutional. This decision dealt a death blow to metropolitan-wide solutions to segregated and underfunded inner-city school systems.[51] Meanwhile, notes Gary Orfield, whites in the urban South discovered "that they could bend segregation laws and adopt neighborhood schools while leaving segregation largely untouched." The effects of white flight to the suburbs, combined with the growth of racially exclusive private schools, made public school desegregation a "bittersweet victory" in many southern cities.[52] Throughout the United States, despite an increase in the average years of schooling completed by black students since the 1960s, there has been little improvement in the quality of the education they receive in the inner city.[53]

To a considerable extent, the difficulties of achieving integrated urban schools were the by-product of continuing racial divisions in housing patterns as American cities moved into the postindustrial era. Most central cities saw no appreciable change in segregation levels after 1970, although in many cases predominantly black areas expanded as older inner-city neighborhoods either gentrified or became depopulated due to abandoned housing.[54] In many cities, the increasing suburbanization of African Americans resulted in only a marginal reduction in levels of residential segregation. Many blacks moved into older suburbs, often in areas that were extensions of central-city ghettos. A good example was East Cleveland, where in the 1960s "block after block underwent rapid transition" from white to black.[55] By the 1980s, such blockbusting practices were illegal. Similar results were obtained at a slower pace, however, through the widespread practice by real estate agents of "steering" prospective black homeowners or renters to racially mixed or predominantly black areas.[56] In the post-civil rights era, most cities could point to the emergence of at least one "model" integrated area, such as Mt. Airy in Philadelphia or the Chicago suburb of Oak Park, where racial cooperation and shared middle-class lifestyles made it possible for blacks and whites to maintain relatively stable, integrated communities. More typical, unfortunately, was Long Island, where, by the 1990s, blatant discrimination in apartment rentals had forced 95 percent of the black residents into 5 percent of the census tracts, creating what one housing administrator called "a Mason-Dixon line in Nassau County."[57]

Whatever its negative consequences, residential segregation, coupled with the continuing movement of whites to the suburbs, helped lay the groundwork for an upsurge of black political power in urban centers beginning in the late 1960s. In many smaller cities, African Americans were able to elect representatives to city councils for the first time.[58] More significantly, blacks were elected to the mayor's office in a number of cities, including Cleveland and Gary (1967), Newark (1970), Detroit, Atlanta, and Los Angeles (1973), New Orleans (1977), Birmingham (1979), Chicago and Philadelphia (1983), Baltimore (1987), New York (1989), and Denver (1991). Beginning in the 1970s, many smaller communities, especially in the South, also elected black mayors. By 1990, almost 300 black mayors held office throughout the United States.[59] Although black candidates benefited from an increase in the black voter base, however, the election of black mayors was far from a simple case of one racial group succeeding another through numerical dominance. Among major cities, only Detroit, Atlanta, New Orleans, Baltimore, and Washington, D.C.[60] had black majorities when they first elected an African American mayor. (In 1990, of the thirty largest cities with black mayors, only twelve had black majorities.[61]) Elsewhere, black candidates needed white votes—in some cases many white votes—in order to win, and their victories should more properly be described as examples of coalition politics than the result of racial solidarity alone.[62] The premier example is Thomas Bradley, who was elected to five terms (1973-93) as mayor of Los Angeles, a city where blacks comprised only about 17 percent of the population. Bradley effectively built a coalition that reached out from his base in the black community to gain support from white liberals and, to a lesser extent, Latino and Asian voters.[63]

Unfortunately, black political gains at the local level occurred during a period of reduced federal spending on domestic programs, which limited black politicians' ability to respond to the needs of their often impoverished constituents. Institutional and financial restraints, as well as resistance from white politicians or business leaders, made it almost impossible for these mayors to bring about basic improvements in jobs and housing in the inner city. The rapid decline of Newark and Gary, two of the most glaring case studies, coincided almost exactly with the rise of blacks to power in those once vital industrial communities.[64]

In some ways, however, it may be quite inaccurate to describe black power at the local level, as one writer predicted in 1969, as a "hollow prize."[65] Although the election of black mayors has had limited benefits for the black poor in Los Angeles and Atlanta, those metropolises can hardly be considered examples of economic failure. Even in "Rust Belt" cities that have suffered decline since the 1960s, black mayors have often been able to provide moderate improvements in urban services (including parks and recreational facilities) in traditionally neglected black neighborhoods. In addition, in many cities headed by black mayors there has been a considerable increase in the number of African American police (including police chiefs) and municipal employees. In one city, Atlanta, these gains were quite dramatic following the election of the first black mayor, Maynard

TABLE 11.1 Black Municipal Employment in Atlanta, 1970-1978 (in percentages)

	Before Black Mayor (1970)	*After Black Mayor (1978)*
Total work force	38.1	55.6
Administrators	7.1	32.6
Professionals	15.2	42.2

SOURCE: Peter K. Eisinger, "Black Employment in Municipal Jobs: The Impact of Black Political Power," *American Political Science Review* 76 (1982), 385.

Jackson, in 1973 (Table 11.1). Increases in black municipal employment in other cities at this time were smaller, but still substantial. In his study of racial politics in Los Angeles, Raphael Sonenshein found that (with the exception of the police department) total black representation in city jobs increased little during the Bradley years, but many more African Americans received political appointments or filled higher-status positions in government. Modest moves toward redistributing resources were carried out during the 1973-1980 period, when Bradley was successful in obtaining federal support for maintaining jobs and services in low-income communities. During the Reagan era, however, "left without federal aid and with little public pressure," the Bradley administration and its allies became "supportive of the status quo, pro-business, and not very innovative."[66]

Expansion of black municipal employment and services in black neighborhoods was not limited to cities with black mayors. One case study found that, between 1966 and 1978, there was a steady growth in black employment in city jobs in a variety of communities where African Americans gained significant representation on city councils but did not elect black mayors. In Philadelphia, as Carolyn Teich Adams has noted, black members of the city council were actually more militant in representing neighborhood interests than was the city's first African American mayor, W. Wilson Goode. Led by black City Council President John Street, they repeatedly blocked approval of a downtown convention center endorsed by Goode and major business interests until affirmative action goals were included in the plan.[67]

In evaluating the efficacy of black power at the local level, however, it would be a mistake to focus exclusively on statistically measurable material gains. One should not underestimate the intangible benefits derived from the election of African Americans to local office, which has greatly increased the access of minorities to the political system and, in the process, reduced a major grievance of the period preceding the riots of the 1960s. Like the increased number of black police officers, African American mayors and councilmembers serve as representative authority figures, not only in the black community but for whites as well.[68] "The mere presence of black and Hispanic council-members," concludes one study of urban politics in the 1970s,

> has increased minority access to councils and changed decision-making processes. We were told repeatedly that minority councilmembers were important in linking minorities to city hall, in providing role models, and in sensitizing white colleagues to minority concerns. Blacks and Hispanics talked about how different it was to be "on the inside," to attend council meetings and see minority representatives sitting on the council, and to be able to call that councilmember on the phone to raise an issue. Even where minorities were not strongly incorporated, councilmembers talked about a new atmosphere and new pressure on the council once minorities were members. One official said, "When minorities talk to the city council now, councilmembers nod their heads rather than yawn."[69]

Two political scientists, using a national sample of voters in 1987, found that the election of a black mayor also had a positive impact on the attitudes of African Americans toward political participation, increasing their trust in and knowledge of local government. "Empowerment," they concluded, "increases black participation. It appears to do so because it increases attentiveness to politics and because it contributes to a more engaged orientation to politics."[70] In the long run, such changed attitudes may be the most important legacy of the new black officeholders of the late twentieth century.

The Underclass Thesis

Regardless of their race or ideological orientation, the mayors of cities undergoing the transition to a postindustrial order have operated under constraints that did not burden their predecessors of the industrial era. Among the most important of these constraints, sociologist William Julius Wilson has argued, are fundamental economic changes that have had an impact on black urban communities since 1970. In his controversial books *The Declining Significance of Race* (1978) and especially *The Truly Disadvantaged* (1987), Wilson has documented the increasing levels of poverty, unemployment, crime, and family instability among inner-city ghetto residents. This black underclass[71] group, he maintains, has become increasingly isolated at the very time when the black middle class has grown in size and moved out of the inner city as a result of new opportunities for advancement that have opened up since the civil rights era. While not denying the underlying importance of "historic discrimination," Wilson places most of the blame for the growth of the so-called underclass on such "impersonal" factors as the decline of manufacturing jobs and the shift to a high-tech, suburban-based economy demanding higher skill levels—changes that have disproportionately hurt blacks living in the inner cities. Wilson and other underclass theorists (especially sociologist John Kasarda) argue that these spatial and educational "mismatches" have undercut employment opportunities for the black poor at the same time that the exodus of the black middle class has removed an element of stability and community leadership from the inner city.[72] The growth rate and age structure of black communities, according to Wilson, have also played an important role in the creation of an underclass. The arrival of large numbers of migrants in a short time span "made it much more difficult for blacks to follow the path of both the new Europeans and the Asian-Americans in overcoming the negative effects of discrimination by finding special occupational niches." At the same time, "the sheer increase in the number of young people, especially young minorities," has exacerbated the problems of crime and unemployment in the inner city.[73] Most scholars agree with Wilson's contention that poverty and female-headed households have increased among poor blacks since the early 1970s and that violent crime, although it may be declining to some extent, remains a serious problem in the inner city, particularly for children growing up in the ghetto.[74] They also find much value in Wilson's emphasis on the negative effects of deindustrialization and the rise of the service economy on African Americans living in cities.[75] However, as recent case studies of Philadelphia and Atlanta indicate, neither deindustrialization per se nor even economic decline are necessary preconditions for the emergence of the kind of racial differences in employment and urban social conditions described by Wilson and many others. As manufacturing declined in Philadelphia after 1960, the percentage of blacks in such industrial work actually rose, but at the same time, black workers found themselves relegated to the lowest-paying manufacturing jobs. In Atlanta, the economic boom of the 1970s and 1980s had minimal effect on black unemployment rates, and high levels of job growth coincided with a rapid increase of income inequality between blacks and whites and continuing high levels of black poverty.[76]

It is also probable that racial discrimination in hiring practices plays a substantially greater role in creating black unemployment, especially among people with low levels of formal education, than is acknowledged by Wilson and others who stress the impact of nonracial factors. Since the 1960s, such discrimination has become more subtle, but it is still widespread.[77] As far as the underemployment of black men is concerned, there is little indication that race has declined as a significant factor. The same may be true for persistent income inequality in the postindustrial city. As Stephanie Coontz has pointed out, racial differences in poverty and unemployment rates are, to a significant extent, independent of family type. Regression analysis of family income for Philadelphia in 1987 suggests that only about 60 percent of the difference between white and black family income can be accounted for by factors like age, education level, and family structure; the remaining 40 percent is the result of "racism and racial discrimination in employment."[78] Perhaps the most glaring evidence of this is the dramatic growth in the homeless population among African Americans since the late 1970s, a phenomenon that curiously has attracted little attention from either underclass theorists or ethnographers.[79]

Although it seems self-evidently valid, there are also serious flaws in the "spatial mismatch" theory linking increases in inner-city unemployment with the expansion of jobs in the suburbs, far removed from where most blacks live. Historically, the "job ceiling," not geography, has been the primary factor limiting black employment opportunities. Urban blacks have generally not benefited from living near major sources of employment. Prior to 1920, in fact, African Americans often lived close to industrial jobs from which they were usually excluded for racial reasons.[80] Did spatial considerations suddenly become more salient in the decentralizing metropolis of the postwar era? The exodus of whites from the central cities in the late 1940s and 1950s, it should be pointed out, did not prevent one group of underpaid black workers—female domestics—from gaining employment in the suburbs or outlying central-city neighborhoods.[81] Nor, at that time, were the job opportunities of black industrial workers necessarily dictated by geography. In metropolitan Detroit, the racial attitudes of auto plant managers were usually much more important than factory location in determining the number of blacks hired. In 1960, one General Motors plant in an all-white suburb had almost no black workers, whereas another in the same community had a workforce that was 41 percent black.[82] The issue is not whether a spatial mismatch exists today (to a considerable extent it does), but how we got there. And although much historical work on this subject remains to be done, it seems clear that the causes of differential access of blacks to expanding work in the suburbs cannot be ascribed wholly, or even perhaps primarily, to the decentralization of jobs.

To a much greater extent than the "spatial mismatch" theory, the argument that a flood of migrants has undercut black economic opportunity is contradicted by the historical record. The eras of heaviest black migration to the urban North—during and immediately after World Wars I and II—were the same periods when industrial work was most available to African Americans. On the other hand (as Wilson himself admits), black unemployment has increased most rapidly since 1970, a period when black migration to the cities fell off sharply and few newcomers had to be absorbed. Wilson is nevertheless buoyed by the fact that, whatever urban racial problems remain at the end of the twentieth century, "one of the major obstacles to urban black advancement—the constant flow of migrants—

has been removed." He suggests that economic conditions for urban blacks may improve of their own accord as the latest wave of migrants, from Mexico and Southeast Asia, replaces African Americans at the bottom of the economic ladder.[83] This "last of the immigrants" thesis is similar to that espoused by neoconservative intellectuals in the 1960s, who argued that blacks would eventually experience the same upward mobility that the children and grandchildren of European immigrants have supposedly attained.[84] However, although many Latinos and some Asian groups have suffered grievous economic problems in recent decades, there is no evidence that blacks are "benefiting" in some perverse way from the arrival of Hispanic or Asian newcomers on the urban scene.[85]

The central thesis of *The Truly Disadvantaged*—that the underclass is growing and is becoming increasingly isolated in inner-city areas —is based primarily on data from Chicago and several other Rust Belt cities for the 1970-1980 decade. Recent studies, however, indicate that these trends are not equally evident in all cities. Nationally, the number of poor blacks living in areas of extremely high poverty (40 percent or more of the census tract below the poverty line) increased by 27 percent between 1970 and 1980. However, almost as many metropolitan centers had declines in the number of black poor in high-poverty districts as experienced increases, and most of the growth of such underclass areas occurred in five declining industrial areas. New York and Chicago alone accounted for half the national total increase. Furthermore, only 21 percent of all poor blacks (26 percent of those living in cities) resided in such areas of dense poverty.[86] If a less-rigorous definition of poverty areas (20 percent or more poor) is used, there was no statistically significant increase in the concentration of the poor in central cities between 1972 and 1989.[87] To understand the evolution of socioeconomic conditions in black communities since the 1960s, it is necessary to differentiate the growth of black poverty and continuing high levels of residential segregation during this period from an increasingly concentrated underclass.[88] It is probably true that underclass areas (using the 40 percent definition) have grown, at least in some cities. But a theory that explains the plight of only one quarter of the urban black poor is of limited value at best. Historically, as well as today, the debilitating effects of poverty have never been limited to an arbitrarily defined underclass.

As Michael Katz has recently pointed out, much scholarship on the underclass suffers from narrow conceptualization.[89] As a means of understanding urban racial conditions during the postwar era, carving off an element of the black urban community for special scrutiny may hide as much as it reveals. Specifically, it diverts attention from changes in urban racial conditions that affect the majority of blacks who do not fall into the underclass category at all. Shifting the focus from the underclass to the other part of William Julius Wilson's thesis—the exodus of the black middle class from inner-city areas and the supposed growing class divisions within the black community—can help to place this issue in a broader context. Since the passage of civil rights legislation in the 1960s, racial differentials in income have declined, but not disappeared. The black middle class made significant gains from the mid-1960s to the late 1970s, and the number of blacks attending college at that time expanded dramatically.[90]

Since the late 1970s, however, the relative economic position of the black middle class when compared with whites has, at best, stagnated.[91] In the 1980s, black access to higher education declined, and for many blacks the economic value of being college educated became more uncertain. Between 1979 and 1987, the number of college-educated African Americans earning poverty-level wages actually increased faster than those receiving high wages.[92] Furthermore, to a much greater degree than whites, the newly enlarged black managerial and professional group is dependent on public or publicly funded employment, which makes them vulnerable during an era of fiscal conservatism and budgetary restraints.[93] Data for the 1970s and 1980s reveal some polarization of income levels within black communities, but the overall income range for African Americans remains much narrower than that of whites and much more weighted toward the lower end of the income spectrum.[94] Finally, it is probable that many blacks who have recently risen into middle-class status have little savings at their disposal. One 1983 survey found that an astonishing one third of all blacks (in contrast to only 7 percent of whites) had no liquid assets at all, and the median assets held by nonwhites and Hispanics was a meager $1,000. Summarizing economic conditions in black communities in the 1980s, Jacqueline Jones notes that "even middle-class families remained just a paycheck away from financial disaster."[95]

A shared perception by the black middle and lower classes of economic uncertainty is reinforced by social conditions imposed on all groups in the black community to some extent. The class segregation within black communities recently discovered by underclass theorists is nothing new; it has been a feature of black urban life for at least seventy years. As early as the 1920s, the oldest sections of black communities experienced high levels of poverty, crime, and delinquency, while the black middle class typically moved into "transitional" areas on the outskirts of the expanding ghetto.[96] Residential class segregation, however, has traditionally operated differently for African Americans than for whites. Whites have long been able to separate themselves out residentially by class. Today, as in the past, most African Americans live in racially circumscribed areas, and this places distinct limitations on the ability of the middle class to separate themselves from the black poor.[97] Although little hard research has been done on the topic, it is clear that middle-class blacks who have fled the inner city have not necessarily moved into homogeneous middle-class neighborhoods. Rather, they frequently have been channeled into areas that contain considerable variation in incomes and lifestyles, especially of people poorer than themselves. In 1967, almost one third of blacks living in urban poverty areas were white-collar workers; 57 percent worked at a variety of semiskilled or unskilled jobs. In a study of Buffalo's African American community in 1980, Henry L. Taylor found some segregation by income, but "no 'exclusive' high or low-income residential sections on the [predominantly black] East Side."[98] The smaller size and later development of Buffalo's black community may have retarded class differentiation to a greater extent than in some larger northern cities, but similar tendencies were evident elsewhere. By the 1980s, black poverty in the suburbs, although remaining significantly lower than

TABLE 11.2 Poverty Rates, Philadelphia and Suburbs, by Race and Hispanic Origin, 1980-1990

	Poverty Rate			
	Total	*Whites*	*Blacks*	*Hispanics*
1980				
Philadelphia	20.6	11.9	32.2	46.3
Suburbs	6.0	5.1	21.0	22.8
1990				
Philadelphia	20.3	11.1	29.0	45.3
Suburbs	4.8	3.7	17.5	15.8

SOURCE: U.S. Census data, courtesy of Professor William Yancey.

central-city levels, far exceeded that of whites, as data from Philadelphia indicate (Table 11.2).

Nationally, by 1992, 23.2 percent of all African Americans in suburban areas had incomes below the poverty level, compared to only 6.3 percent of whites and 19.3 percent of Hispanics.[99] Despite a much higher poverty rate for Hispanics[100] than for blacks in the central city, the lower rate for this group in the suburbs is one indication that movement out of underclass areas creates greater class separation among the new Latino groups than it does for African Americans.[101] The residential patterns of people with more than a high school education also demonstrates this trend (Table 11.3). Blacks who had attended college were much more likely to live in areas having substantial numbers of poor people than were the college educated of other groups. Separation by class in urban America during the post-civil rights era, then, is much less advanced for African Americans than is imagined by those who focus primarily on the exodus of the middle class from deteriorated inner-city neighborhoods.[102]

Even successful blacks who have managed to find housing in an area suitable to their income and lifestyle are unable to escape many aspects of racism that remain embedded in the social and cultural framework of the emerging postindustrial order. In-depth interviews with thirty-seven college-educated African Americans residing in a dozen cities during 1988-1990 led sociologist Joe R. Feagin to

TABLE 11.3 College-Educated Persons Living in Poverty Areas, by Race and Hispanic Origin, 1992 (in percentages)

	Whites	*Blacks*	*Hispanics*
Some college	6.8	31.4	16.9
B.A. degree or more	4.7	21.9	10.3

SOURCE: U.S. Bureau of the Census, *Poverty in the United States: 1992* (Washington, D.C., 1993), Table 9, 60-65.
NOTE: Some college = high school graduates who have attended college but not received a B.A. degree; poverty areas = census tracts containing 20 percent or more people living below the poverty line. Figures are for the entire United States, metropolitan and nonmetropolitan areas combined.[103]

argue for "the continuing significance of race" in the postindustrial city. Although incidents of verbal racial insult or even physical attack by whites were not unknown to the blacks interviewed, far more common were "avoidance" or "rejection actions, such as poor service" in restaurants or stores. Ignored in some stores by sales clerks, blacks on other occasions reported a "problem of burdensome visibility" in dealing with shopkeepers who perceived them as potential shoplifters. "The excessive policing of black shoppers and the discourtesy of clerks illustrate the extra burden of being black in public places," Feagin concluded.[104] Individually, such incidents seem trivial. Cumulatively, however, they reinforce a sense of shared experience among blacks regardless of their economic class.

Three decades after the passage of the Civil Rights Act of 1964, racism in the postindustrial city has neither disappeared nor been restricted in its effects on an impoverished and isolated underclass group. Although operating in a much more subtle manner today than during the era of legal segregation, it affects all classes in the black community but in different ways and to different degrees, creating crosscurrents of racial and class identification in black urban America at the end of the twentieth century. The sometimes contradictory tendencies of race and class were strikingly evident in the conflict between the W. Wilson Goode administration in Philadelphia and the radical black group called MOVE, which had armed and barricaded itself in a home in a black middle-class neighborhood. Despite MOVE's extremist rhetoric and overt hostility to the mayor and other black leaders (as well as to the white police), the Goode administration tried for some time to ignore the black-on-black community conflict engendered by the group. Then, in 1985, the administration drastically reversed its course by ordering an assault on the MOVE compound and allowing a bomb to be dropped on the roof of the building. The resulting fire killed eleven MOVE members and destroyed sixty homes—most of which, ironically, were owned by middle-class blacks who were Goode's core political constituency.[105] The MOVE tragedy represented a failure of personal leadership by Goode, but in a larger sense it exemplified the potential for disaster that existed when conflict within the black community could not be resolved except through force. Although Goode narrowly won reelection in 1987, the assault on MOVE undercut the mayor's support among white liberals, revealing the fragility of the biracial coalition that had elected him four years before.[106] In 1991, the black community was unable to unite around a black candidate to succeed Goode, and the mayor's office reverted to a white Democrat, Ed Rendell.

Conflicting loyalties of class and race also help explain why middle-class blacks in Los Angeles were deeply ambivalent about the trials of Damian Williams and other poor blacks accused of assaulting white truck driver Reginald Denny during the riot of April, 1992.

> "Emotionally we are tugged," said Harold C. Hart-Bibbrig, a black lawyer who grew up only a few streets from the intersection where Mr. Denny was attacked. "I never would have thrown that brick. Most of the people who are screaming for amnesty [for the defendants] would never have thrown that brick."
>
> But, he added: "When I'm out of my community I feel I am treated with suspicion. When I am in my community I feel there is no distinction being made by the police between me and the perpetrators of crime. That's a commonality between me and Damian Williams."[107]

Such "commonalities" help to explain distinctive voting patterns of urban African Americans, who have generally not taken part in the conservative shift that has dominated much of American politics since the late 1960s. Regardless of education or income levels, African Americans have consistently supported liberal issues and black candidates for local office. No economic subgroup among black voters gave Harold Washington less than 95 percent of the vote when he was elected mayor of Chicago in 1983, and blacks in Los Angeles regularly put aside factional differences to help elect Thomas Bradley mayor five times. A careful study of black voting behavior in Boston in the late 1970s concluded that blacks, to a much greater extent than whites, were "voting in ways that suggest similar political perceptions regardless of socioeconomic background."[108] The continuing support of middle-class African Americans for liberal candidates, whether black or white, may be partly a defensive response to perceived threats to their own uncertain economic status, which to a greater extent than for whites relies on government-funded employment. It also demonstrates, however, a degree of racial solidarity with the poor, who constitute a much larger group in the black community than among whites. In politics, as in other aspects of black communities of the postindustrial era, race still often supersedes class.

Beyond the Underclass Debate: Work and Community

The themes and issues raised by underclass theories have expanded our knowledge of urban racial conditions in the late twentieth century. It is important to recognize the limitations of such an analysis, however. The history of African American communities during the post-World War II era (or any other time period) is more than the story of segregation and racism, however profound these may be in their effects. Too much emphasis on the debilitating effects of unemployment, for example, detracts from a broader understanding of the black working class during the transition to a postindustrial economy. To a surprising extent, the history of the black working poor—or even the working class as a whole—during the past fifty years remains unwritten. This is especially true for black male workers. Since the late 1940s, despite substantially lower average wages and higher rates of unemployment than whites, most adult black men have regularly been part of the labor force. A preliminary statistical analysis reveals that the sharp decline in factory work for blacks evident in some major industrial centers during the 1960s and 1970s was not duplicated everywhere. Between 1960 and 1980, at a time when the number of white males working as semiskilled operatives declined from 19 to 16 percent, the proportion of black men in such jobs remained fairly steady at about a quarter of the work force. Meanwhile, the rate of black males in unskilled labor jobs, which had once been a mainstay for less educated men, declined from 23 to 12 percent, and employment of black men in the emerging service occupations increased only marginally, from 15 to 16 percent. The strongest advancement for black men came in white-collar work, which overall increased from 14 to 28 percent of the workforce, but they continued to lag far behind whites in higher income professional and managerial occupations.[109] The occupational position of black men in the postindustrial city remained shaky at best. In contrast to whites, a higher percentage of jobs held by blacks were in low-wage industries or service occupations. It was not deindustrialization alone that weakened the general economic status of black men during the 1960s and 1970s. Rather, the loss of high-paying industrial work in some locales, a general decline in unskilled labor jobs, stricter educational requirements for newer high-tech manufacturing work, and slow movement into white-collar occupations all played a role.[110]

Black women workers, in some respects, appear to have made a more successful transition to the postindustrial economy. The 1960-1980 period saw a dramatic decline (from 35 to 7 percent) in their employment as domestics, poorly paid work traditionally associated with black women. At the same time, African American women made rapid strides in clerical work (9 percent in 1960; 29 percent in 1980) and in professional occupations (6 percent in 1960; 15 percent in 1980), if not managerial positions. The proportion of black women working in personal service occupations increased from 21 to 24 percent, and a surprising 15 percent worked as factory operatives during this period.[111] The movement of black women into clerical work and even service jobs represented real progress compared to the drudgery and irregular nature of domestic service employment.

However, as Bette Woody has explained, black women in manufacturing were more likely than white women to do the heavier or more physically demanding work in such industries as dry cleaning, food processing, and mail-order retailing. The same is true in personal services, where they were often employed in hotels or hospitals as maintenance or food service workers. Even in the newer clerical occupations, where their employment is quite diversified, "they continue to dominate low-status traditional job groups in most business establishments."[112] Furthermore, most of the new jobs that opened up to black women offered few benefits or opportunities for advancement. Nor, despite modest wages, is such employment always secure. Some of these jobs became expendable in the 1980s and 1990s as a result of mechanization, corporate "restructuring," or downsizing.[113]

The general pattern of African American employment in the postindustrial city is clear. Despite some gains in higher-status work, the main tendency is the creation of a new proletariat, based on low-skill (and often nonunion) manufacturing, service, and clerical work. Labor historians, however, have yet to investigate the nature of such work in the manner suggested by Joe W. Trotter and Peter Gottlieb for the industrial period.[114] It is especially important to analyze the changing work experiences of black women in the new economy, because they contribute a substantially greater percentage of total family income than do white women who work. We need a historical approach to contemporary black women's work similar to that provided for the 1880-1930 period by Sharon Harley, who goes beyond economic statistics to explore the relationships between gender, work, and domestic roles in the African American community.[115]

Equally important for a more comprehensive understanding of African American communities since World War II is the history of urban communal activities, including interracial human relations groups, organized protest activities, and distinctive social, cultural, and religious institutions. Local organizations promoting interracial cooperation and equal access to jobs, public services, and housing initially came into being in many northern cities in response to the race riots and tension resulting from renewed black migration during

World War II.[116] Throughout the postwar period, local commissions on human relations worked with black and white homeowners to prevent panic selling or racial violence when blacks began to move into previously all-white neighborhoods. Usually, such activities merely slowed the process of ghetto expansion, but in the 1970s, there were examples of stable, integrated suburban neighborhoods achieved through the concerted efforts of local interracial groups.[117] The role of such organizations in ameliorating racial conflict, or of local branches of the Urban League and the NAACP, has received little attention from historians. If Nina Mjagkij's case study of the Cincinnati Urban League is a guide, the league's behind-the-scenes approach to discrimination by employers did open up skilled jobs for African Americans in the 1950s, but the league was most successful when its "quiet negotiations" were combined with the threat of direct action by more militant civil rights organizations.[118]

At the same time the civil rights movement was reaching its apogee in the South, there was an upsurge in protest activities among blacks in the urban North. Beginning with massive boycotts against segregated and underfunded schools in Chicago and New York in 1963 and 1964, there were numerous protests over such issues as inner-city education, the police, housing, and unresponsive or discriminatory health care or welfare institutions. The struggle over schooling, culminating in the busing controversy in Boston and the Ocean Hill-Brownsville conflict in New York, became the centerpiece of black protest in the North in the late 1960s and early 1970s. Although such actions resulted in greater black representation on school boards and the hiring of more black teachers and administrators, it had relatively little impact on the overall structure of the urban educational bureaucracy.[119] Insurgent black protest organizations were often quite short-lived, but occasionally some, such as the Woodlawn Organization in Chicago, became ongoing community action organizations.[120] In contrast to the South, such activities in the urban North have not generally been subjected to careful historical analysis.[121] Case studies of the goals, methods, and community context of protest movements in the North are essential to a broader understanding of black communities in the postindustrial era.

The same can be said for black urban cultural and religious institutions. The forthcoming history of the Howard Theater and the black theater movement by Bettye Collier-Thomas augments the traditional approach to such cultural activities by placing them in the economic and social context of the black communities in which they flourished and eventually, in the 1960s, declined.[122] The history of black music also has an important local context. It is well known that jazz and the blues, as musical forms, have important roots in the urban experience of black musicians in New Orleans, Memphis, and Chicago.[123] Although the "urban" themes of black music are important, the economic and social function of musicians and musical performance in African American communities also deserves attention. Partly as a result of discrimination in many other professions, a higher percentage of blacks than whites have traditionally worked as musicians in major urban centers.[124] A recent study of the black musicians' union in twentieth-century Philadelphia shows that black musicians were more than entertainers who carried forward an important African American cultural tradition. Through their own labor organization (until they reluctantly merged with the white local in 1971), they also attempted to define and defend their position as workers and artists, responding to changes in the music industry as well as to handicaps imposed on them because of their race.[125]

The church is by far the most important African American urban institution, with important connections to politics, economics, social welfare, and women's activities. As C. Eric Lincoln and Lawrence H. Mamiya have noted, endemic racism has forced the black church to take on a far broader range of functions than most white churches, and ghettoization has anchored the majority of black urban churches in inner-city neighborhoods.[126] The religious institutions and activities of African Americans in specific cities reveal much about the black communities of which they are a vital part. It is evident, for example, that black churches played an important role in the mobilization of the black vote that elected Harold Washington mayor of Chicago in 1983 and reelected him four years later. The study of election statistics and political strategies alone does not do justice to those remarkable political events.[127] Washington's campaigns, one observer noted, "were steeped in religious symbolism. Both campaigns were launched with ministers flanking the candidate, prayers were issued, and the candidate was blessed as he went off to battle. The veterans did not know what to make of the strange mix, one saying it more resembled a Baptist revival meeting than a political event."[128] In New York City during the 1980s, a new generation of black clergy, influenced by black liberation theology, became "particularly sensitive to the question of community empowerment" and started to forge linkages with black secular organizations and African American political leaders.[129] The interaction of the black church with migrants may also explain much about the process of social adaptation of southern newcomers to the urban North during the postwar era.[130] Of all major black institutions, it is primarily churches like the St. Paul Community Baptist Church in East Brooklyn that have been most successful in bringing some measure of hope to the most impoverished inner-city neighborhoods of the postindustrial era.[131] In addition to churches, black secular institutions, including recreational associations and black hospitals, deserve more than the passing mention they are given in most community studies.[132] Nor should informal group life, expressed through family and neighborhood interaction, be neglected.[133] Common patterns of interpersonal communication, mutual self-help, and a shared sense of identity in facing an often hostile outside world underlie and inform day-to-day life in many black neighborhoods, including underclass areas.[134] Even inner-city gangs, despite their use of violence, are more than mere symptoms of social pathology. Recent ethnographic studies have shown they have a variety of organizational structures and ways of interacting with the communities of which they are a part.[135]

Toward a Comparative Perspective

Perhaps because the dominant tendency of black migration in the twentieth century has been toward the urban North, most historical and sociological studies of black urban conditions have focused on that region. In many ways, this bias in the literature would seem to be less significant for the postindustrial period than for the industrial era. Especially since the 1960s, urban centers in the South and West have exhibited many of the same tendencies in general economic conditions and black community development as have cities in the North, including ghettoization and increasing levels of poverty among blacks, the expansion of suburbs, and the decentralization of many economic activities.[136]

One significant difference, however, is the varying effect of Hispanic and Asian minorities on urban centers in different regions. In major northern cities, with the exception of New York and, to a lesser extent, Chicago, blacks are by far the largest nonwhite group. That is also true of Atlanta and New Orleans, but not of other sunbelt metropolises. By 1990, Latinos outnumbered blacks in Miami, San Antonio, Houston, Phoenix, Los Angeles, and San Diego; and Asians constituted a significant minority group as well in San Francisco, Seattle, and other West Coast cities. Much of this growth of nonblack minority populations in southern and western cities has taken place during the past thirty years.[137]

The effects of these demographic changes have been especially profound in the realm of urban politics. Although blacks in the urban North have gained political power during a period of industrial decline, their counterparts in growing Sunbelt metropolises have often been stymied as a result of the annexation of suburban areas or the rapid influx of other racial minorities, both of which have diluted the black vote.[138] In Miami, blacks remained at about 18 percent of the population between 1940 and the mid-1980s, whereas the Hispanic group (mostly Cubans) grew exponentially from 4 percent to over 60 percent.[139] "In the urban South," says Ronald Bayor in summarizing this trend, "it is no longer possible to focus on white or black categories alone since other groups are now legitimized and a new awareness of coalition politics exists."[140] Bayor maintains that cities like Atlanta, Miami, San Antonio, and Houston have adopted a "northern" style of multiethnic coalition politics. His analysis could be applied to Los Angeles and other western cities as well.[141] The interracial violence that has plagued Miami and Los Angeles in the 1980s and 1990s, however, is one indication that the new political configuration of Sunbelt cities has not yet created an adequate framework for addressing—perhaps even acknowledging—the grievances of African Americans. In the last decade of the twentieth century, it is evident that the problems of economic inequality, poor housing, and inadequate schools continue to affect African Americans to some degree in virtually all major urban areas. Competition with other minority groups for scarce economic and political resources during an era of fiscal restraint has sharpened the edge of black discontent in cities like Los Angeles and Miami as well as in New York, where the narrow defeat of black Mayor David Dinkins in his bid for reelection in 1993 left blacks with a painful sense of betrayal.[142]

One challenge to historians of urban racial conditions in the postwar period will be to evaluate the impact of emergent multiracial (as opposed to multiethnic) cities on African American communities and race relations. Historical comparisons between blacks and white European ethnic groups for the late nineteenth and early twentieth century have clarified the ways in which an earlier generation of black migrants both shared and, to a significant degree, remained outside the mainstream of industrial society. To properly assess the history of black urban America in the postindustrial period will require genuinely comparative studies, not only of blacks and whites but also between nonwhite racial minorities, both within the United States and in "global cities" in other parts of the world.[143]

Conclusion

A comprehensive understanding of the history of urban racial conditions requires an assessment of the shifting effect of external, internal, and structural factors over time.[144] In contrast to the industrial period, scholars of the postwar era have focused heavily on the impact of structural forces on the black community, especially the role of the state. In light of the enlarged place of government in American society since the 1930s, such an approach has considerable value, but it would be a mistake to underestimate the power of either the external effects of racism or the internal dynamics of social or cultural forces originating in the black community itself. The role of governmental or economic policies are important, but if their effects on African Americans are harmful, that is as much a product of the racial attitudes of those who approve or carry them out as of the bureaucratic or technological framework in which they are formulated. Likewise, the evolution of African American communities is not simply a response to external or structural factors. The aspirations and struggles of urban African Americans, individually and collectively, have continually shaped and reshaped the social, cultural, and political life of black communities. Underclass theorists are prone to emphasize "impersonal forces" when discussing the social conditions they deplore. Historians—who take a longer view and are perhaps better attuned to the vagaries of human behavior—know better. Although it is necessary to generalize to make sense of historical change, at bottom there are no impersonal forces at work in history—only personal forces writ large. Social change, whether beneficial or injurious, is ultimately the product of millions of decisions made by countless individuals over many years. A history of black urban America during the postwar era that proceeds from this basis will, hopefully, do justice to the complexity of that story.

Notes

1. Studies of black urban communities during the industrial period include W. E. B. Du Bois, *The Philadelphia Negro: A Social Study* (Philadelphia, 1899); St. Clair Drake and Horace R. Cayton, *Black Metropolis: A Study of Negro Life in a Northern City* (New York, 1945); Gilbert Osofsky, *Harlem: The Making of a Ghetto, 1890-1930* (New York, 1966); Allan H. Spear, *Black Chicago: The Making of a Negro Ghetto, 1890-1920* (Chicago, 1967); William Tuttle, Jr., *Race Riot: Chicago in the Red Summer of 1919* (New York, 1970); Neil Betten and Raymond A. Mohl, "The Evolution of Racism in an Industrial City, 1906-1940: A Case Study of Gary, Indiana," *Journal of Negro History* 59 (1974), 51-64; Bettye Collier-Thomas, "The Baltimore Black Community, 1865-1910" (Ph.D. diss., George Washington University, 1974); Zane Miller, "Urban Blacks in the South, 1865-1920," in Leo Schnore, ed., *The New Urban History* (Princeton, 1975), 184-204; Kenneth L. Kusmer, *A Ghetto Takes Shape: Black Cleveland, 1870-1930* (Urbana, 1976); James Borchert, *Alley Life in Washington: Family, Community, Religion, and Folklife in the City, 1850-1970* (Urbana, 1980); Kenneth L. Kusmer, ed., *From Reconstruction to the Great Migration, 1877-1917* (New York, 1991); Kenneth L. Kusmer, ed., *The Great Migration and After, 1917-1930* (New York, 1991); Elizabeth Pleck, *Black Migration and Poverty: Boston, 1865-1900* (New York, 1979); August Meier and Elliott Rudwick, *Black Detroit and the Rise of the UAW* (New York, 1979); Joe W. Trotter, *Black Milwaukee: The Making of an Industrial Proletariat, 1915-45* (Urbana, 1985); George C. Wright, *Life Behind*

a Veil: Blacks in Louisville, Kentucky, 1865-1930 (Baton Rouge, 1985); Peter Gottlieb, *Making Their Own Way: Southern Blacks' Migration to Pittsburgh, 1916-1930* (Urbana, 1987); Darrel E. Bigham, *We Ask Only a Fair Trial: A History of the Black Community of Evansville, Indiana* (Bloomington, 1987); Lawrence B. De Graaf, "The City of Black Angels: Emergence of the Los Angeles Ghetto, 1890-1930," *Pacific Historical Review* 39 (1970), 323-52; Quintard Taylor, "The Emergence of Black Communities in the Pacific Northwest, 1890-1930," *Journal of Negro History* 64 (1979), 342-54; Taylor, "Black Urban Development—Another View: Seattle's Central District, 1910-1940," *Pacific Historical Review* 58 (1989), 429-48; Lillian S. Williams, "Afro-Americans in Buffalo, 1900-1930: A Study in Community Formation," *Afro-Americans in New York Life and History* 8 (1984), 7-35; Richard Alan Ballou, "Even in 'Freedom's Birthplace'! The Development of Boston's Black Ghetto, 1900-1940" (Ph.D. diss., University of Michigan, 1984); James R. Grossman, *Land of Hope: Chicago, Black Southerners, and the Great Migration* (Chicago, 1990); Roger Lane, *Roots of Violence in Black Philadelphia, 1860-1900* (Cambridge, 1986); Lane, *William Dorsey's Philadelphia and Ours: The Past and Future of the Black City in America* (Cambridge, 1991), chaps. 1-11; Laurence Glasco, "Double Burden: The Black Experience in Pittsburgh," in Samuel P. Hays, ed., *City at the Point: Essays on the Social History of Pittsburgh* (Pittsburgh, 1989), 69-110; Charles Ashley Hardy III, "Race and Opportunity: Black Philadelphia During the Era of the Great Migration, 1916-1930" (Ph.D. diss., Temple University, 1989); Eric Brian Halpern, " 'Black and White Unite and Fight': Race and Labor in Meatpacking, 1904-1948" (Ph.D. diss., University of Pennsylvania, 1989); Earl Lewis, *In Their Own Interests: Race, Class, and Power in Twentieth-Century Norfolk, Virginia* (Berkeley, 1991); Joe W. Trotter, ed., *The Black Migration in Historical Perspective: New Dimensions of Race, Class, and Gender* (Bloomington, 1991); Peter Rachleff, *Black Labor in Richmond, 1865-1890* (Urbana, 1988); Eric Arnesen, *Waterfront Workers of New Orleans: Race, Class, and Politics, 1863-1923* (New York, 1991); Michael K. Honey, *Southern Labor and Black Civil Rights: Organizing Memphis Workers* (Urbana, 1993); Albert S. Broussard, *Black San Francisco: The Struggle for Racial Equality, 1900-1954* (Lawrence, 1993). On black women's organizations, see Gerda Lerner, "Early Community Work of Black Club Women," *Journal of Negro History* 59 (1974), 158-67; Adrienne Lash Jones, *Jane Edna Hunter: A Case Study in Black Leadership, 1905-1950* (New York, 1989); and Dorothy Salem, *To Better Our World: Black Women and Organized Reform* (New York, 1990). Two important studies that compare African Americans and urban white ethnic groups are John Bodnar, Roger Simon, and Michael P. Weber, *Lives of Their Own: Blacks, Italians and Poles in Pittsburgh, 1900-1960* (Urbana, 1983), and Joel Perlmann's study of Providence, *Ethnic Differences: Schooling and Social Structure Among the Irish, Italians, Jews, and Blacks in an American City, 1880-1935* (Cambridge, England, 1988).

2. See Dominic J. Capeci, Jr., *The Harlem Riot of 1943* (Philadelphia, 1977); Harvard Sitkoff, "The Detroit Race Riot of 1943," *Michigan History* 53 (1969), 183-206; Dominic J. Capeci, Jr., and Martha Wilkerson, *Layered Violence: The Detroit Rioters of 1943* (Jackson, 1991); Cheryl Greenberg, "The Politics of Disorder: Reexamining Harlem's Riots of 1935 and 1943," *Journal of Urban History* 18 (1992), 395-441.

3. On socioeconomic conditions in the black North during the Depression, see E. Franklin Frazier, "Some Effects of the Depression on the Negro in Northern Cities," *Science and Society* 2 (1938), 489-99; Drake and Cayton, *Black Metropolis,* parts 2 and 3; Cheryl Lynn Greenberg, *Or Does It Explode? Black Harlem in the Great Depression* (New York, 1991).

4. See Herbert Garfinkle, *When Negroes March: The March on Washington Movement in the Organizational Politics for FEPC* (New York, 1959).

5. Drake and Cayton, *Black Metropolis*, chap. 12; Horace R. Cayton and George S. Mitchell, *Black Workers and the New Unions* (Chapel Hill, 1939); Christopher G. Wye, "The Black Worker and the Labor Movement in Cleveland: Forging a New Relationship," in Kenneth L. Kusmer, ed., *Depression, War, and the New Migration, 1930-1960* (New York, 1991), 166-91; Trotter, *Black Milwaukee*, chap. 5; William H. Harris, *The Harder We Run: Black Workers Since the Civil War* (New York, 1982), chap. 5; Lewis, *In Their Own Interests*, 137-43.

6. William J. McKenna, "The Negro Vote in Philadelphia Elections," *Pennsylvania History* 32 (1965), 406-15; Charles Russell Branham, "The Transformation of Black Political Leadership in Chicago, 1864-1942" (Ph.D. diss., University of Chicago, 1981); Christopher Robert Reed, "Black Chicago Political Realignment During the Great Depression and New Deal," *Illinois Historical Journal* 78 (1985), 242-56; Harvard Sitkoff, *A New Deal for Blacks: The Emergence of Civil Rights as a National Issue* (New York, 1978); Nancy Weiss, *Farewell to the Party of Lincoln: Black Politics in the Age of F. D. R.* (Princeton, 1983).

7. Christopher G. Wye, "The New Deal and the Negro Community: Towards a Broader Conceptualization," *Journal of American History* 59 (1972), 621-39; Kenneth T. Jackson, "Race, Ethnicity, and Real Estate Appraisal: The Home Owners Loan Corporation and the Federal Housing Administration," *Journal of Urban History* 6 (1980), 419-52; Raymond A. Mohl, "Trouble in Paradise: Race and Housing in Miami During the New Deal Era," *Prologue* 19 (1987), 7-21; Ronald H. Bayor, "Urban Renewal, Public Housing and the Racial Shaping of Atlanta," *Journal of Policy History* 1 (1989), 419-39; John F.Bauman, *Public Housing, Race, and Renewal: Urban Planning in Philadelphia, 1920-1974* (Philadelphia, 1987).

8. An impressionistic view is provided by Nicholas Lemann, *The Promised Land: The Great Black Migration and How It Changed America* (New York, 1991). A study of the impact of black migration during World War II alone would be valuable. On the impact of black migration to the West, see Gerald D. Nash, *The American West Transformed: The Impact of World War II* (Bloomington, 1985), 88-107.

9. Lemann, *The Promised Land*, 3-7, 48-50.

10. Nicholas Lemann's contention that the "underclass" problem of the urban North is traceable directly to the black rural culture of the sharecropper South (Lemann, "The Origins of the Underclass, Pt. I," *Atlantic Monthly* 257 [June 1986], 31-61; Lemann, *The Promised Land*, 31) is ahistorical and flies in the face of the well-known fact that black migrants have usually been more successful economically than northern-born African Americans. See Larry H. Long, "Poverty, Welfare and Migration," *American Sociological Review* 39 (1974), 46-56, and Jacqueline Jones, "Southern Diaspora: Origins of the Northern 'Underclass,' " in Michael B. Katz, ed., *The "Underclass" Debate: Views From History* (Princeton, 1993), 27-30.

11. Richard M. Bernard, ed., *Snowbelt Cities: Metropolitan Politics in the Northeast and Midwest Since World War II* (Bloomington, 1993), appendix, 272-73; Richard M. Bernard and Bradley R. Rice, eds., "Introduction," in *Sunbelt Cities: Politics and Growth Since World War II* (Austin, 1990), 24-25.

12. For an overview of these changes, see Raymond A. Mohl, "The Transformation of Urban America Since World War II," *Amerikastudien* 33 (1988), 53-71, and Howard P. Chudacoff and Judith E. Smith, *The Evolution of American Urban Society*, 4th ed. (Englewood Cliffs, 1994), 255-87.

13. Arnold R. Hirsch, *Making the Second Ghetto: Race and Housing in Chicago, 1940-1960* (Cambridge, England, 1983); Charles F. Casey-Leininger, "Making the Second Ghetto in Cincinnati: Avondale, 1925-70," in Henry Louis Taylor, Jr., ed., *Race*

and the City: Cincinnati, 1820-1970 (Urbana, 1993), 232-57; Carolyn T. Adams et al., *Philadelphia: Neighborhoods, Division, and Conflict in a Postindustrial City* (Philadelphia, 1991), 73-81; Adrienne Lash Jones, "The Central Avenue Community, 1930-1983" (unpublished paper, 1984); author's personal experience of Hough Avenue area in the 1950s.

14. Hirsch, *Making the Second Ghetto*, 28-39; Gregory D. Squires et al., *Chicago: Race, Class, and the Response to Urban Decline* (Philadelphia, 1987), chap. 4; Thomas J. Sugrue, "The Origins of the Urban Crisis: Race, Industrial Decline, and Housing in Detroit, 1940-1960" (Ph.D. diss., Harvard University, 1992), 150-94; Broussard, *Black San Francisco*, 172-73; Casey-Leininger, "Making the Second Ghetto in Cincinnati," 239-40.

15. Broussard, *Black San Francisco*, 240.

16. Hirsch, *Making the Second Ghetto*, 40-99. The history of urban racial conflict has tended to focus on large-scale incidents that resulted in fatalities. One implication of Hirsch's study is that the dividing line between such events and incidents that did not escalate into full-scale racial warfare is much less clear than previously understood.

17. F. James Davis, "The Effects of Freeway Displacement on Racial Housing Segregation in a Northern City," *Phylon* 26 (1965), 301-08; Joe T. Darden et al., *Detroit: Race and Uneven Development* (Philadelphia, 1987), 158-70; Beth Anne Shelton et al., *Houston: Growth and Decline in a Sunbelt Boomtown* (Philadelphia, 1989), 73; Bayor, "Urban Renewal," 423-35; Ronald Bayor, "Roads to Racial Segregation: Atlanta in the Twentieth Century," *Journal of Urban History* 15 (1988), 3-21; Christopher Silver, "The Changing Face of Neighborhoods in Memphis and Richmond, 1940-1985," in Randall M. Miller and George E. Pozzetta, eds., *Shades of the Sunbelt: Essays on Ethnicity, Race, and the Urban South* (Boca Raton, 1988), 109-10.

18. Kenneth T. Jackson, *Crabgrass Frontier: The Suburbanization of the United States* (New York, 1985), 190-230; Gregory D. Squires and William Velez, "Insurance Redlining and the Transformation of an Urban Metropolis [Milwaukee]," *Urban Affairs Quarterly* 23 (1987), 63-83; Bauman, *Public Housing, Race, and Renewal;* Bayor, "Urban Renewal, Public Housing, and the Racial Shaping of Atlanta"; John F. Bauman, Norman P. Hummon, and Edward K. Muller, "Public Housing, Isolation, and the Urban Underclass: Philadelphia's Richard Allen Homes, 1941-1965," *Journal of Urban History* 17 (May 1991), 264-92; Todd Swanstrom, *The Crisis of Growth Politics: Cleveland, Kucinich, and the Challenge of Urban Populism* (Philadelphia, 1985), 67-68; David W. Bartelt, "Housing the 'Underclass,' " in Katz, ed., *The "Underclass" Debate*, 145-54.

19. Charles S. Johnson, *Patterns of Negro Segregation* (New York, 1943), chap. 1; Karl Taueber and Alma Taueber, *Negroes in Cities: Residential Segregation and Neighborhood Change* (Chicago, 1965), 189-92; Abram L. Harris, *The Negro Population of Minneapolis: A Study of Race Relations* (Minneapolis, 1927), 13-14; Lawrence B. De Graff, "City of Black Angels: Emergence of the Los Angeles Ghetto, 1890-1930," 345-50; Trotter, *Black Milwaukee: The Making of an Industrial Proletariat, 1915-45*, 178; Taylor, "Black Urban Development—Another View," 429-48; Douglas Massey and Nancy Denton, *American Apartheid: Segregation and the Making of the Underclass* (New York, 1993), 26-42; Kenneth L. Kusmer, "The Origin of Black Ghettos in the North, 1870-1930: A Study in Comparative History" (unpublished paper).

20. Massey and Denton, *American Apartheid*, 57 and 42-57. Massey and Denton provide the best summary statement on the rise of the black ghetto in the twentieth century. Their comparison of northern and southern cities is excellent, but they include data on only one western city, Los Angeles, which they categorize as a northern city. Especially prior to World War II, western cities may well have represented a third pattern of racial segregation, distinctive from most northern and southern cities.

21. Numan V. Bartley, *The Rise of Massive Resistance: Race and Politics in the South in the 1950s* (Baton Rouge, 1969); David R. Goldfield, *Black, White, and Southern: Race Relations and Southern Culture, 1940 to the Present* (Baton Rouge, 1990), 63-117. The historical literature on the civil rights movement is enormous, but surprisingly few studies focus on local communities. Those that do so examine relatively small cities. Two valuable studies are William Chafe, *Civilities and Civil Rights: Greensboro, North Carolina, and the Black Struggle for Freedom* (New York, 1980) and Robert J. Norrell, *Reaping the Whirlwind: The Civil Rights Movement in Tuskegee* (New York, 1985), which covers the period from the 1930s to the 1970s. In-progress studies of Atlanta by Ronald Bayor and of Miami by Raymond A. Mohl will provide a much-needed assessment of changing race relations in a big-city context.

22. In most northern states, segregation in, or exclusion from, public accommodations was not legal. It nevertheless existed and varied widely from one community to another prior to World War II. Even in a city like Cleveland, with a "liberal" reputation, racial discrimination in restaurants, theaters, and hotels was widespread in the 1920s and 1930s. See Kusmer, *A Ghetto Takes Shape*, 174-89; Christopher G. Wye, "Midwest Ghetto: Patterns of Negro Life and Thought in Cleveland, 1929-1945" (Ph.D. diss., Kent State University, 1973), chap. 2.

23. Glasco, "Double Burden," in Hays, ed., *City at the Point*, 93; Emma Lou Thornbrough, "Breaking Racial Barriers to Public Accommodations in Indiana, 1935-1963," *Indiana Magazine of History* 83 (1987), 301-43. On the integration of public accommodations and schools in Washington in the 1950s, see Steven J. Diner, "The Black Majority: Race and Politics in the Nation's Capital," in Bernard, ed., *Snowbelt Cities*, 250-51.

24. Nina Mjagkij, "Behind the Scenes: The Cincinnati Urban League, 1948-63," in Taylor, ed., *Race and the City*, 282-89.

25. Goldfield, *Black, White, and Southern*, 47-48, 93; Ronald H. Bayor, "Race and City Services: The Shaping of Atlanta's Police and Fire Departments," *Atlanta History: A Journal of Georgia and the South* (Fall 1992), 19-27.

26. August Meier and Elliott Rudwick, *From Plantation to Ghetto*, 3rd ed. (New York, 1976), 268-69.

27. Swanstrom, *The Crisis of Growth Politics*, 101; Martin Kilson, "Political Change in the Negro Ghetto, 1900-1940s," in Nathan Huggins et al., eds., *Key Issues in the Afro-American Experience* (New York, 1971), part 2, 167-85.

28. Paul Kleppner, *Chicago Divided: The Making of a Black Mayor* (De Kalb, Ill., 1985); William J. Grimshaw, *Bitter Fruit: Black Politics and the Chicago Machine, 1931-1991* (Chicago, 1992). In addition to white hostility, Lawrence Glasco stresses the negative impact of geographic dispersal of blacks in several ghettos in Pittsburgh as an explanation of the weak political influence of blacks there. Glasco, "Internally Divided: Class and Neighborhood in Black Pittsburgh," *Amerikastudien* 34 (1989), 223-30.

29. Kusmer, *A Ghetto Takes Shape*, 183-84; Michael W. Homel, *Down From Equality: Black Chicagoans and the Public Schools, 1920-41* (Urbana, 1984); Grossman, *Land of Hope*, chap. 8; Bigham, *We Ask Only a Fair Trial*, 124-29.

30. Kleppner, *Chicago Divided*, 50-54; Darden et al., *Detroit*, 220-25, 230-31; Richard M. Bernard, "Milwaukee: The Death and Rebirth of a Midwestern Metropolis," in Bernard, *Snowbelt Cities*, 177-79. For an overview of the emerging school crisis in the North that relates it to structural changes in the postwar city, see Harvey Kantor and Barbara Brenzel, "Urban Education and the 'Truly Disadvantaged': The Historical Roots of the Contemporary Crisis, 1945-1990," in Katz, ed., *The "Underclass" Debate*, 369-73.

31. Mark I. Gelfand, "Boston: Back to the Politics of the Future," in Bernard, ed., *Snowbelt Cities,* 50-51; Glasco, "Double Burden," in Hays, ed., *City at the Point,* 80, 90; Raphael J. Sonenshein, *Politics in Black and White: Race and Power in Los Angeles* (Princeton, 1993), 33-35; Rufus P. Browning, Dale Rogers Marshall, and David H. Tabb, *Protest Is Not Enough: The Struggle of Blacks and Hispanics for Equality in Urban Politics* (Berkeley, 1984), 20.

32. See especially William Julius Wilson, *The Truly Disadvantaged: The Inner City, the Underclass, and Public Policy* (Chicago, 1987), 3-106.

33. Jon Teaford, *The Rough Road to Renaissance: Urban Revitalization Movements in America, 1940-1985* (Baltimore, 1990), 10-43.

34. Carl Abbott, *The New Urban America: Growth and Politics in Sunbelt Cities* (Chapel Hill, 1981), 98-119; Nash, *The American West Transformed,* 17-36.

35. Squires et al., *Chicago,* 25-29; Adams et al., *Philadelphia,* 30-39; Sugrue, "The Origins of the Urban Crisis," 94-148. For a macroeconomic view of the history of this period, see Barry Bluestone and Bennett Harrison, *The Deindustrialization of America: Plant Closings, Community Abandonment, and the Dismantling of Basic Industry* (New York, 1982), 111-39.

36. John T. Cumbler, *A Social History of Economic Decline: Business, Politics, and Work in Trenton* (New Brunswick, 1989), 173-74; Sugrue, "The Origins of the Urban Crisis," 45-80. As a result of legal action, many unions were forced to hire blacks in the late 1960s and early 1970s, but in 1977 the pace of integration in the better-paying trades was described by one observer as "exceedingly slow." William B. Gould, *Black Workers in White Unions: Job Discrimination in the United States* (Ithaca, 1977), 282 and 281-89. See also Herbert Hill, "Black Workers, Organized Labor, and Title VII of the 1964 Civil Rights Act: Legislative History and Litigation Record," in Herbert Hill and James E. Jones, Jr., eds., *Race in America: The Struggle for Equality* (Madison, 1993), 314-24.

37. Sugrue, "Origins of the Urban Crisis," 126; Robert J. Norrell, "Caste in Steel: Jim Crow Careers in Birmingham, Alabama," *Journal of American History* 73 (1986), 690-93. For basic data on unemployment trends, see Reynolds Farley and Walter R. Allen, *The Color Line and the Quality of Life in America* (New York, 1987), 213-18.

38. Mark J. Stern, "Poverty and Family Composition Since 1940," in Katz, ed., *The "Underclass" Debate,* 222-23.

39. Herbert Hill, "Demographic Change and Racial Ghettos: The Crisis of American Cities," *Journal of Urban Law* 44 (1966), 231-85, remains the best general introduction to the historical background to the riots. On the riots themselves, see *The Report of the National Advisory Commission on Civil Disorders (Washington, D.C., 1968);* Joseph Boskin, *"The Revolt of the Urban Ghettos, 1964-1967," Annals of the American Academy of Political and Social Science* 382 (1969), 1-14; Robert M. Fogelson, *Violence as Protest: A Study of Riots and Ghettos* (Garden City, 1971); John S. Adams, "The Geography of Riots and Civil Disorders in the 1960s," *Economic Geography* 48 (1972), 24-42; Sidney Fine, *Violence in the Model City: The Cavenaugh Administration, Race Relations, and the Detroit Riot of 1967* (Ann Arbor, 1989).

40. See Robert M. Fogelson, "White on Black: A Critique of the McCone Commission Report on the Los Angeles Riot," *Political Science Quarterly* 82 (1967), 337-67.

41. On the history of race riots generally, see Allen D. Grimshaw, *Racial Violence in the United States* (Chicago, 1969). The Harlem riots of 1935 and 1943 were early examples of "commodity riots," and the Detroit riot of 1943 contained elements of both the World War I-style "communal" riot and the 1960s "commodity" riot. It is not

surprising that the early examples of 1960s-style riots occurred in cities where the black population was both large and already fairly isolated at an early stage.

42. *The Kerner Report: The Report of the National Advisory Commission on Civil Disorders* (New York, 1988), 111. Note the specific rejection of an emergent underclass as a cause of the riot.

43. Jerome L. McElroy and Larry D. Singell, "Riot and Non-Riot Cities: An Examination of Structural Contours," *Urban Affairs Quarterly* 8 (1973), 281-302.

44. Paula B. Johnson, David O. Sears, and John B. McConahay, "Black Invisibility, the Press, and the Los Angeles Riot," *American Journal of Sociology* 76 (1971), 698-721. On black attitudes toward the riot, see also David O. Sears and T. M. Tomlinson, "Riot Ideology in Los Angeles: A Study of Negro Attitudes," *Social Science Quarterly* 49 (1968), 485-503. Gelfand, "Boston: Back to the Politics of the Future," in Bernard, ed., *Snowbelt Cities*, 51, also comments on "the almost total invisibility of blacks in the pages of Boston's newspapers" prior to the 1960s.

45. The almost total exclusion of blacks from television programming during the 1950s and early 1960s may have played an even more important role than lack of press coverage in reinforcing a sense of alienation among African Americans during the period preceding the riots. For a general discussion of African American stereotypes that later developed in commercial television, see Jeannette L. Dates, ed., *Split Image: African Americans in the Mass Media* (Washington, D.C., 1990), 253-302.

46. Godfrey Hodgson, *America in Our Time* (New York, 1976), 180.

47. Swanstrom, *The Crisis of Growth Politics*, 101.

48. The only significant study that attempts to place the riot in this broader context is Cumbler, *A Social History of Economic Decline*, 174-79.

49. Abbott, *The New Urban America*, illustrates the uneven economic growth of the "Sunbelt" region; Goldfield, *Promised Land*, places recent southern urbanization in a regional context. See also Randall M. Miller, "Migrations and Economic Growth in the Metropolitan Southern Sunbelt," *Perspectives on the American South* 4 (1986), 163-86.

50. In most northern cities in the 1970s, the increase in the percentage of the population that was black was only about half what it had been during the 1960s. Typical was Philadelphia, which went from 26 to 34 percent black between 1960 and 1970, but increased only to 38 percent black in 1980. In the South, the increase was even slower, and in western cities almost nonexistent, primarily because the growth of the Hispanic population outstripped that of African Americans. For data, see Bernard, ed., *Snowbelt Cities*, 272-73; Bernard and Rice, "Introduction," in Bernard and Rice, eds., *Sunbelt Cities*, 24-25.

51. *Bradley v. Milliken* U.S. 418 (1974). For the background to this case, see Darden et al., *Detroit*, 225-30.

52. Gary Orfield, "School Desegregation After Two Generations: Race, Schools, and Opportunity in Urban Society," in Hill and Jones, eds., *Race in America*, 239; Roger Biles, "A Bittersweet Victory: Public School Desegregation in Memphis," *Journal of Negro Education* 55 (1986), 470-83. For an excellent case study of Atlanta schools, see Gary Orfield and Carole Ashkinaze, *The Closing Door: Conservative Policy and Black Opportunity* (Chicago, 1991), 103-48.

53. Kantor and Brenzel, "Urban Education," 373-92, and Orfield, "School Desegregation After Two Generations," 234-62, provide valuable overviews of race and schooling in the postindustrial era.

54. David W. Bartelt, "Housing the 'Underclass,' " 126-27; Robert D. Bullard, *Invisible Houston: The Black Experience in Boom and Bust* (College Station, Tex., 1987), 32-39.

55. Arthur D. Little, Inc., *East Cleveland: Response to Urban Change* (East Cleveland, 1969), quoted in Teaford, *Cities of the Heartland,* 233; John R. Logan and Mark Schneider, "Racial Segregation and Racial Change in American Suburbs, 1970-1980," *American Journal of Sociology* 89 (1984), 874-88; John F. Kain, "Housing Market Discrimination and Black Suburbanization in the 1980s," in Gary A. Tobin, ed., *Divided Neighborhoods: Changing Patterns of Racial Segregation* (Newbury Park, Calif., 1987), 87. As Nancy Denton and Douglas Massey point out (*American Apartheid,* 69-70) the rate of suburbanization of blacks in the South is sometimes much higher than the North. In some southern cities, this may be due to the traditional regional urban pattern of blacks living in outlying areas, but to some extent it reflects the statistically artificial incorporation of older black rural populations into metropolitan areas as Census Bureau designations have changed in the post-World War II period.

56. George Galster, "Racial Steering by Real Estate Agents: Mechanisms and Motives," *Review of Black Political Economy* 57 (Summer 1990), 51-62; Bullard, *Invisible Houston,* 50-59.

57. Diana Jean Schemo, "L. I. Apartment Service Accused of Bias in Rental Offers," *New York Times,* February 15, 1994.

58. Browning, Marshall, and Tabb, *Protest Is Not Enough,* 19-22.

59. Georgia A. Persons, "Racial Politics and Black Power in the Cities," in George C. Galster and Edward W. Hill, eds., *The Metropolis in Black and White: Place, Power and Polarization* (New Brunswick, 1992), 173-75.

60. Washington, D.C. was already two thirds black at the time its first black mayor, Walter Washington, was appointed by President Lyndon Johnson in 1967. Washington was elected in his own right in 1974 after Congress enacted home-rule legislation.

61. Persons, "Racial Politics," table 10.1, 175.

62. In the case of Chicago, a sharp division between two white candidates for the Democratic Party nomination also played an important role in Harold Washington's victory in 1983. See Kleppner, *Chicago Divided,* 134-85.

63. Sonenshein, *Politics in Black and White.*

64. Studies stressing the limitations on the power of the new black mayors include Mack H. Jones, "Black Political Empowerment in Atlanta: Myth and Reality," *Annals of the American Academy of Political and Social Science* 439 (September 1978), 90-117; James Button, "Southern Black Elected Officials: Impact on Socioeconomic Change," *Review of Black Political Economy* 12 (Fall 1982), 29-45; and Robert A. Beauregard, "Tenacious Inequalities: Politics and Race in Philadelphia," *Urban Affairs Quarterly* 25 (1990), 420-34. On Newark and Gary, see also Edmund J. Keller, "The Impact of Black Mayors on Urban Policy," *Annals of the American Academy of Political and Social Science* 439 (September 1978), 40-52, and Persons, "Racial Politics," 179-83.

65. H. Paul Friesema, "Black Control of the Central Cities: The Hollow Prize," *Journal of the American Institute of Planners* 35 (1969), 75-79.

66. Persons, "Racial Politics," 183; Kenneth R. Mladenka, "The Distribution of an Urban Public Service: The Changing Role of Race and Politics," *Urban Affairs Quarterly* 24 (1989), 556-83; Peter K. Eisinger, "Black Employment in Municipal Jobs: The Impact of Black Political Power," *American Political Science Review* 76 (1982), 383 and 380-87; Sonenshein, *Politics in Black and White,* 174, 139-75.

67. Browning, Marshall, and Tabb, *Protest Is Not Enough,* 184-87, 195-98; Carolyn Teich Adams, "Philadelphia: The Private City in the Postindustrial Era," in Bernard, ed., *Snowbelt Cities,* 218-20. The study by Browning et al. found that black gains in municipal employment during this period significantly exceeded that of Hispanics.

68. See James Button, "Southern Black Elected Officials: Impact of Economic Change," *Review of Black Political Economy* 12 (Fall 1982), 43. Monte Piliawsky, "The Impact of Black Mayors on the Black Community: The Case of New Orleans' Ernest Morial," *Review of Black Political Economy* 13 (1985), 5-23, and some other scholars have stressed the "symbolic" significance of black elected officials. Regardless of the limitations placed on such officials, I question whether *symbolic* is the best word to describe something as fundamental as access to the political process.

69. Browning, Marshall, and Tabb, *Protest Is Not Enough,* 141.

70. Lawrence Bobo and Franklin D. Gilliam, Jr., "Race, Sociopolitical Participation, and Black Empowerment," *American Political Science Review* 84 (1990), 383 and 380-85.

71. When loosely applied to life in the contemporary inner city, the term *underclass* is inherently vague and has much in common with the pejorative nineteenth-century concept of the "undeserving poor." See Herbert Gans, "Deconstructing the Underclass: The Term's Danger as a Planning Concept," *Journal of the American Planning Association* 56 (1990), 271; Michael B. Katz, "The Urban 'Underclass' as a Metaphor of Social Transformation," in Katz, ed., *The "Underclass" Debate,* 3-23. In the discussion that follows, I avoid behavioralist definitions and use underclass or underclass areas to denote a high level of the geographic concentration of poverty.

72. Wilson, *The Truly Disadvantaged,* 39-62. See also John D. Kasarda, "Urban Change and Minority Opportunities," in Paul E. Peterson, ed., *The New Urban Reality* (Washington, D.C., 1985); Kasarda, "Urban Industrial Transition and the Underclass," *Annals of the American Academy of Political and Social Science* 501 (January 1989), 26-47; Kasarda, "Jobs, Migration, and Emerging Urban Mismatches," in Michael G. McGeary and Lawrence E. Lynn, eds., *Urban Change and Poverty* (Washington, D.C., 1988), 148-98. Although modifying his thesis in response to criticism, Wilson has retained the basic argument of *The Truly Disadvantaged.* See William Julius Wilson, "Public Policy Research and *The Truly Disadvantaged,*" in Christopher Jencks and Paul E. Peterson, eds., *The Urban Underclass* (Washington, D.C., 1991), 463-76; Wilson, "Studying Inner-City Social Dislocations: The Challenge of Public Agenda Research," *American Sociological Review* 56 (1991), 1-14.

73. Ibid., 34, 37, and 33-39.

74. Christopher Jencks, "Is the Underclass Growing?" in Jencks and Peterson, eds., *The Urban Underclass,* 28-100; Robert J. Sampson, "Urban Black Violence: The Effect of Male Joblessness and Family Disruption," *American Journal of Sociology* 93 (1987), 348-82. For a sensitive appreciation of the impact of violence on inner-city "kids" (as he affectionately calls them), see Carl H. Nightingale's ethnographic/historical study, *On the Edge: A History of Poor Black Children and Their American Dreams* (New York, 1993), 19-51.

75. See, for example, Squires et al., *Chicago,* 29-30; Thomas Sugrue, "The Structures of Urban Poverty: The Reorganization of Space and Work in Three Periods of American History," in Katz, ed., *The "Underclass" Debate,* 105-10.

76. Adams et al., *Philadelphia,* 50-52; Orfield and Ashkenaze, *The Closing Door,* 48-54, 55-56, 66-67.

77. Joleen Kirschenman and Kathryn M. Neckerman, " 'We'd Love to Hire Them, but . . . ': The Meaning of Race for Employers," in Christopher Jencks and Paul E. Peterson, eds., *The Urban Underclass* (Washington, D.C., 1991), 203-33; Jomills Henry Braddock II and James M. McPartland, "How Minorities Continue to Be Excluded From Equal Employment Opportunities: Research on Labor Market and Institutional Barriers," *Journal of Social Issues* 43 (1987), 5-39.

78. Douglas T. Lichter, "Racial Differences in Underemployment in American Cities," *American Journal of Sociology* 93 (1988), 771-92; Stephanie Coontz, *The Way We Never Were: American Families and the Nostalgia Trap* (New York, 1992), 252-53; Adams et al., *Philadelphia*, appendix C, 187-88. The average black family income in Philadelphia in 1987 ($19,229) was only 52 percent that of white families ($36,964).

79. Kim Hopper, "Marginalia: Notes on Homelessness in the United States, 1992," in M. Jarvinen and C. Tigerstedt, *Hemloshet i Norden* (Helsinki, 1992), 132. One study of the homeless population of Pennsylvania calculated that 64.7 percent of the individuals using shelters were African Americans. Phyllis Ryan, Ira Goldstein, and David Bartelt, *Homelessness in Pennsylvania: How Can This Be* (Philadelphia, 1989), 26.

80. Theodore Hershberg et al., "A Tale of Three Cities," in Hershberg, ed., *Philadelphia: Work, Space, Family, and Group Experience in the Nineteenth Century* (New York, 1982), 461-91; Kusmer, *A Ghetto Takes Shape*, 42-52; Eugene P. Ericksen and William L. Yancey, "Work and Residency in Industrial Philadelphia," *Journal of Urban History* 5 (1979), 147-82.

81. Because of the ghettoization of many black communities, as early as 1920 most domestic workers in the urban North probably had to travel considerable distances from their homes to their place of employment. During World War II, many of these women were able to leave domestic service for good-paying industrial jobs (which often may have been closer to home than their previous work), but as Jacqueline Jones notes, reconversion at the end of the war forced many black women into domestic service. In 1950, a third of all black women were employed in household service. This figure declined steadily during the next twenty years, but, in 1970, 14 percent still worked as domestics. Jacqueline Jones, *Labor of Love, Labor of Sorrow: Black Women, Work, and the Family From Slavery to the Present* (New York, 1985), 256-60; Bette Woody, *Black Women in the Workplace: Impact of Structural Change in the Economy* (Westport, 1992), 66.

82. Sugrue, "The Origins of the Urban Crisis," 66-73.

83. Wilson, *The Truly Disadvantaged*, 35 and 33-36.

84. For a discussion of the "last of the immigrants" thesis, as applied to black migrants, see Kenneth L. Kusmer, "The Black Urban Experience in American History," in Darlene Clark Hine, ed., *The State of Afro-American History: Past, Present, and Future* (Baton Rouge, 1986), 111-12. Wilson does not cite conservative theorists like Irving Kristol, but his argument is similar.

85. In Miami, the influx of refugees from Cuba in the 1960s and 1970s has had the exact opposite effect as that envisioned by Wilson in *The Truly Disadvantaged*. The movement of Cubans into service jobs, construction, and small businesses undercut black employment in those areas. See Raymond A. Mohl, "Miami: The Ethnic Cauldron," in Bernard and Rice, eds., *Sunbelt Cities*, 86-87.

86. Paul A. Jagorsky and Mary Jo Bane, "Ghetto Poverty in the United States, 1970-1980," in Jencks and Peterson, eds., *The Urban Underclass*, 251-59; the figure of 26 percent is computed from table 5, 251.

87. Mark S. Littman, "Poverty Areas and the 'Underclass': Untangling the Web," *Monthly Labor Review* (March 1991), 25-27.

88. Misuse of the term *ghetto* has contributed to misunderstanding of the connection between black segregation, poverty, and the underclass. Wilson, in *The Truly Disadvantaged*, uses *ghetto*, *inner city*, and *underclass areas* interchangeably; and Jagorsky and Bane, in "Ghetto Poverty in the United States," 235-73, specifically refer to high-poverty areas as *ghetto areas*. This is confusing, because traditionally the term ghetto has been used to designate *racially* segregated areas of cities, regardless of the

social class of the inhabitants. Ghettos are not necessarily slums, and black ghettos have always had sections where middle-class residents predominate.

89. Michael B. Katz, "Conclusion: Reframing the Underclass Debate," in Katz, ed., *The "Underclass" Debate,* 442.

90. Gary Puckrein, "Moving Up," *Wilson Quarterly* 8 (1984), 75.

91. Norman Fainstein, "The Underclass/Mismatch Hypothesis as an Explanation for Black Economic Deprivation," *Politics & Society* 15 (1986-1987), 403-51.

92. Orfield and Ashkenaze, *The Closing Door,* 149-73; Bennett Harrison and Lucy Gorham, "What Happened to African-American Wages in the 1980s?" in Galster and Hill, eds., *The Metropolis in Black and White,* 66. Harrison and Gorman define "high wages" as three times the poverty rate, or about $35,000 in 1987.

93. Fainstein, "The Underclass/Mismatch Hypothesis," 441.

94. Adams et al., *Philadelphia,* 48-52; Harrison and Gorman, "What Happened to African-American Wages in the 1980s?" 57-58.

95. Harrison and Gorham, "What Happened to African American Wages in the 1980s?" 58; Jacqueline Jones, *Labor of Love, Labor of Sorrow,* 305.

96. E. Franklin Frazier, *The Negro Family in Chicago* (Chicago, 1932); Drake and Cayton, *Black Metropolis,* 174-213; Kusmer, *A Ghetto Takes Shape,* 209-14. Middle-class enclaves, dating to the pre-World War I era, sometimes existed outside of the main area of black concentration, but the number of blacks living in such areas was usually quite small. As in the case of Cleveland's Hough Avenue district, these enclaves sometimes formed the basis for the development of the new ghettos of the post-World War II period.

97. See Denton and Massey, *American Apartheid,* 144-45.

98. William H. Harris, *The Harder We Run: Black Workers Since the Civil War* (New York, 1982), 154; Henry L. Taylor, Jr., "The Theories of William Julius Wilson and the Black Experience in Buffalo, New York," in Taylor, ed., *African Americans and the Rise of Buffalo's Postindustrial City, 1940 to Present* (Buffalo, 1990), part 2, 77; Susan and Norman Fainstein, "The Racial Dimension of Black Political Economy," *Urban Affairs Quarterly* 25 (1989), 188-89.

99. U.S. Bureau of the Census, *Poverty in the United States: 1992* (Washington, D.C., 1993), table 21, 137. See also the data on work experience by racial groups, 1950-1980, in Stern, "Poverty and Family Composition," table 7.2, 225. For the entire postwar period, the rate of chronic unemployment for blacks living in suburbs differed little from those residing in central cities. In both cases, the rates were substantially higher for blacks than whites.

100. The Census Bureau defines Hispanics as people of Hispanic origin, regardless of race.

101. In a study of metropolitan areas throughout the United States in 1980, the Hispanic population exhibited a strong positive correlation between income levels and degree of residential integration. There was no such correlation for blacks. See Douglas S. Massey and Nancy A. Denton, "Trends in Residential Segregation of Blacks, Hispanics, and Asians, 1970-1980," *American Sociological Review* 52 (1987), 802-25. In a fascinating study, these scholars have also shown that Caribbean Hispanics of noticeably African ancestry are more highly segregated than other members of this group. Massey and Denton, "Racial Identity Among Caribbean Hispanics: The Effect of Double Minority Status on Residential Segregation," *American Sociological Review* 54 (1989), 790-808.

102. For a roseate view that class has largely superseded racial attachments among blacks in the new suburban areas, see Joel Garreau, *Edge City: Life on the New Frontier* (New York, 1988), 148-69.

103. Although the Census Bureau does not provide a breakdown of these statistics for metropolitan areas alone, it is highly unlikely that such data would vary much from data presented here.

104. Joe R. Feagin, "The Continuing Significance of Race: Antiblack Discrimination in Public Places," *American Sociological Review* 56 (1991), 102, 107, and 101-16. See also Joe R. Feagin and Melvin P. Sikes, *Living With Racism: The Black Middle Class Experience* (Boston, 1994).

105. Although lacking in historical depth, the best journalistic account of the MOVE tragedy is Michael Boyette, with Randi Boyette, *Let It Burn! The Philadelphia Tragedy* (Chicago, 1989).

106. Adams, "Philadelphia," 218.

107. "City Splits in Its View of Accused in Beatings," *New York Times*, October 5, 1993.

108. Kleppner, *Chicago Divided*, 217, 227; Harlan Hahn et al., "Cleavages, Coalitions, and the Black Candidate: The Los Angeles Mayoralty Elections of 1969 and 1973," *Western Political Quarterly* 29 (1976), 521-30; Sonenshein, *Politics in Black and White*, 134, 202-03; James Jennings, "Race, Class, and Politics in the Black Community of Boston," *Review of Black Political Economy* 10 (Fall 1982), 60-61 and 47-63.

109. The source for this data is Farley and Allen, *The Color Line and the Quality of Life in America*, table 9.1, 264-65.

110. The decline in unskilled labor jobs probably played a large role in the dramatic decline in employment among those not graduating from high school. In 1970, in the Philadelphia metropolitan area, 68 percent of individuals with less 10 years of schooling and 76 percent of those with 10 or 11 years of schooling were employed; ten years later the figures were only 18 and 35 percent. Because black adults were much more likely than whites (35 versus 20 percent) to have dropped out of high school before graduating, they were disproportionately hurt by the elimination of low-skill jobs. Adams et al., *Philadelphia*, 59.

111. Farley and Allen, *The Color Line and the Quality of Life in America*, 264-65. Stern, "Poverty and Family Composition," table 7.4, 228-29, using a large sample from the Census Bureau, provides data showing similar trends in occupations by race.

112. Woody, *Black Women in the Workplace*, 71-72 and 71-91. Many of Woody's generalizations are based on 1982 census data.

113. Ibid., 3, 87-88.

114. Trotter, *Black Milwaukee*; Gottlieb, *Making Their Own Way*. One valuable study of a labor group with a significant proportion of black workers is Sue Cobble, *Dishing It Out: Waitresses and Their Unions in the Twentieth Century* (Urbana, 1991).

115. Sharon Harley, "For the Good of Family and Race: Gender, Work, and Domestic Roles in the Black Community, 1880-1930," *Signs* 15 (1990), 336-49. See also Jacqueline Jones, *Labor of Love, Labor of Sorrow*, chaps. 5-8, and the pioneering essay by Darlene Clark Hine, "Black Migration to the Urban Midwest: The Gender Dimension, 1915-1945," in Trotter, ed., *The Great Migration*, 127-46.

116. For examples of human relations groups formed during the war, see Robert A. Burnham, "The Mayor's Friendly Relations Committee: Cultural Pluralism and the Struggle for Black Advancement," in Taylor, ed., *Race and the City*, 258-79; and Broussard, *Black San Francisco*, 193-204.

117. Darden et al., *Detroit*, 138-40, 145-49.

118. Mjagkij, "Behind the Scenes," 280-94.

119. Kleppner, *Chicago Divided*, 50-63; Louis Kushnick, "Race, Class, and Power: The New York Decentralization Controversy," *Journal of American Studies* 3 (1969), 201-19; Ronald P. Formisano, *Boston Against Busing: Race, Class, the Ethnicity in the*

1960s and 1970s (Chapel Hill, 1991); Marjorie Murphy, *Blackboard Unions: The AFT and the NEA, 1900-1980* (Ithaca, 1990), 232-51; Kantor and Brenzel, "Urban Education," 389-90.

120. John Hall Fish, *Black Power/White Control: The Struggle of the Woodlawn Organization in Chicago* (Princeton, 1973).

121. For a perceptive overview, see Thomas F. Jackson, "The State, the Movement, and the Urban Poor: The War on Poverty and Political Mobilization in the 1960s," in Katz, ed., *The "Underclass" Debate*, 422-30.

122. Bettye Collier-Thomas, *The Howard Theater and the National Black Theater Movement* (New York, forthcoming).

123. The literature on the history of African American music is enormous. Works specifically relating black music to urban themes include Charles Keil, *Urban Blues* (Chicago, 1966); Marshall W. Stearns, *The Story of Jazz* (New York, 1970), 37-75 (on New Orleans); Joseph Thomas Hennessey, "From Jazz to Swing: Black Jazz Musicians and Their Music, 1917-1935" (Ph.D. diss., Northwestern University, 1973), which deals mostly with New York and Chicago; and Lawrence Levine, *Black Culture and Black Consciousness: Afro-American Folk Thought From Slavery to Freedom* (New York, 1977), 217-39, which relates the rise of the blues and jazz to migration, urbanization, and the secularization of African American culture in the early twentieth century. Michael W. Harris, *The Rise of Gospel Blues: The Music of Thomas Andrew Dorsey in the Urban Church* (New York, 1992), imaginatively explores the connections between African American music and urban religious institutions.

124. For example, in 1910 and 1920 black men made up 3.7 and 4.7 percent of the total male workforce in Chicago, but constituted 6.3 and 7.4 percent of the musicians. Spear, *Black Chicago*, 32-33, 152-53.

125. Diane Delores Turner, "Organizing and Improvising: A History of Philadelphia's Black Musicians' Protective Union Local 274, American Federation of Musicians" (Ph.D. diss., Temple University, 1993). On Chicago's black musicians' union, which merged with the white local in 1966, see Clark Halker, "Banding Together," *Chicago History* 18 (Summer, 1989), 41-59.

126. C. Eric Lincoln and Lawrence H. Mamiya, *The Black Church in the African American Experience* (Durham, 1990), 162-63.

127. One weakness of Paul Kleppner's otherwise excellent history of racial politics in Chicago, *Chicago Divided*, is that it does not address the social basis of black mobilization for Washington.

128. William J. Grimshaw, "Unravelling the Enigma: Mayor Harold Washington and the Black Political Tradition," *Urban Affairs Quarterly* 23 (1987), 203-04.

129. Basil Wilson and Charles Green, "The Black Church and the Struggle for Community Empowerment in New York City," *Afro-Americans in New York Life and History* 12 (1988), 75, 51-79.

130. For the industrialization period, see the excellent case study by Robert S. Gregg, *Sparks From the Anvil of Oppression: Philadelphia's African Methodists and the Great Migration, 1890-1930* (Philadelphia, 1993). Unfortunately, no similar study exists for the postindustrial era.

131. See the account of the St. Paul Church and its charismatic minister, Johnny Ray Youngblood, in Samuel Freedman, *Upon This Rock: The Miracles of a Black Church* (New York, 1993).

132. On black hospitals during the industrial era, see Vanessa Northington Gamble, "The Negro Hospital Renaissance: The Black Hospital Movement, 1920-1945," in Diana Elizabeth Hall and Janet Golden, eds., *The American General Hospital: Communities and Social Contexts* (Ithaca, 1989), 82-105. A forthcoming book

by Gamble will explore the history of black hospitals in great detail. For a case study of one city, see David McBride, *Integrating the City of Medicine: Blacks in Philadelphia's Health Care, 1910-1965* (Philadelphia, 1988).

133. For a theoretical statement of the importance of the historical study of informal group life, see James Borchert, "Urban Neighborhood and Community: Informal Group Life, 1850-1970," *Journal of Interdisciplinary History* 11 (1981), 607-31. See also Borchert, *Alley Life in Washington.*

134. For a perceptive study by a geographer, see David Ley, *The Black Inner City as Frontier Outpost: Images and Behavior of a Philadelphia Neighborhood* (Washington, D.C., 1974). For the post-World War II period, the only specifically historical study that explores black family life and informal group interaction at the neighborhood level is Nightingale, *On the Edge.* Nightingale's study is not a holistic study of a community, however, but focuses largely on the causes and consequences of violence and deprivation.

135. See Martin Sanchez Jankowski, *Islands in the Street: Gangs and American Urban Society* (Berkeley, 1991).

136. For the South, see Miller, "Migrations and Economic Growth in the Metropolitan Southern Sunbelt," 165-70.

137. For urban population data for 1940-1980, see Bernard, *Snowbelt Cities,* 273-74; Rice and Bernard, "Introduction," *Sunbelt Cities,* 24-25.

138. David Goldfield, *Black, White, and Southern: Race Relations and Southern Culture, 1940 to the Present* (Baton Rouge, 1990), 227-41, provides a regional context for the development of black voting and racial politics since 1976.

139. Raymond A. Mohl, "Ethnic Politics in Miami, 1960-1986," in Miller and Pozzetta, eds., *Shades of the Sunbelt,* 144-45.

140. Ronald Bayor, "Race, Ethnicity, and Political Change in the Urban South," in Miller and Pozzetta, *Shades of the Sunbelt,* 130.

141. Ibid., 132-39.

142. Raymond A. Mohl, "On the Edge: Blacks and Hispanics in Metropolitan Miami Since 1959," *Florida Historical Quarterly* 78 (1990), 37-56; Seth Mydans, "Blacks Complaining of Neglect as Los Angeles Is Rebuilt," *New York Times,* August 30, 1992; Larry Rohter, "As Hispanic Presence Grows, So Does Black Anger," *New York Times,* June 20, 1993.

143. Kenneth L. Kusmer, "The Next Agenda in American Ethnic History," *Reviews in American History* 20 (1992), 580-84; Saskia Sassen, *The Global City: New York, London, Tokyo* (Princeton, 1991), 245-320.

144. Kusmer, "The Black Urban Experience," 105-8.

Index

About the Authors

Elsa Barkley Brown teaches in the Center for Afro-American and African Studies and the Department of History at the University of Michigan. She is associate editor of the two-volume *Black Women in America: An Historical Encyclopedia* (1993). Her articles have appeared in *Signs, Sage, History Workshop,* and *Feminist Studies.*

Kenneth W. Goings is Professor of History at the University of Memphis. His most recent book is *Mammy and Uncle Mose: Black Collectibles and American Stereotyping* (1994). He is also author of *The NAACP Comes of Age: The Defeat of Judge John J. Parker* (1990). He is collaborating with Gerald L. Smith on a study of African American Memphis, 1862-1920.

Darlene Clark Hine is the John A. Hannah Professor of History at Michigan State University. She is author and editor of numerous books in African American history, including *Black Victory: The Rise and Fall of the White Primary in Texas* (1979), *Black Women in White: Racial Conflict and Cooperation in the Nursing Profession, 1890-1950* (1989), *Hine Sight: Black Women and the Reconstruction of American History* (1994), and the highly acclaimed *Black Women in America: An Historical Encyclopedia* (1993).

Tera W. Hunter teaches southern and labor history at the University of North Carolina at Chapel Hill, where she is Assistant Professor of History. She is currently working on a study of the politics and culture of wage household labor in Atlanta between 1865 and 1920.

Robin D. G. Kelley is Professor of History and Africana Studies at New York University. He is author of *Hammer and Hoe: Alabama Communists During the Great Depression* (1990) and *Race Rebels: Culture, Politics, and the Black Working Class* (1994) and coeditor of *Imagining Home: Class, Culture, and Nationalism in the African Diaspora* (1994).

Gregg D. Kimball, historian at the Valentine Museum in Richmond, Virginia, is currently working on his Ph.D. dissertation in history at the University of Virginia. His exhibitions at the Valentine Museum include "Bondage and Freedom: Antebellum Black Life in Richmond, Virginia" (with Marie Tyler-McGraw), "The Working People of Richmond: Life and Labor in an Industrial City," and "Shared Spaces, Separate Lives" (with Barbara C. Batson), all funded by the National Endowment for the Humanities. He is currently working on Valentine Riverside, a history park on the James River in Richmond, formerly the site of the Tredegar Iron Works.

Kenneth L. Kusmer is Professor of History at Temple University. He has also taught at the University of Pennsylvania and at Göttingen University in Germany. He is author of *A Ghetto Takes Shape: Black Cleveland, 1870-1930* (1976) and editor of *Black Communities and Urban Development in America, 1720-1990* (1991). His current research deals with African American urban history since 1940 and with the history of homeless men in the United States, 1870-1940.

Earl Lewis is Associate Professor of History and Afro-American and African Studies at the University of Michigan. He has published articles on African American migration and urbanization, the black family, race, and identity. He is author of *In Their Own Interests: Race, Class, and Power in Twentieth-Century Norfolk, Virginia* (1991), and, with Robin D. G. Kelley, general editor of a forthcoming eleven-volume history of African Americans for young adults.

Raymond A. Mohl is Professor of History and Chair of the Department of History at the University of Alabama at Birmingham. He is author of *The New City: Urban America in the Industrial Age, 1860-1920*

(1985). His most recent books include *Searching for the Sunbelt: Historical Perspectives on a Region* (1990) and, with Arnold R. Hirsch, *Urban Policy in Twentieth-Century America* (1993). He is currently completing a revision of his urban history reader, *The Making of Urban America,* as well as a history of race relations in twentieth-century Miami.

Gerald L. Smith is Associate Professor of History at the University of Kentucky. He is author of *A Black Educator in the Segregated South: Kentucky's Rufus Atwood* (1994). He is working with Kenneth W. Goings on a study of African American Memphis from 1862 to 1920.

Joe W. Trotter is Professor of History at Carnegie Mellon University in Pittsburgh. He is editor of *The Great Migration in Historical Perspective: New Dimensions of Race, Class, and Gender* (1991) and author of *Coal, Class, and Color: Blacks in Southern West Virginia, 1915-32* (1990) and *Black Milwaukee: The Making of an Industrial Proletariat, 1915-45* (1985). He is currently conducting research on African Americans in the urban deep South.

Shane White is Senior Lecturer in the Department of History at the University of Sydney, Australia. He is author of *Somewhat More Independent: The End of Slavery in New York City, 1770-1810* (1991).